Climatic change

Climatic change

EDITED BY

JOHN GRIBBIN

Visiting Fellow, Science Policy Research Unit,
University of Sussex

CAMBRIDGE UNIVERSITY PRESS

CAMBRIDGE
LONDON · NEW YORK · MELBOURNE

Published by the Syndics of the Cambridge University
    Press
The Pitt Building, Trumpington Street, Cambridge CB2 1RP
Bentley House, 200 Euston Road, London NW1 2DB
32 East 57th Street, New York, NY 10022, USA
296 Beaconsfield Parade, Middle Park, Melbourne 3206,
    Australia

© Cambridge University Press 1978

First published 1978

Printed in Great Britain by
Cox & Wyman Ltd
London, Fakenham and Reading

Library of Congress Cataloguing in Publication Data
Main entry under title:

Climatic change.

1. Climatic changes. I. Gribbin, John R.
QC981.8.C5C55 551.5 76–52185
ISBN 0 521 21594 3 hard covers
ISBN 0 521 29205 0 paperback

FOR NAN AND VAL –

'There are holes in the sky,
where the rain gets in;
but the holes are small –
that's why rain is thin!'

Spike Milligan

Contents

Contents

List of contributors

Dr T. P. Barnett, Scripps Institute of Oceanography, University of California, San Diego, P.O. Box 1529, La Jolla, California 92037, USA

Professor M. I. Budyko, State Hydrological Institute, Basil Island, 23 Line 2, Leningrad 199053, USSR

Dr R. M. Chervin, National Center for Atmospheric Research,* P.O. Box 3000, Boulder, Colorado 80303, USA

Dr J. C. Duplessy, Centre des faibles Radioactivités, Laboratoire mixte CNRS-CEA, 91190 Gif-sur-Yvette, France

Professor H. Flohn, Department of Meteorology, University of Bonn, Federal Republic of Germany

Dr J. Gribbin, Science Policy Research Unit, Mantell Building, University of Sussex, Falmer, Brighton BN1 9RF, England

Dr W. W. Kellogg, National Center for Atmospheric Research,* P.O. Box 3000, Boulder, Colorado 80303, USA

Dr G. J. Kukla, Lamont-Doherty Geological Observatory, Palisades, New York 10964, USA

Professor H. H. Lamb, Climatic Research Unit, University of East Anglia, Norwich, England

Dr S. H. Schneider, National Center for Atmospheric Research,* P.O. Box 3000, Boulder, Colorado 80303, USA

Dr D. H. Tarling, Department of Geophysics, The University, Newcastle upon Tyne NE1 7RU, England

Dr R. L. Temkin, National Center for Atmospheric Research,* P.O. Box 3000, Boulder, Colorado 80303, USA

Dr T. A. Wijmstra, Hugo de Vries Laboratorium, Universitiet van Amsterdam, Afdeling Palynologie, Sarphatistraat 221, Amsterdam 4, The Netherlands

Dr J. Williams, International Institute for Applied Systems Analysis, Schloss Laxenburg, 2361 Laxenburg, Austria

*The National Center for Atmospheric Research is sponsored by the National Science Foundation.

Preface

During the early 1970s a series of climate-related disasters – droughts in the Sahel and Ethiopia, failure of the Indian monsoon in several successive years, bad harvests in the USSR, floods in Brazil and so on – helped to focus attention on the possibility that climate can and does change rapidly enough and to a large enough extent to affect the activities of our global society. In particular, it became clear that at a time of rapidly growing population with only slender reserves of food there now exists a greater danger than ever before that a small shift in, say, rainfall distribution, could have widespread and devastating implications. Having trained as an astrophysicist, my own interest in these problems was as a journalist, reporting and interpreting the relevant specialist work in different climate-related disciplines to a wide audience of scientists, chiefly through the pages of *Nature*. To do so effectively, I needed to develop some understanding of the basics of climatic change, and I sought unsuccessfully for a 'standard text' that would encapsulate up-to-date thinking on the subject and provide the necessary background. The failure to find such a book led directly to the idea of the present volume as an overview of the basics of climatic change, intended for any scientifically literate person with an interest in climate. In essence, this is the book I needed in 1973, with the bonus that more recent developments are now also included.

With the benefit of the knowledge I have gained through editing the book, it seems clear to me that our understanding of climatic change is likely to improve rapidly over the next few years, partly through the development of better computer models, partly through the availability of better data from instruments flown above the atmosphere in artificial satellites, and partly through the synthesis of ideas from different disciplines which are all relevant to the climate problem – or problems. Although the analogy is not exact, it seems that the study of climatic change is in some ways developing as the 1970s' 'revolution in the Earth sciences', following the revolution in our understanding of the solid Earth brought about in the 1960s through the development of the concepts of plate tectonics. I am grateful to all the researchers who agreed so readily to contribute to this volume and have provided me with an inside view of the development of this new revolution in our understanding of Planet Earth. The importance of the subject is now beyond doubt, and I hope that in this book we have retained also some flavour of the excitement of working in such a field as a comprehensive overall picture begins to emerge. There is still room for a great deal more development, from many disciplines, and I hope that this volume will help to stimulate some of that development.

March 1977   John Gribbin

SECTION 1

Studying the climates of the past

I

The geological—geophysical framework of ice ages

D. H. TARLING

Perhaps the most well-known evidence for the reality of climatic change is that for the successive advances and retreats of immense ice sheets over temperate latitudes during the past million years or so. The occurrence of this Late Cenozoic Ice Age at the same time as Man's discovery of tools, fire and so on may be no coincidence, since the drastic changes in climatic zonation, with associated changes in flora and fauna, resulted in a strong evolutionary bias towards any organisms which could adapt quickly to such changes and which could modify their effects.

Direct records of climatic change are only available for the past few hundred years (Lamb, 1972) and it is difficult to isolate 'natural' climatic trends from those due to other influences. In any case, the recognition of climatic cycles does not necessarily indicate their cause because of the strong interactive nature of most climatic controls. The most fundamental need is, of course, to fill the gap in knowledge between climatic changes over geological time-scales and meteorological records (Lamb, 1975). However, geologically controlled climatic changes are still occurring and their effects must be isolated from other factors if any understanding of climatic change is to result. It is probable, for example, that a much better evaluation of the effects of increased dust and carbon dioxide in the atmosphere or the diversion of freshwater rivers away from the Arctic, can be made from studying the past effect of similar situations (Gates & Imbrie, 1975) than from computer simulations (Barry, 1975) – at least for some years to come. Even more significant is the possibility that certain geological factors may operate much more rapidly than previously thought. If ice sheets disappear rapidly, as is indicated by the geological record, then major ice surges from Antarctica may be possible and almost instantaneous (Hughes, 1975) with resultant rapid rise in sea-level in days rather than over several thousand years. Similarly, it is possible that the unstable ice sheets which covered much of northern Europe and North America grew rapidly to a significant extent and quickly gave rise to surrounding periglacial conditions – even though it took several thousand years for them to reach their maximum development. The evidence for or against the reality of such mechanisms must lie within the geological records and a knowledge of the development and recession of past ice sheets may then provide clues by which relatively minor, microclimatic controls could be used to prevent or reduce their speed of development. Indeed, the geological dictum 'the present is the key to the past' could well be converted, for climatology, into 'the past is the key to the present and future'. After all, weather forecasting still depends largely on comparison of past and present atmospheric conditions in order to predict over only a few days.

In the following sections an attempt is made to identify some of the characteristics of previous ice ages and thus to evaluate the significance of various geological and geophysical factors that have influenced past climatic changes and may be responsible for current change. It is clear, however, that no one factor alone is responsible and it is only by comparison of the past records that it becomes more evident how these different climatic controls interact with each other.

Table 1.1 gives a summary of the radiometric

3

age of Palaeozoic stages involved in glacial episodes for readers unfamiliar with these.

Table 1.1. *Approximate radiometric age of Palaeozoic stratigraphic stages involved in glacial episodes*

	Stages	× 10⁶ yr ago
		———245
	Kazanian	
		———250
	Kungarian	
Permian		
		———255
	Artinskian	
		———270
	Sakmarian	
		———280
	Stephanian	
		———290
	Westphalian	
Carboniferous		
		———310
	Namurian	
		———325
	Visean	
		———335
	Tournasean	
		345
		———480
	Caradocian	
		———450
Ordovician		
	Llandeilo	
		———465
	Llanvirn	
		———475
	Arenig	

Previous Ice Ages

For well over 200 yr, the sediments and morphological features bordering the Alps have been recognised as indicating the former extent of glaciation in Europe. Subsequent studies and bibliographies for such sediments and morphological features in Europe and North America, in particular, are now extensive (Embleton & King, 1968; Péwé, 1969; Flint, 1971; Washburn, 1973) and key papers have recently been collected into one volume (Goldthwait, 1975). But although such features undoubtedly accompanied earlier glaciations, most of them are likely to have been eroded subsequently, particularly where mountain glaciation was involved, and it is the detritus from such later glaciation that is most likely to be preserved. Even so, most deposits are likely to have been reworked by river action which would have been stronger following the melting of the ice sheets. Nonetheless, certain characteristics can be recognised (Flint, 1961, 1975) as strongly indicative of a glacial origin, although each of these features can be created by other means (Schermerhorn, 1975), particularly by deposition from turbidity currents (Carter, 1975). It is, therefore, the morphological features of striated surfaces, roches moutonnées and so on that are usually considered the most diagnostic. Although even these rarely preserved features can be created by other means, such as landslips, if several such glacial features can be discovered over an extensive area, with associated periglacial sediments and low fauna and floral speciation, then the evidence becomes very convincing. To be climatically significant, however, such evidence must indicate the presence of ice sheet activity, rather than highland glaciation which, though locally important, may not be diagnostic of the regional climate. The occurrence of such indicators over a very wide area of low relief and elevation can, therefore, unambiguously indicate the presence of an ice sheet with the concomitant implications of major climatic changes.

Precambrian Glaciations

At least four major ice ages have been recognised within Precambrian times (Fig. 1.1), although rocks older than 2800 Myr tend to be metamorphosed to the extent that identification of an original glacial origin is now unlikely. The most striking of these is the youngest, the Varangian, which occurred some 660 to 680 Myr ago; glacial deposits of this age have been recognised in North America, Greenland, Spitsbergen, Scandinavia, the British Isles, France, USSR, China, India, Australia, Africa and South America (Harland, 1972; Harland & Herod, 1975). The most detailed and thorough studies so far available are for the deposits in Scotland (Spencer, 1971, 1975) where most glacial features, other than striated pavements, are present. Striated pavements of this age are, in fact, only known in northern Norway and Sweden. It has been suggested that many of these deposits are not truly glacial, but owe their glacial characteristics to other causes, such as turbidite deposition in active tectonic environments (Schermerhorn, 1975). Palaeomagnetic evidence seems to indicate that these deposits in Greenland, Spitsbergen and Britain did not form in high latitudes (Girdler, 1964; Tarling, 1974) as is also indicated by the presence in most areas of interbedded, well-developed limestones, dolomites and stromatolites. Nonetheless the glaciogenic features of these sediments are extremely striking and it seems probable that some of these are indeed of glacial origin, but further research is still necessary before final conclusions can be drawn.

The Sturtian Ice Age (Harland & Herod, 1975) has been recognised some 750 Myr ago in Australia, China, southwest Africa and Scandinavia, and the Gnejsö Ice Age occurred some 950 Myr ago in Greenland, Spitzbergen and Norway. The Huronian Ice Age, some 2300 Myr ago, is strongly indicated by sedimentary (Gowgandan Tillite) and morphological evidence in Canada and may correlate with other, possibly glacial, debris in South Africa and India (Harland & Herod, 1975). The Canadian deposits seem to have accumulated at high latitudes (Symons, 1975) and therefore are probably derived from a polar ice-cap. It seems probable that the reality of these ancient ice ages will be confirmed by further study and it may eventually be possible to establish their palaeogeographical situation. At this stage, however, the data available are inadequate to throw significant light on the factors associated with the occurrence of glaciation, other than their possible chronological relationship with galactic rotations and so on (see Chapter 7).

Lower Palaeozoic Glaciations

The most striking and well-known glacial deposits of this age are in Ordovician rocks in the Sahara (Fairbridge, 1974; Allen, 1975). These occur through much of North Africa (Fig. 1.2) where they have mainly been researched by Beuf and his colleagues (Beuf, Biju-Duval, Mauvier & Legrand, 1968a; Beuf et al., 1968b; Gariel, de Charpal & Beenacef, 1968; Rognon, de Charpal, Biju-Duval & Gariel, 1968). In some areas, U-shaped palaeo-valleys are present with polished and striated floors overlain by tills, some of which are water-laden and include dropstones. Graded sediments and outwash sediments are also present with marginal permafrost characteristics bordering them which include polygons, kettle-hole structures and so on. The age of these deposits is not known precisely but they are overlain by Silurian

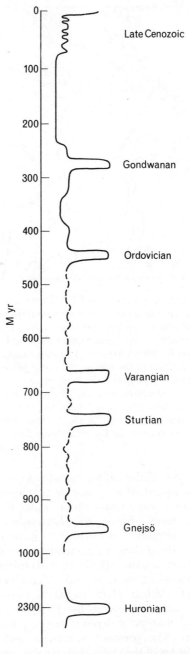

Fig. 1.1 Ice ages through geological time. The right hand side of the curve corresponds to periods of major ice sheet formation with each period including several glacial and interglacial stages. The left hand side of the graph corresponds to periods with no known glaciation, with intermediate position indicating the possible extent of mountain glaciation.

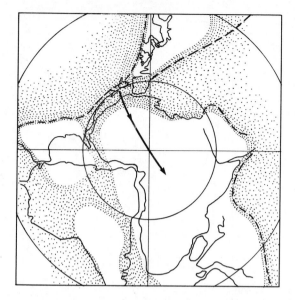

Fig. 1.2. The palaeogeography of the Ordovician (Saharan) glaciation. The continental boundaries at this time are very uncertain – and present-day coastlines are only indicated for general location purposes. The polar movement relative to the continent for this period (broad arrow) is still badly defined, that is, it is uncertain within some 20° of its indicated position, but must travel from the top left hand side of the map to the bottom right hand side during the Lower Palaeozoic.

deposits and one of the tillites includes Upper Ordovician fauna; it seems probable that most of the glaciation was during the Caradocian, with some activity during the Llandeilian. Attempts to determine directly the palaeomagnetic latitude of these sediments have been largely unsuccessful (Abou-Deeb, 1976) but it is clear from studies of younger and older rocks from the North African platform that the area was in high latitudes during the Ordovician (Fig. 1.2) and that these deposits are of polar ice-cap origin. The presence of marine sediments within some of the deposits shows that the ice was at or close to sea-level, although most of the deposits were laid down in unfossiliferous waters. Glacial deposits of Ordovician age have also been recognised in various parts of South America, South Africa (although these may be Cambrian), West Africa, Newfoundland and possibly Britain (Harland & Herod, 1975) but

the glacial origin and age of most of these deposits are uncertain.

Upper Palaeozoic Glaciations

With the obvious exception of those of the Late Cenozoic Ice Age, the glacial deposits of Permo-Carboniferous age (Fig. 1.3) in all of the Gondwana continents (South America, Africa, India, Australia and Antarctica) are the best documented and studied of all glaciations (Crowell & Frakes, 1970, 1975). Most of the discoveries, except in Antarctica, were made in the nineteenth century and were possibly the major factor influencing geologists from these areas in accepting the concept of continental drift during times when most 'northern' geologists were unwilling to consider the theory.

(a) South America

Late Palaeozoic glacial rocks were not reported from South America until the beginning of this century (White, 1907) but have now been found over some 1 500 000 km² in the Parana Basin alone. Although much of the glacial material is reworked, particularly in the outcrops along the edge of the Andes, all of the classical features of glacial activity have been recognised, including tills, striated platforms, glacially cut and filled valleys, dropstones, faceted pebbles, roches moutonnées, eskers and other features, and it seems certain that the deposits of the Parana Basin are of ice sheet derivation, although those of the Andes and further south may be associated with mountain or piedmont glaciation (Frakes & Crowell, 1969). The oldest glacial deposits of this Gondwanan sequence have been found in the Paganzo Basin where the flora indicate a probable pre-Westphalian age (Kemp, 1975), and glacial deposits in the Rio Blanco basin are thought to be of Early Carboniferous age (Frakes & Crowell, 1969). Glacials deposits of Middle and Upper Carboniferous age are also known, although the maximum glacial extent appears to have been confined to the very latest Carboniferous and earliest Permian (Fig. 1.2). Glaciation, possibly mountainous and on a very local scale, seems to have persisted locally in Brazil until the Artinskian or possibly slightly later (McClung, 1975). This record of glaciation is the longest for all of the Gondwanan continents for this period and it may be significant that the estimate of a minimum of 17 advance/

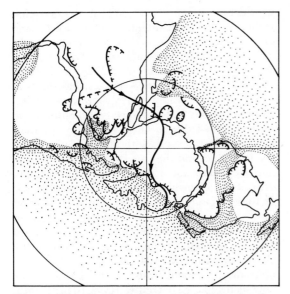

Fig. 1.3. The Gondwanan glaciation. The distribution of ice sheets at this time is indicated by the barbed line. Although there is some doubt about the exact synchroneity, it seems probable that most of these ice sheets were active simultaneously in the very Early Permian and disappeared almost simultaneously from all of these continents as soon as the Gondwanan continent had drifted so that the pole was approaching the 'Pacific' Ocean. The polar path, although better defined than for the Ordovician (Fig. 1.2), can only be broadly indicated by the arrowed line with a probable uncertainty of some 10° as it moves from its Lower Carboniferous to Middle Permian position.

retreat cycles (Rocha-Campos, 1967) is by far the largest determination of the number of glacial fluctuations.

(b) Southern Africa

Since the original description of glacial deposits of Late Palaeozoic age (Sutherland, 1870), there has been little doubt about the glacial origin of the Dwyka tillite with its associated striated pavements, glacial valleys infilled with glacial debris and other features. Most geologists who have visited outcrops have been convinced of its glacial origin (Frakes & Crowell, 1970a; Crowell & Frakes, 1972), although there is increasing recognition of the extent of fluvial reworking (Elliott, 1975). Several ice lobes have been recognised, some deriving from the south-

east of the continent and others passing from the continent onto the then adjacent South American continent (Fig. 1.3). It has often been assumed that these lobes are of very similar age, with different directions of motion indicating the local effect of topography (Matthews, 1970; Rust, 1975) but at least four major advance/retreat cycles have been recognised (Theron & Blignault, 1975). The flora within the various outcrops indicate glaciation spanning much of the Carboniferous and Early Permian (Plumstead, 1973), although recent studies indicate that the maximum glaciation was largely confined to the uppermost Carboniferous and earliest Permian (Kemp, 1975). This age seems to be confirmed by the Early Permian marine fauna near the top of the glacial sequence in the Karroo Basin (McLachlan & Anderson, 1975) which also suggests that much of the glaciation occurred at low levels and was of a polar ice-cap form, although undoubted mountain glacial features can be recognised on a local scale. The youngest glacial deposits seem to be those of probable highland glaciation in the Congo, where an Artinskian age is suggested (McClung, 1975).

(c) The Indian subcontinent

The Talchir glaciation was the first of the Late Palaeozoic Gondwanan glaciations to be recognised (Blandford, Blandford & Theobald, 1859) and was of critical significance in persuading Earth scientists about the reality of continental drift, since these glacial deposits implied that the Gondwanan glaciation extended well past the equator and therefore implied world-wide glaciation if the continents were then in their present-day positions. The fact that the remaining northern continents were mostly characterised by equatorial conditions at this time therefore provided very strong evidence for the continents being in different positions relative both to the rotation poles and to each other at this time. Glacial beds have been recognised in many areas (Frakes, Kemp & Crowell, 1975) and include diamictites, tills, glaciated pavements, etc. (Fig. 1.3). The flora generally indicate a Westphalian to Sakmarian age (Frakes *et al.*, 1975) although there is increasing floral and faunal evidence for glaciation being mostly restricted to latest Carboniferous and earliest Permian, mainly Sakmarian (McClung,

1975; Kemp, 1975; Shah & Sastry, 1975). The glaciation, which includes at least three major episodes (Casshyap & Qidwai, 1974), was during a time of low relief and close to sea-level, as indicated by the series of extensive marine transgressions in various places. The final ice recession, accompanied by minor fluctuations, appears to have been very rapid (Ghosh & Mitra, 1970).

(d) Australia

Palaeozoic glacial deposits were recognised (Selwyn, 1859) near Adelaide at almost the same time as the first glacial deposits of similar age in India and these have subsequently been identified over most of the Australian continent (Crowell & Frakes, 1971). These deposits include all features associated with both mountain and ice-cap glaciation. The glaciation seems to have commenced by at least Westphalian time within mountains bordering the Tasman geosyncline and remained active until the end of the Carboniferous, by which time the mountains were largely eroded. Faunal and sedimentological evidence strongly indicates drastic climatic change which commenced in the earliest Namurian, marked by a strong decrease in the diversity of floras (Morris, 1975), changes in marine faunas and the disappearance of true reefs and the subsequent paucity of significant limestone development (Runnegar & Campbell, 1976). Mountain glaciation probably ceased before the formation of ice sheets, in southeastern Australia, in Sakmarian times, although ice sheets may have been present in western areas towards the end of Stephanian times. The ice sheets largely disappeared at the end of the Sakmarian, but some areas retained some glaciers, possibly as late as Kazanian, as evidenced by rafted debris of this age in Queensland, and possibly Tasmania. Marine levels are quite common and it is thought that a direct marine connection existed between northwestern Australia and the Tasman geosyncline for much of the glacial period. At least three phases of glaciation can be recognised, but this is probably a gross underestimate of the number of glacial episodes during this period.

(e) Antarctica

Although Palaeozoic glaciation had long been suspected in this continent by proponents of continental drift, it was not until 1960 that evidence was first found (Long, 1962) and deposits are now known along most of the length of the Transantarctic Mountains of eastern Antarctica and the Ellsworth Mountains of western Antarctica (Frakes, Matthews & Crowell, 1971; Elliot, 1975). However, difficulties of access and exposure have continued to restrict knowledge of these deposits and their terrestrial nature and, with a lack of diagnostic fossils, means that they are still only known to be post-Devonian and to be overlain by Permian deposits laid down when glaciation had ceased in these regions. Frakes et al. (1971) suggested that the glaciation was of different ages in different localities, on the basis that Antarctica drifted through polar regions during the Carboniferous and Lower Permian, but an uppermost Carboniferous–earliest Permian age now seems likely (Barrett & Kyle, 1975; Kemp, 1975), although not proved. For palaeogeographic considerations (Fig. 1.3) it is important to note that there is evidence for a Jurassic–Cretaceous strike-slip motion of Antarctica relative to East Antarctica so that at this time cratonic blocks lay alongside the Transantarctic Mountains (Barrett & Kohn, 1975; Elliot, 1975;) and much of West Antarctica was the site of geosynclinal accumulations.

(f) Elsewhere

Only two other localities are considered to show evidence for glacial activity at or about the same time as the glaciation of the Gondwanan continents, as the Wajid Sandstones of Saudi Arabia are now known to be neither of glacial origin nor of Permo-Carboniferous age (Hadley & Schmidt, 1975).

(i) The North American Permo-Carboniferous Squantum 'tillite' of Massachusetts seems to be a very localised deposit (Newell, 1957; Dott, 1961); and it has been argued that it may not even be glacial (Frakes et al., 1975), although a glacial origin would be consistent with its grain textures (Rehmer & Hepburn, 1974). It certainly does not seem to be associated with ice-cap glaciation, if only because of its restricted extent, and is not therefore considered a major indication of the palaeoclimate of North America which was almost certainly equatorial to tropical from the studies of fauna, flora and nature of sediments of this period (such as those of Chaloner & Meyen, 1973 and Milner & Panchen, 1973) and from palaeomagnetic studies (Turner & Tarling, 1975).

(ii) Glacial sediments within marine strata have been recorded in the Omolon River, east of the Verkhoyansk Mountains and northeast of the Okhotsk Sea (Mikhazlov, Ustritshii, Chernyak & Yavshits, 1970). These are well dated as Kazanian and therefore later than all of the Gondwanan glaciations, except for isolated deposits in southeast Australia, the Congo (Kemp, 1975) and Brazil (McClung, 1975). These deposits are not on the main Siberian platform and, at that time, this Angaran area was probably isolated from the West Siberian Shield. Although further information on these deposits is needed, it seems probable that they are glacial since the Angaran flora, which were also originally confined to this block (Meyen, 1970; Chaloner & Meyen, 1973), and the close parallism of development of the Angaran and Gondwanan flora, suggest similar climatic environments. It is quite probable that this part of the Siberia Shield lay in high northern latitudes and was not finally welded onto the main Eurasian block until later.

Late Cenozoic Ice Age

More is known about the latest and the maximum phases of the Late Cenozoic Ice Age than about any other, and it is the deposits associated with these and present-day glaciations that provide the diagnostic properties for recognising previous ice ages. The features of these deposits are comprehensively covered in many text books, especially Flint (1971), and such details will not be repeated here. However, there has been a considerable expansion in knowledge about the record for such glaciations from studies of deep-sea sediments, in particular, and it is worth reviewing, briefly, the evidence for the timing of the onset of glaciation changes.

The Late Cenozoic Ice Age was originally labelled the Great Ice Age (Geikie, 1874) and the first appearance of its glacial deposits were used to define the start of the Pleistocene. The Plio-Pleistocene boundary is still defined on the first appearance of cold conditions in the Mediterranean basin (Gignoux, 1913) some 1.7 Myr ago but glaciation was clearly initiated in the northern hemisphere some 3 to 4 Myr ago.

(Fig. 1.4) and at least 10 glacial cycles have been recognised in sedimentary sequences in Iceland (Einarsson, Hopkins & Doell, 1967) during the past 3.1 Myr. Much earlier glaciation is indicated in the southern hemisphere (Fig. 1.5 and 1.6). Comparison of the fauna in Australia and South America from Permian times to the present (Colbert, 1973; Cox, 1973, 1974 Denton, Armstrong & Stuiver, 1971; Keast, 1973;) indicates that the climate along the East Antarctic migration route must have been at least mild, probably warm, from Middle Permian times right through to the end of the Cretaceous, at which time it was the formation of new seaways, rather than climatic change, that prevented further migration along this route. Sedimentological studies of rocks in Antarctica and Australia, then still attached to Antarctica, support the evidence for this prolonged equitable climatic situation and indicate that mild climatic conditions persisted in parts of Antarctica through much of the Tertiary period, with drastic faunal and floral changes only commencing in the Miocene, at a similar but probably earlier time than the initiation of glaciation in the northern hemisphere. Examination of the Southern Ocean deep-sea sediments for the presence of glacial sands, cold-water foraminifera, studies of clay mineralogy and palaeotemperatures (Emiliani, 1954; Denton *et al.*, 1971; Margolis & Kennett, 1971; Anon 1973*a, b*; Jacobs, 1974; Blank & Margolis, 1975) indicate that glaciation in Antarctica was active in the Late Cretaceous, Early Eocene, late Middle Oecene, Oligocene, Lower Miocene and Late Miocene, with particularly detailed climatic oscillations recorded for the past 5 Myr (Fig. 1.6)

There is no unequivocal evidence for the nature of the Antarctic glaciation during the earlier Tertiary, but it seems probable that the pre-Miocene evidence is largely associated with local, mountain glaciation as the deep-sea data suggest that glacial conditions were not as severe as in the Late Miocene. Furthermore, glaciation in South America only occurred after major oceanic cooling some 3.5 Myr ago (Mercer, Fleck, Mankinen & Sander, 1975) and only this Miocene cooling is strongly marked by floral and faunal changes in Australia. It seems probable that it was only in Late Miocene times that major ice-cap glaciation commenced in the

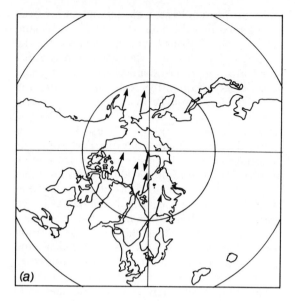

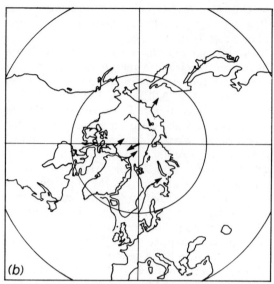

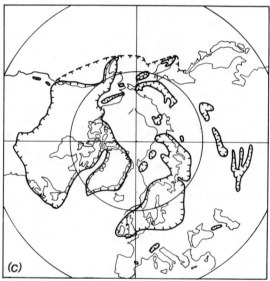

Fig. 1.4. The palaeogeographic changes in the Late Cenozoic for the northern hemisphere. For convenience, present-day coastlines are indicated, although these would be strongly modified during the waxing and waning of the ice sheets in both the northern and southern hemispheres. The 1000-m bathymetry lines for Greenland and Eurasia are indicated in order to illustrate the process of deep-water opening in the northern Atlantic. These reconstructions are for (*a*) Miocene times, (*b*) Pleistocene times and (*c*) for the Quaternary. The arrow from the north pole indicates the movement of the pole, relative to North America, from (*a*) Miocene to Pliocene and (*b*) Pliocene to the present. Similarly the relative movements of the continents, relative to the pole, are indicated on each map. The maximum distribution of ice sheets for the Quaternary is based on Flint (1971). The distribution of the ice sheets is indicated by the barbed line.

southern hemisphere and earlier Tertiary glaciations were of mountain glacial type and not ice sheets (Drewry, 1975). One major difference between the two hemispheres has been that ice seems to have disappeared off the land in the northern hemisphere during the interglacials (although an ice-free or frozen Arctic Ocean at such times is a subject of dispute) while glaciation on Antarctica has been continuous (Markov 1969; Mercer, 1972; Keany, Ledbetter, Watkins & Huang, 1976) with the ice front advancing and retreating so that there was an oscillation of the area of maximum glacial deposition in the surrounding oceans. Two types of Quaternary ice sheet are recognised (Andrews, 1975), with the present Greenland and Antarctic ice-caps being examples of stable conditions, while the extensive Laurentian and Scandinavian ice sheets were unstable and possibly formed and disappeared rapidly, probably in response to only small changes in environmental conditions (Flint, 1971; Hughes, 1975).

Unfortunately, there is little information on

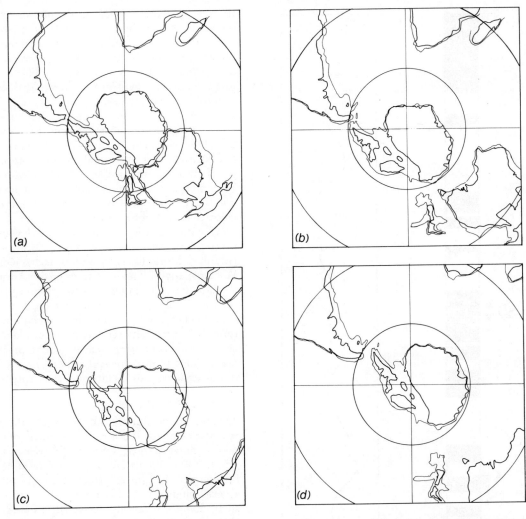

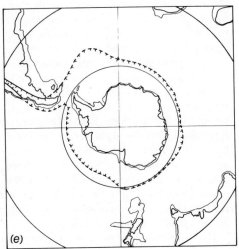

Fig. 1.5. The Cenozoic palaeogeographic evolution of the polar southern hemisphere. These reconstructions are for the (a) Eocene, (b) Oligocene, (c) Miocene, (d) Pliocene and (e) Quaternary. The coastline for western Antarctica is very approximately based on the −800-m surface of the land below the present ice-cap and therefore assumes this degree of isostatic readjustment. No attempt is made to adjust for tectonic activity along the Pacific margins of South America, within western Antarctica or New Zealand. The maximum extent of Quaternary glaciation (barbed line) is following Hughes (1975).

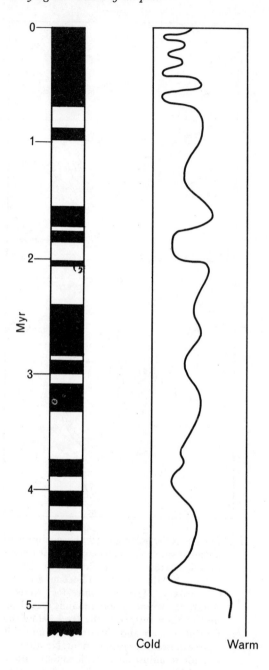

Fig. 1.6. Generalised temperature trends for the past
5 Myr. The age and polarity scale (Tarling &
Mitchell, 1976) has been combined with
curves of Theyer (1972) and Blank &
Margolis (1975). It is emphasised that these
are only generalised trends which cannot yet
be adequately quantified.

the actual characteristics and evolution of ice-
sheets in either hemisphere. Most described
evolutionary sequences are in fact based more on
theoretical models for ice-sheet growth and
decay than on actual observations. There seems
to be increasing evidence for a rapid decay of
unstable ice sheets, which possibly take some
20 000 yr to disappear completely, but considerable argument still exists about their rate of
initial development. Deep-sea sediments show
that considerable ice-rafted material appeared
well before oceanic cooling (Blank & Margolis,
1975) and that periods of maximum ice development are associated with strongly increased
velocity of Antarctic and Indian Ocean circulation, with oceanic cooling largely developing
from the increase in Antarctic bottom water
(Blank & Margolis, 1975; Fillon, 1975; Kennett
Watkins, 1976). The formation of the Antarctic
ice-cap therefore appears to be responsible for
oceanic cooling and circulation rather than
cooler oceanic waters initiating glaciation. The
time difference between the onset of glaciation
and oceanic cooling is unclear, but it probably
takes about 1000 yr for Antarctic bottom waters
to upwell in equatorial latitudes (see Appendix).

Quaternary Climatic Changes
The record of climatic change since the most
recent glacial advance forms the subject of many
books (Butzer, 1964; Embleton & King, 1968;
Péwé, 1969; Evans, 1971; Flint, 1971; Washburn, 1973; Wright & Moseley, 1975) and also
several chapters in this book. The setting for
such changes is largely derived from studies of
climatic indicators in cores of deep-sea sediments (Berger & Roth, 1975) spanning the past
5 Myr (Fig. 1.6); these only seem to reflect
changes with a periodicity of greater than
100 000 yr in sediments older than some 500 000
yr, although 20 000-yr periodicites are recorded
in more recent sediments, with as many as 20
possible cycles within the past 1.7 Myr (Damuth,
1975). Such cycles are probably astronomically
controlled and the data are reviewed in detail
by Evans (1971) and in Section 3 of this book,
so the nature and reliability of these cycles
will not be discussed further here. These deep-
sea sedimentary climatic records are based on
studies of lithology, palaeontology and oxygen
isotope studies of microfossils.

High-latitude, deep-sea sediments show clear

signs of glaciation by the appearance of ice-rafted sand grains (Kellog, 1975; Keany *et al.*, 1976), while at lower latitudes there is generally an increase in the coarseness of the detrital grains, probably reflecting the increased periglacial area exposed to aeolian transport. Changes in the clay mineralogy (Jacobs, 1974) and particularly of the percentage calcium carbonate content (Damuth, 1975) are also indicators of the changing temperature of the surface oceanic waters, although there is always the possibility of later dissolution of these components during diagenetic changes or in response to changes of the bottom waters. Similar considerations also apply to the palaeontological evidence, for, as in lithological studies, it is always difficult to eliminate the possibility of reworked material being involved in the analyses. The composition of the flora and fauna in the upper oceanic waters largely depends on its temperature, although such influences as salinity and predators may cause local differences (Ramsay, 1972). As these organisms die and accumulate at the bottom of the ocean, the fossil content of the deep-sea sediments will contain a filtered reflection of the surface composition because of partial or total solution of certain tests as they sink or lie on the bottom. The recognition of a particular species, such as *Globorotalia menardii*, within the sediments can therefore be a strong indication of warm water at the ocean surface at that time, while its absence from the microfossil content will suggest cold conditions (Corliss, 1975); similarly, with the identification of left-coiling and right-coiling attitudes of the skeletal tests (Ericson & Wollin, 1956). Possibly better quantitative temperature estimates can be obtained by studying the total fauna (Corliss, 1975) whereby small changes in the fossil community can be used to indicate relatively small temperature fluctuations which would not be so well reflected by examination for the relative concentration of only a few species.

The oxygen isotope ratio O^{18}/O^{16} in skeletal remains within sediments or within ice cores varies according to the temperature at which the skeleton or ice is formed, as long as there has been no subsequent chemical change (Emiliani, 1954; Bowen, 1966; Shackleton & Opdyke, 1973; Emiliani & Shackleton, 1974). Studies of the variation in this ratio in the Greenland and Antarctic ice-caps (Gow, 1968; Dansgaard, Johnson, Clausen & Langway, 1971; Dansgaard & Hammer, 1974) are particularly valuable in determining temperature fluctuations during the past 8 to 15 000 yr, but dating becomes increasingly difficult for earlier times and uncertainties about the effects of recrystallisation at greater depths increase and these also obscure the annual laminations used for dating. Dating in deep-sea cores tends to be even less precise but isotopic studies can be extended through tens of millions of years (Saito & Van Donk, 1974; Savin, Douglas & Stehli 1975). Detailed temperature fluctuations are best revealed during the past few thousand years, before recrystallisation of the skeletal remains causes increased uncertainty about the significance of the ratio. The problem of chemical change is present within individual specimens and it is also generally necessary to ensure that foraminifera tests, for example, are all of the same size as there is a tendency for the ratio to vary in different parts of the same organism. The oxygen isotope ratio is also highly sensitive to changes in the salinity of the water in which the organism was living. Consequently, variations in the isotopic ratio can only be used as a measurement of temperature trends and these are difficult to relate to absolute temperatures.

Perhaps the most important studies for determining the cause of climatic change during the past few millenia are those attempting to explain variations in the production of ^{14}C in the upper atmosphere. This can be closely and precisely monitored by combined carbon isotope studies of tree-ring samples which have been dated by dendrochronology (Damon, Long & Grey, 1970; Suess, 1970*a*). The observed differences between the observed and expected ^{14}C content may partially reflect changes in the effectiveness of exchange between the atmospheric and oceanic reservoirs, but more probably reflects either variations in the solar radiation 'constant' or of the effectiveness of the ionosphere as a barrier to solar radiation. At the moment it is difficult to evaluate these different sources although there seems to be a clear correlation with the strength of the geomagnetic field (Bucha, 1970) which suggests that fluctuations in the effectiveness of the ionosphere may be the first order control on this variation. It is, however, difficult to see how short-term fluctuation are accounted for by such changes and the total interaction is clearly more complex (Suess, 1970*b*). Further clarification may come if correlations can also be obtained between the magnetic properties of varved sediments (Nöel & Tarling, 1975) and their thickness, as changes in the varve thickness on a regional scale can also reflect temperature changes.

The major problem of most existing techniques is that most are merely temperature indicators, which is only a part, albeit an important part, of the total climatic scenario. Most of the evidence for rainfall must come from prolonged, tedious geomorphological, sedimentological and biological studies of terrestrial deposits (Newell, Herman, Gould-Steward & Tanaka, 1975; Williams, M. A. J., 1975). Similarly, it will always be important to distinguish between local climatic effects and those of regional or world-wide significance so that regional studies for the same time, such as those currently being made for the Caribbean area some 18 000 yr ago (Brunner & Cooley, 1976), will become increasingly essential to evaluate the total climatic picture. In the short term, however, the major difficulty is likely to be the development of circular arguments, in view of suggestions that climatic cycles should form the basis of dating for the Pleistocene in particular (Evans, 1971). This will make it extremely difficult to determine the preciseness of the synchroneity of climatic change in the different hemispheres (Williams, M. A. J., 1975) or even within the same continent. It is, however, symptomatic of the increasing evidence for the dominant climatic influence of the astronomical variations in the Earth's orbit and the tilt of its axis (Milankovich, 1938; Evans, 1971; Kukla, 1975) that it is climatic variations which are now forming the basis for geological dating.

The Causes of Ice Ages

No one factor is solely responsible for causing ice ages but it is generally accepted that certain factors must be operative in order for otherwise minor climatic influences to become sufficiently amplified to account for the marked climatic variability which characterises all ice ages. Apart from the difficulty in isolating different interactive causes, the evidence is always complicated by the strong climatic influences exerted by the ice sheets themselves, as these locally increase the Earth's albedo and create their own atmospheric-pressure zones, with resultant equatorial displacement of pre-existing climatic belts. However, such world-wide changes are often obscured by more local effects resulting from changes in the oceanic and atmospheric circulation systems, the gradual cooling of equatorial oceans, changes in geography and albedo in response to sea-level changes (about 8% more land than at present was exposed during the maximum Cenozoic glaciation and about 20% less land when there were no ice sheets) and so forth. Nonetheless, examination of the record of previous ice ages appears to indicate the significance of different effects, not only in terms of ice ages but also in terms of their general climatic influence.

Galactic factors

Although it is sometimes claimed that there has only been one ice age during the history of the Earth (Cornwall, 1970), there now seems no doubt about the existence of several ice ages, even though controversy exists about some of the evidence in Precambrian times. A possible explanation could therefore be that there was cooling as the solar system passed through 'stationary' galactic dust clouds, which would therefore reduce the amount of solar radiation reaching the Earth. At the moment, there seems to be no obvious periodicity in the occurrences of known ice ages (Fig. 1.1) as is indicated by comparing the claims of Williams, G. E., (1975*a*) who estimates a periodicity of some 155 Myr and Steiner & Grillmair (1973) who suggest a decreasing periodicity from 400 Myr in the Precambrian to 275 Myr now. Even more significant is the fact that the Gondwanan Ice Age does not seem to be associated with world-wide cooling, as the continents which were then in low latitudes (North America, Greenland, Europe and parts of Asia) show typical equatorial and tropical environments, the only northern Asiatic glaciation occurring at a time when the Gondwanan ice sheets had disappeared. And it is difficult to explain by this hypothesis the sequence of glacials and interglacials or the fact that the appearance or disappearance of at least some ice sheets seems to have been very rapid.

Williams, G. E., (1975*a*, *b*) has suggested that the Earth's axis of rotation could have been in the plane of the ecliptic during the Varangian Ice Age so that each year would then be characterised by six months daytime and six months night. Such conditions could make the Earth highly sensitive to other climatic variables, such as the eccentricity of the Earth's orbit, thus explaining the interjacent occurrence of sedimentary evidence for both high- and low-latitude environments. It seems impossible, however, to invoke this situation for all ice ages.

Volcanic dust

A sudden, prolonged increase in volcanic activity would result in an increase in temperature, associated with the carbon dioxide added to the atmosphere, and cooling, associated with the increased atmospheric albedo due to atmospheric dust, particularly if this dust provides nuclei for the condensation of clouds. There appears to be increasing evidence of the reality of both of these effects (Bray, 1974) and the net effect of volcanic activity appears to be mainly associated with cooling. The persistence of dust in the atmosphere generally seems to be up to about 13 yr (Lamb, 1972), so that very prolonged volcanic activity would be required, certainly over 1000 yr or more, in order to initiate and maintain an ice age, even if other palaeogeographic considerations were favourable. In general, andesitic volcanic activity is likely to be the most important, as the explosive, gaseous nature of such volcanoes gives rise to a much more rapid transfer of particles and vapour into the stratosphere, in contrast with that of the less explosive basaltic volcanoes. The Late Miocene was a period of considerable tectonic activity along the Alpine–Himalayan belt and the so-called Pacific Girdle, some of which activity was accompanied by andesitic volcanic activity. However, the rate of sea-floor spreading is probably the major control on the rate of andesitic eruption along subduction zones. The Miocene seems to be marked by a reduction in the rate of spreading as the Tethys Ocean became closed and sea-floor spreading in the Early Tertiary was generally faster than in the Late Tertiary, implying greater andesitic activity at a time which was not characterised by ice sheet formation. The Eocambrian Ice Age does not seem to be associated with significant volcanic activity, although this may not be a typical ice age. The Ordovician glaciation was approximately contemporaneous with considerable andesitic volcanic activity along the closing Appalachian–Caledonian Ocean (Iapetus Ocean), but this activity was not significantly greater in Caradocian times than for other times in the Lower Palaeozoic, although study of records of volcanic activity in other parts of the world are still necessary. The Permo-Carboniferous glaciations occurred during a period of relative igneous quiescene, with major basaltic flood basalts occurring much later in the Triassic–Jurassic, associated with the break-up of Gondwanaland.

Examination of the various ice ages, therefore, suggests that while increased volcanic activity can give rise to short-term cooling which could be periodic if volcanic activity was periodic, its effects are more likely to be both random and small. It certainly seems to be unlikely that increased volcanic activity is the actual triggering mechanism for the initial development of an ice age and is certainly inadequate to explain sequences of advances and retreats (Chappell, 1975).

Carbon dioxide in the atmosphere

An increase of carbon dioxide in the atmosphere produces a 'greenhouse effect' in which incoming solar radiation is largely unaffected, but reflected heat rays are retained within the atmosphere, resulting in an increase in the temperature of the atmosphere. This effect is currently the subject of considerable concern in view of the increase in atmospheric carbon dioxide resulting from Man's burning of fossil fuels. However, while not dismissing these effects as real and of major concern to Man's future (see Broecker, 1975), it is hard to see how variations in the carbon dioxide content of the atmosphere have been the cause of previous ice ages. The main source of carbon dioxide before Man must have been volcanic activity and this does not seem to be in antiphase with the formation of ice ages.

Nor does the withdrawal of carbon dioxide from the atmosphere by plants and skeletal organisms seem to be related to the cessation or appearance of ice ages. Land plants appeared in the Devonian, for example, and would have major, but local, climatic effects in reducing the rate of run-off of rainfall, reducing rates of erosion because of the binding effects of their roots, reducing the albedo of previously barren rock areas and so on. However, no world-wide climatic effect, such as the appearance or destruction of ice sheet formation, appears at this time or when photosynthesis commenced some 3000 Myr ago. Similarly, the major calcium carbonate depositions of Cretaceous times must have reduced the atmospheric carbon dioxide content, yet this does not appear to have had any significant climatic effect, unless the Tertiary cooling of the oceans can somehow be related to a somewhat delayed effect.

Mountain building at subduction and collision zones
Clearly, land above the snow line is a source for potential ice sheets if accumulation exceeds ablation. As the glaciers grow, they spread out as piedmont glaciers, locally modifying the climate and, as in the case of the Cordilleran ice sheet in the Cenozoic, eventually coalesce to form an extensive ice sheet with its own climatic regime. The regional cooling could then result in neighbouring plateau areas being sufficiently cooled that vast areas are brought below the equilibrium line and, if precipitation remains adequate, large, unstable ice sheets may form. Such models can be applied to several areas in the Cenozoic, for example, in Canada (Flint, 1971) and Antarctica (Drewry, 1975; Hughes, 1975). The Miocene was certainly a period of strong tectonic activity in many parts of the world and it is tempting to associate the appearance of the large Late Cenozoic ice sheets with this event (Flint, 1971). However the correlation does not seem to exist for earlier ice ages and is not always evident during the Late Cenozoic.

The Varagian Ice Age, in the North Atlantic, seems to have taken place close to sea-level over an extensive area which was not characterised by significant mountain belts. However, in view of uncertainties about the nature of this ice age, it may not be diagnostic. The Ordovician Saharan glaciation took place well after the Anti–Atlas Mountains had been formed and eroded, although during earlier periods (Cambrian) there were minor mountain glaciations which apparently did not expand into ice sheets. The Gondwanan glaciation appears to have been preceded by mountain glaciation in several areas, particularly in South America and southeastern Australia, but these Carboniferous mountain glaciers did not themselves generate the ice sheets of the earliest Permian and it

appears that the mountains in southeastern Australia were completely eroded before the sudden appearance of the ice sheets. In other areas, such as parts of India, western and northern Australia, there seem to be no mountain belts as source areas for piedmont glaciers, although clearly the formation of ice sheets elsewhere could have resulted in sufficient cooling of other areas without being in direct contact. In the Cenozoic, however, ice sheets on the Tibetan plateau did not expand outside the area, and the growth of Alpine glaciers did not result in a southern European ice sheet. In the Antarctic, the orogenic events in western Antarctica are still unclear, although major activity certainly took place in Cretaceous and Early Tertiary times, with no apparent formation of extensive ice sheets.

There is clearly a sound logical argument for the development of ice sheets in response to increased elevation, but it seems that there is little direct connection with mountain building episodes for many ice ages.

Isostatic uplift
The growth and decay of ice sheets has an obvious isostatic response which is delayed by the slow relaxation time of mantle rocks (Andrews, 1975) and this can be of major importance in allowing plateau areas, as they become free of ice, to rise to levels where the increase in elevation takes them above the snow line. The elevation will continue to increase after the initial accumulation of ice has commenced until the mantle rocks readjust to the newly imposed load, after which isostatic sinking will commence. But while such considerations are critical to any concepts of the evolution of ice sheets, after their initiation, such considerations are not relevant to the first formation of a sheet. Flint (1971), Drewry (1975) and others have suggested that an initial isostatic uplift of continental edges in high latitudes, such as the Transantarctic Mountains or Scandinavia during the Late Cenozoic, may have resulted in response to continental separation as the North Atlantic and the Southern Ocean formed. However, the uplift of the Transantarctic Mountains seems to have been mainly in pre-Oligocene times (Drewry, 1975) and the opening of the Norwegian Basin commenced in Eocene times (Fig. 1.5) with major isostatic adjustments

taking place within a few million years of separation. Little is known about the reasons for vertical movements within continents and it is conceivable that such events could be of major significance in the Late Cenozoic; however, most previous ice sheets were not associated with mountain formation or continental separation, and were usually very extensive and close to sea-level during the earlier phases of ice sheet formation, for example in the Gondwanan glaciation. In fact, the Gondwanan continents seems to have been remarkably stable until the initiation of continental separation well over 100 Myr after their glaciation.

Palaeogeographic changes in response to continental movements

The movements of the continents, with the associated formation of new oceans and combination of pre-existing continents into new blocks, clearly must have major effects on both oceanic and atmospheric circulation. At the moment, only highly simplified estimates can be made of these effects on previous climatic patterns (Frakes & Crowell, 1970b); Frakes & Kemp, 1972, 1973; Robinson, 1973) but it is difficult to underestimate their effects on regional climates. Such changes could thus create a sensitivity to minor influences in medium to high latitudes, precipitating the onset of glaciation in reference to minor factors such as astronomical changes. Ewing and Donn (Ewing & Donn, 1956; Ewing, 1960, 1971; Donn & Ewing, 1968) have suggested that, in the Late Cenozoic, oceanic exchange between the Arctic and Atlantic may have been the critical factor in explaining glacials and interglacials while the rotational pole lay within the Arctic Basin, with the implication that the Bering Straits and northern Atlantic connections to the Arctic are critical areas. Their proposed mechanisms do offer interesting explanations for the different phases of the Late Cenozoic Ice Age, but the occurrence of a land-locked polar ocean seems to be unique in the available geological record. Previous ice ages are often associated with shallow marine incursions in polar regions (Fig. 1.2–1.5) and quite deep waters must have existed alongside the Transantarctic Mountains before their glaciations (Fig. 1.5). However, the rotational pole has lain in the Arctic Basin from at least Cretaceous

times; the uncertainty of its relationship in earlier times reflecting different hypothesis for the origin of the Basin itself.* In the Late Cretaceous, with the rotational pole in the Arctic Basin near Alaska, connections to the Caribbean existed along the Mid-Continental Seaway through central North America, and the Turgai Sea linked it with the closing Tethys Ocean. During the Tertiary, these seaways closed and the Bering Straits offered a discontinuous, shallow connection with the Pacific (Hopkins, 1967). On this basis, ice sheets should have been initiated at different times throughout the Late Cretaceous and Tertiary, rather than in the Late Miocene. It is possible, of course, that the critical factor could be the opening of a deep oceanic link between the Arctic and Atlantic in post-Oligocene times, but although this deep-water connection did not form to any significant amount until Miocene times (Fig. 1.4), the Icelandic Ridge still prevented the free interchange of deep oceanic waters and restricted the interchange to relatively shallow current exchanges except where narrow deep channels were cut across the ridge, and that exchange was already in progress by Early Miocene times (Anon, 1975), well before the initiation of ice sheet glaciation. The combined evidence from this and earlier glaciations therefore suggest that the occurrence of a partially land-locked polar ocean is not prime factor in initiating an ice age.

Continental movements in the southern hemisphere, during the Tertiary, must have had a major effect on the oceanic and atmospheric circulation (Frakes & Kemp, 1972, 1973), with the final creation of continuous circum-Antarctic circulation by the Upper Oligocene (Anon, 1974; Foster, 1973b; Jenkins, 1974) following the separation of Australia from Antarctica and the opening of the Drake Passage between the Antarctic Peninsula and Patagonia. However there is no spectacular lateral Miocene tectonic event which could be associated with even more drastic climatic changes, although on a large

*It should, perhaps, be stressed here that in referring to movement of the pole, or polar wandering, we are really talking about drift of continental features relative to the pole of rotation. It is from the viewpoint of a being dwelling on the (moving) surface of the Earth that the poles appear to wander, and therefore natural that historically this relative movement of land and pole should have been called polar wandering.

scale the final collision of Africa with Europe and India with Asia marked the final disappearance of a persistent, longitudinal, low latitude ocean, the Tethys, and, at a similar time but on a smaller scale, the closure of the Panama isthmus prevented further longitudinal exchange between the Atlantic and Pacific (Luyendyk, Forsyth & Phillips, 1972). Such events created radically different geographic conditions by not merely removing a low-latitude ocean, but by replacing it with a huge land mass comprising Europe, Asia, India and Africa. It may be, however, that relatively small-scale changes, such as that in the Caribbean area, may have as much climatic effect as the major movement of the continents. Relatively small changes in eastern Asia during the past 7000 yr seem to have had major effects on the Kuroshio current (Taira, 1975).

Such changes in the palaeogeography, on both a large and small scale, could easily create the conditions in high and intermediate latitudes where comparatively small factors, such as those due to astronomical factors, could have major climatic effects. Not only are the consequences of such changes almost impossible to evaluate, but there are also major problems in determining when the Tethys was sufficiently closed to restrict longitudinal exchange of oceanic waters. It seems quite probable that much of the exchange was already restricted well before the onset of ice sheet glaciation and that relatively minor changes in the geography of critical areas such as the Caribbean may have been more influential than the larger scale changes.

Examinations of earlier palaeogeographic conditions are severely hampered by lack of adequate information. In the Gondwanan glaciation, however, the Tethys was in existence and there seems to have been fairly free oceanic circulation in a longitudinal direction at low latitudes. This would suggest that changes in the degree of longitudinal oceanic exchange in low latitudes may not be significant in creating conditions for the initiation of ice ages. It may, however, be relevant that in all ice ages marine incursions occur during glaciation, but seem to be somewhat less common shortly afterwards, in spite of the rise in sea-level that must have occurred as the ice sheets melted. These marine incursions would certainly provide the moisture that is necessary to initiate and maintain the

growth of ice sheets and such conditions may be necessary to allow the development of full ice sheets when other factors are fulfilled.

Location of the rotational poles

With the arguable exception of the Late Precambrian Ice Age(s), the most notable feature of all ice ages is that they only occur at high latitudes, mostly within 30° of the pole – although the Late Cenozoic glaciation shows that ice sheets can extend up to 50° away from the rotational pole and, conversely, some regions near the poles, such as Alaska north of the Brooks Ranges and much of Siberia, may not be glaciated at all (Fig. 1.4). It seems necessary that the pole lies sufficiently inland before ice sheets can be developed. Throughout the Mesozoic, the South Pole was close to Antarctica, but it was only after it became located inland* that ice sheet formation commenced. The drift of the Gondwanan continent to the situation where the pole, by Middle Permian times, was merely on the edge of the land mass seems to coincide with the rapid, simultaneous disappearance of ice sheets throughout the Gondwanan continent, leaving only isolated mountain glaciers until the continent drifted sufficiently far back under the pole in Miocene times. It seems, therefore, that it is essential for the pole to lie well within a continental block before ice sheets can form, but this is insufficient, on its own, to result in ice sheet formation. During the Silurian, Devonian and Lower Carboniferous, only isolated patches of mountain glaciation occurred in western Gondwanaland (South America and Africa) as this continent drifted so that the pole apparently moved from the Sahara to southern Africa (Fig. 1.3). It seems probable that the frigid zone was too far from adequate water supplies at this period for the formation of ice sheets and it was only when the movements had carried the pole near to shallow marine waters off southwestern Africa that an adequate moisture supply became available for ice sheet formation on the low-lying polar plateau areas.

The situation seems to be that when the pole only lies on the edge of a continental block oceanic circulation is adequate to maintain

*That is, after the continent drifted over the pole.

equable conditions on that continent in spite of the angle of the Sun's rays and the long, polar nights. There seems to be some optimum distance from the ocean when the polar coldness becomes adequate to initiate a permanent snow cover and thereby its own climatic regime. But such conditions also entail sufficient sources of moisture that adequate precipitation can still be carried into this high pressure zone but without causing a permanent reduction of the pressure system. In the case of Antarctica during the Late Cenozoic, such moisture sources were clearly available from the variable-depth seas in the area now occupied by western Antarctic ice (Fig. 1.5); these could, in fact, have acted as channels controlling the direction of water-laden winds onto the Antarctic plateau (Hughes, 1975).

The Late Cenozoic situation in the Arctic area seems to be unique in geological history as this is the only ice sheet glaciation episode that has so far been recorded in which a land-locked ocean has formed in polar regions. Presumably the highly restricted circulation in the Arctic could then result in similar conditions arising, once thoroughly cooled, as would happen in very low-relief plateau situations in terrestrial polar areas. However, this supposition is not very satisfactory and it is important to note that models based on this Arctic Ocean situation alone cannot be applied to glaciations at other times or other areas. It seems that the Antarctic situation in Miocene times was the controlling factor that strongly influenced oceanic circulation systems. One result of this circulation pattern was that the medium latitude, northern lands were much more sensitive to astronomically controlled climatic variations than those of the southern hemisphere. This difference in sensitivity resulted in the total disappearance of ice in the northern hemisphere while glaciation persisted during interglacials in Antarctica.

Geomagnetic effects
Reversals of the geomagnetic field have occurred on average at least three times per million years for the last 70 Myr. The change of polarity of the geomagnetic field seems to take some 2000 yr, but is preceded and followed by periods of some 3000 yr of reduced intensity of the field. Studies of such reversals in deep-sea sediment cores indicate that the reversals seem to coincide

with periods when index fossils tend to appear or become extinct and it was suggested that this may be due to changes in the protectiveness of the Van Allen radiation belts against incoming, harmful radiation during the period of a weak geomagnetic field (Black, 1967). This mechanism is thought to be inadequate in its direct effect but may operate by causing climatic change (Hays & Opdyke, 1967). There seem to be some degree of correlation between summer temperatures, in particular, and the change in the local strength of the geomagnetic field (Wollin *et al.*, 1973) for the past 50 yr or so, although the data for such correlations seem to have been somewhat selective. There also seems to be some correlation between 'anomalous' geomagnetic field behaviour and climatic change, such as the Laschamp event (Nöel & Tarling, 1975) some 12 400 yr ago and changes in lake levels, and other climatic indicators, in Britain (Clapperton, Gunson & Sugden, 1975), Africa, India, Australia, and elsewhere (Williams, 1975) at this time.

If there is a strong climatic effect due to polarity changes, then there should be an association between ice ages and the rate of polarity change. During the Cenozoic, polarity changes were much more frequent than during the preceding Cretaceous and Jurassic, but, conversely, the maximum Gondwanan glaciation was during one of the longest periods of constant polarity so far identified. More fundamentally, there seems to be no reason why regional weather trends should be linked to regional geomagnetic field trends as most proposed mechanisms involve upper atmosphere changes which should result in world-wide changes. Nonetheless, the observed correlations seem to exist, although requiring more study, and suggest geomagnetic effects may have some as yet unqualified influence on climate, accounting for some of the 'noise' on the astronomically controlled cycles. At the moment, however, it appears unlikely to be a significant cause for major climatic changes.

Conclusions
Studies of previous ice age situations strongly suggest that two main factors are involved, the occurrence of a continental block in polar regions so that the pole lies away from oceanic influences, with shallow, epicontinental seas

sufficiently near to provide a moisture supply. Under such conditions, the middle latitudes become particularly sensitive to relatively minor changes in solar radiation, such as those due to astronomical variations in the Earth's orbit and tilt of its axis. The duration of these sensitive conditions seems to be of the order of 10 Myr with only mountain glaciation being present in high latitudes during intervening periods. The duration of any one ice age, however, probably only reflects the rate of motion of the polar continent relative to the rotational pole. The evidence for the rate of formation of stable or unstable ice sheets is particularly unclear, but there is increasing evidence for their rapid disappearance, with the possible implication that their formation may have been rapid. Many of the climatic changes away from high to medium latitudes seem to be largely controlled by the effects of increased ocean circulation and cooling as a result of iceberg calving from already developed ice sheets. Within this general pattern, however, local changes will result from the changing land–sea distribution, but local tectonic events may be more crucial to dramatic changes in oceanic circulation than the gradual, persistent motion of the continents, although it is only when such movements allow oceanic waters sufficiently close to the pole that the Earth's sensitivity to climatic changes ceases and more clement climatic conditions, typical of most geological time, re-establish themselves.

References

Abou-Deeb, J. M. (1976). Palaeomagnetic studies in Algeria and Tunisia Ph.D. thesis, University, Newcastle-upon-Tyne, 464 pp.

Allen, P. (1975). Ordovician glacials of the central Sahara. In *Ice ages: ancient and modern*, ed. E. A. Wright & F. Moseley, pp. 275–85. Seel House Press, Liverpool.

Andrews, J. T. (1975). *Glacial systems*. Duxbury Press, Massachusetts. 191 pp.

Anon (1973a). Leg 28 deep-sea drilling in the Southern Ocean. *Geotimes*, 18(6), 19–24.

Anon (1973b). Deep-sea drilling in the roaring 40s. *Geotimes*, **18**(7), 14–17.

Anon (1975). Leg 38. *Geotimes*, **20**(2), 24–6.

Barrett, P. J. & Kohn, B. P. (1975). Changing sediment transport directions from Devonian to Triassic in the Beacon Super-Group of south Victoria Land, Antarctica. In *Gondwanan geology*,

ed. K.S.W. Campbell, pp. 15–35. ANU Press, Canberra.

Barrett, P. J. & Kyle, R. A. (1975). The Early Permian glacial beds of south Victoria Land and the Darwin Mountains, Antarctica. In *Gondwanan geology*, ed. K. S. W. Campbell, pp. 333–46. ANU Press, Canberra.

Barry, R. G. (1975. Climatic models in palaeoclimatic reconstruction. *Palaeogeogr. Palaeoclimatol. Palaeoecol.*, **17**, 123–37.

Berger, W. H. & Roth, P. H. (1975). Ocean micropaleontology: progress and prospect. *Rev. Geophys. Space Phys.*, **13**(3), 561–85.

Beuf, S., Biju-Duval, B., Mauvier, A. & Legrand, P. (1968a). Nouvelles observations sur le 'Cambro-Ordovician' du Bled el Mass (Sahara central). *Service Geol. Algerie, Bull.* **38**, 39–52.

Beuf, S., Beenacef, A. K., Biju-Duval, B., du Charpal, O., Gariel, O. & Rognon, P. (1968b). Les grands ensembles sedimentaires du Paleozoique inferieur du Sahara. *C. R. Somm Seances Soc. Geol. Fr.*, **8**, 260–3.

Black, D. I. (1967). Cosmic ray effects and faunal extinctions at geomagnetic reversals. *Earth Planet. Sci. Lett.*, **3**, 225–36.

Blandford, W. T., Blandford, H. F. & Theobald, W. (1859). On the geological structure and relations of the Talcheer Coal Field, in the district of Cuttack. *Mem. geol. Surv. India*, **1**, 33–89.

Blank, R. G. & Margolis, S. V. (1975). Pliocene climatic and glacial history of Antarctica as revealed by southeast Indian Ocean deep-sea cores. *Bull. geol. Soc. Am.*, **86**, 1058–66.

Bowen, R. (1966). Oxygen istotopes as climatic indicators. *Earth Sci. Rev.*, **2**, 199–224.

Bray, J. R. (1974). Volcanism and glaciation during the past 40 millenia. *Nature*, **252**, 679–80.

Broecker, W. S. (1975). Climatic change: are we on the brink of a pronounced global warning? *Science*, **189**, 460–3.

Brunner, C. A. & Cooley, J. F. (1976). Circulation in the Gulf of Mexico during the last glacial maximum 18 000 yr ago. *Bull. geol. Soc. Am.*, **87**, 681–6.

Bucha, V. (1970). Evidence for changes in the Earth's magnetic field intensity. *Phil. Trans. R. Soc. A*, **269**, 47–55.

Butzer, K. W. (1964). *Environment and archeology*. Methuen, London. 524 pp.

Carter, R. M. (1975). A discussion and classification of subaqueous mass-transport with particular application to grain-flor, slurry-flow, and fluxoturbidites. *Earth Sci. Rev.*, **11**(2), 145–77.

Casshyap, S. N. & Qidwai, H. A. (1974). Glacial sedimentation of Late Palaeozoic Talchir Diamictite, Pench Valley Coalfield, central India. *Bull. geol. Soc. Am.*, **85**, 749–60.

Chaloner, W. G. & Meyen, S. V. (1973). Carboniferous and Permian floras of the northern continents. In *Atlas of palaeobiogeography*, ed. A. Hallam, pp. 169–84. Elsevier, Amsterdam.

Chappell, J. (1975). On possible relationships between Upper Quaternary glaciations, geomagnetism, and volcanism. *Earth Planet. Sci. Lett.*, **26**, 370–6.

Churkin, M., Jr. (1973). Geologic concepts of Arctic Ocean Basin. *Bull. Am. Ass. Petrol. Geol.*, **19**, 485–99.

Clapperton, C. M., Gunson, A. R. & Sugden, D. E. (1975). Loch Lomond readvance in the eastern Cairngorns. *Nature*, **253**, 710–12.

Colbert, E. H. (1973). Continental drift and the distribution of fossil reptiles. In *Implications of continental drift to the Earth sciences*, ed. D. H. Tarling & S. K. Runcorn, vol. 1, pp. 395–412. Academic Press, London.

Corliss, B. H. (1975). Late Pleistocene palaeoclimatology: planktonic foraminiferal analyses of sediment cores from the central North Atlantic. *Palaeogeogr. Palaeoclimatol. Palaeoecol.*, **18**(1), 45–61.

Cornwall, I. (1970). *Ice ages*. Humanities Press, New York. 180 pp.

Cox, C. B. (1973). Systematics and plate tectonics in the spread of marsupials. In *Organisms and continents through time*, ed. N. F. Hughes, *Spec. Pap. Palaeontol.*, 12, pp. 113–19.

Cox, C. B. (1974). Vertebrate palaeodistributional patterns and continental drift. *J. Biogeogr.*, **1**, 75–94.

Crowell, J. C. & Frakes, L. A. (1970). Phanerozoic glaciation and the causes of ice ages. *Am. J. Sci.*, **268**, 193–224.

Crowell, J. C. & Frakes, L. A. (1971). Late Paleozoic glaciation: Part IV, Australia. *Bull. geol. Soc. Am.*, **82**, 2515–40.

Crowell, J. C. & Frakes, L. A. (1972). Late Paleozoic glaciation: Part V, Karroo Basin, South Africa. *Bull. geol. Soc. Am.*, **83**, 2887–912.

Crowell, J. C. & Frakes, L. A. (1975). The Late Palaeozoic glaciation. In *Gondwanan geology*, ed. K. S. W. Campbell, pp. 313–31. ANU Press, Canberra.

Damon, P. E., Long, A. & Grey, D. C. (1970). Arizona radiocarbon dates for dendrochronologically dated samples. In *Radiocarbon variations and absolute chronology*, ed. I. U. Olsson, pp. 615–18. Wiley, New York.

Damuth, J. E. (1975). Quaternary climate change as revealed by calcium carbonate fluctuations in western equatorial Atlantic sediments. *Deep Sea Res.*, **22**, 725–43.

Dansgaard, W., Johnsen, S. J., Clausen, H. B. & Langway, C. C., Jr. (1971). Climatic record revealed by the Camp Century ice core. In *The Late Cenozoic glacial ages*, ed. K. K. Turekian, pp. 37–56. Yale University Press, New Haven.

Dansgaard, W. & Hammer, C. U. (1974). Stable isotopes and dust in deep ice cores. *Abstr. with Programs geol. Soc. Am.*, **6**(7), 1014.

Denton, G. H., Armstrong, R. L. & Stuiver, M. (1971). The Late Cenozoic glacial history of Antarctica. In *The Late Cenozoic glacial ages*, ed. K. K. Turekian, pp. 267–306. Yale University Press, New Haven.

Donn, W. L. & M. Ewing (1968). The theory of an ice-free Arctic Ocean. *Met. Monogr.*, **8**, 100–5.

Dott, J. H., Jr (1961). Squantun 'tillite', Massachusetts – evidence of glaciation of subaqueous mass movement? *Bull. geol. Soc. Am.*, **72**, 1289–1304.

Drewry, D. J. (1975). Initiation and growth of the East Antarctic ice sheet. *J. geol. Soc. Lond.*, **131**(3), 255–73.

Einarsson, T., Hopkins, D. M. & Doell, R. R. (1967). The stratigraphy of Tjornes, northern Iceland, and the history of the Bering land bridge. In *The Bering land Bridge*, ed. D. M. Hopkins, pp. 312–25. Stanford University Press, California.

Elliot, D. H. (1975). Gondwana basins of Antarctica. In *Gondwana geology*, ed. K. S. W. Campbell, pp. 493–536. ANU Press, Canberra.

Emiliani, C. (1954). Temperatures of Pacific bottom waters during the Tertiary. *Science*, **119**, 853–5.

Emiliani, C. & Shackleton, N. J. (1974). The Brunhes epoch: isotopic paleotemperatures and geochronology. *Science*, **183**, 511–14.

Embleton, C. & King, C. A. M. (1968). *Glacial and periglacial geomorphology*. Arnold, London. 608 pp.

Ericson, D. B. & Wollin, G. (1956). Micropaleontological and isotopic determinations of Pleistocene climate. *Micropaleontology*, **2**, 257–70.

Evans, P. (1971). Towards a Pleistocene time-scale. In *The Phanerozoic time-scale – a supplement*, *Spec. Publ. geol. Soc. Lond.*, 5, pp. 123–356.

Ewing, M. (1960). The ice ages – theory. *J. Alberta Soc. Petrol. Geol.*, **8**, 191–200.

Ewing, M. (1971). The late Cenozoic history of the Atlantic Basin and its bearing on the cause of ice ages. In *The Late Cenozoic glacial ages*, ed. K. K. Turekian, pp. 565–73. Yale University Press, New Haven.

Ewing, N. & Donn, W. L. (1956). A theory of ice ages, 1. *Science*, **123**, 1061–6.

Fairbridge, R. W. (1974). Glacial Grooves and Peri glacial features in the Saharan Ordovocian. In *Glacial geomorphology*, ed. D. R. Coats, pp. 317–27. New York State University Publication in Geomorphology.

Fillon, R. H. (1975). Late Cenozoic paleo-oceanography of the Ross Sea, Antarctica. *Bull. geol. Soc. Am.*, **86**, 839–45.

Flint, R. F. (1961). Geological evidence of cold climate. In *Descriptive palaeoclimatology*, ed. A. E. M. Nairn, pp. 140–55. Interscience, New York.

Flint, R. F. (1971). *Glacial geology and Quaternary geology*, Wiley, New York. 892 pp.

Flint, R. F. (1975). Features other than diamicts as evidence of ancient glaciations. In *Ice ages: ancient and modern*, ed. A. E. Wright & F. Moseley, pp. 121–36. Seel House Press, Liverpool.

Foster, R. J. (1974). Eocene echinoids and the Drake Passage. *Nature*, **249**, 751.

Frakes, L. A. & Crowell J. C. (1969). Late Paleozoic glaciation: I, South America. *Bull. geol. Soc. Am.*, **80**, 1007–42.

Frakes, L. A. & Crowell, J. C. (1970a). Late Paleozo glaciation: II, Africa exclusive of the Karroo Basin. *Bull. geol. Soc. Am.*, **81**, 2261–86.

Frakes, L. A. & Crowell, J. C. (1970b). Glaciation and associated Circulation effects resulting from Late Paleozoic drift of Gondwanaland. *Proc. Papers Second Gondwanan Symp.*, pp. 99–109. CSIRO, South Africa.

Frakes, L. A., Matthews, J. L. & Crowell, J. C. (1971) Late Paleozoic glaciation: Part III, Antarctica. *Bull. geol. Soc. Am.*, **82**, 1581–604.

Frakes, L. A. & Kemp, E. M. (1972). Influence of continental positions on Early Tertiary climates. *Nature*, **240**, 97–100.

Frakes, L. A. & Kemp, E. M. (1973). Palaeogene continental positions and evolution of climate. In *Implications of continental drift to the Earth sciences*, ed. D. H. Tarling & S. K. Runcorn, vol. 1, pp. 539–58. Academic Press, London.

Frakes, L. A., Kemp, E. M. & Crowell, J. C. (1975). Late Paleozoic glaciation: Part VI, Asia. *Bull. geol. Soc. Am.*, **86**, 454–64.

Gariel, O., de Charpal & Beenacef, A. (1968). Sur la sedimentation des gres Cambro-Ordovician (Unite II) dans l'Ahnat et le Mouydir (Sahara central). *Service Geol. Algerie, Bull.* **38**, 7–37.

Gates, W. L. & Imbrie, J. (1975). Climatic Change. *Rev. Geophys. Space Phys.*, **13**, 726–31.

Geikie, J. (1874). *The great ice age and its relation to the antiquity of Man*. W. Isbister, London. 575 pp.

Ghosh, P. K. & Mitra, N. D. (1970). Sedimentary framework of Glacial and periglacial deposits of the Talchir formation of India. *Proc. Papers Second Gondwanan Symp.*, pp. 213–23. CSIRO, South Africa.

Gignoux, M. (1913). Les formations marines Pliocenes et Quarternaires de l'Italie du sud et de la Sicile. *Annls Univ. Lyons*, **36**, 1–693.

Girdler, R. W. (1964). The palaeomagnetic latitudes of possible ancient glaciations. In *Problems in palaeoclimatology*, pp. 155–78. Academic Press.

Goldthwait, R. P. (Ed.) (1975). *Glacial deposits*. Dowden Hutchinson & Ross, Pennsylvannia. 464 pp.

Gow, A. J. (1968). Preliminary analysis of ice cores from Byrd Station. *Antarct. J. US*, **3**(4), 113–14.

Hadley, D. G. & Schmidt, D. L. (1975). Non-glacial origin for conglomerate beds in the Wajid sandstone of Saudi Arabia. In *Gondwana geology*, ed. K. S. W. Campbell, pp. 357–71. ANU Press, Canberra.

Harland, W. B. (1972). The Ordovician Ice Age. *Geol. Mag.*, **109**, 451–6.

Harland, W. B. & Herod, K. N. (1975). Glaciations through time. In *Ice ages: ancient and modern*, ed. A. E. Wright & F. Moseley, pp. 189–216. Seel House Press, Liverpoool.

Hays, J. D. & Opdyke, N. D. (1967). Antarctic radiolaria, magnetic reversals, and climatic change. *Science*, **158**, 1001–11.

Hopkins, D. M. (1967). The Cenozoic history of Beringia – a synthesis. In *The Bering land bridge*, ed. D. M. Hopkins, pp. 451–84. Stanford University Press, California.

Hughes, T. (1975). The West Antarctic ice sheet: instability, disintegration, and initiation of ice ages. *Rev. Geophys. Space Phys.*, **13**(4) 502–26.

Jacobs, M. B. (1974). Clay mineral changes in Antarctic deep-sea sediments and Cenozoic climatic events. *J. sedim. Petrol.*, **44**, 1079–86.

Jenkins, D. G. (1974). Initiation of the proto circum Antarctic current. *Nature*, **252**, 371–3.

Keany, J., Ledbetter, M., Watkins, N. & Huang, T. C. (1976). Diachronous deposition of ice-rafted debris in sub-antarctic deep-sea sediments. *Bull. geol. Soc. Am.*, **87**, 873–82.

Keast, A. (1973) Contemporary biotas and the separation sequence of the southern continents. In *Implications of continental drift to the Earth Sciences* ed. D. H. Tarling & S. K. Runcorn, vol. 1, pp. 309–43. Academic Press, London.

Kellog, T. B. (1975). Late Quaternary climatic changes in the Norwegian and Greenland Seas. In *Climate of the Arctic*, ed. G. Weller & S. A. Bowling, pp. 3–36. Geophysical Institute, University of Alaska, Fairbanks.

Kemp, E. M. (1975). The palynology of Late Palaeozoic glacial deposits of Gondwanaland. In *Gondwana geology*, ed. K. W. S. Campbell, pp. 397–413. ANU Press, Canberra.

Kennett, J. P. & Watkins, N. D. (1976). Regional deep-sea dynamic processes recorded by late Cenozoic sediments of the southeastern Indian Ocean. *Bull. geol. Soc. Am.*, **87**, 321–39.

Kukla, G. J. (1975). Missing link between Milankovitch and climate. *Nature*, **253**, 600–3.

Lamb, H. H. (1972). *Climate: present, past and future*, vol. 1: *Fundamentals and climate now*. Methuen, London. 613 pp. (Vol. 2 published 1977.)

Lamb, H. H. (1975). Changes of climate: the perspective of time scales and a particular examination of recent changes. In *Ice ages: ancient and modern*, ed. A. E. Wright & F. Moseley, pp. 169–88. Seel House Press, Liverpool.

Long, W. E. (1962). Sedimentary rocks of the Buckeye Range, Horlick Mountains, Antarctica. *Science*, **136,** 319–21.

Luyendyk, B. P., Forsyth, D. & Phillips, J. D. (1972). Experimental approach to the paleocirculation of the oceanic surface waters. *Bull. geol. Soc. Am.*, **83,** 2649–64.

Margolis, S. V. & Kennett, J. P. (1971). Cenozoic paleoglacial history of Antarctica recorded in sub-antarctic deep-sea cores. *Am. J. Sci.*, **271,** 1–36.

Markov, K. K. (1969). The Pleistocene history of Antarctica. In *The periglacial environment past and present*, ed. T. L. Peive, pp. 263–9. McGill University Press, Montreal.

Matthews, P. E. (1970). Paleorelief and the Dwyka glaciation in the eastern region of South Africa. *Proc. Papers Second Gondwana Symp.*, 491–9. CSIRO, South Africa.

McClung, G. (1975). Late Palaeozoic glacial faunas of Australia: distribution and age. In *Gondwana geology*, ed. K. W. S. Campbell. pp. 381–90. ANU Press, Canberra.

McLachlan, I. R. & Anderson, A. M. (1975). The age and stratigraphic relationship of the glacial sediments in southern Africa. In *Gondwana geology*, ed. K. W. S. Campbell, pp. 415–22. ANU Press, Canberra.

Mercer, J. H. (1972). Cainozoic temperature trends in the southern hemisphere: Antarctic and Andean glacial evidence. Second SCAR Conference on the Quaternary of the Antarctic, Canberra, 1972.

Mercer, J. H., Fleck, R. J., Mankinen, E. A. & Sander, W. (1975a). Southern Patagonia: glacial events between 4 m.y. and 1 m.y. ago. In *Quaternary studies*, ed. R. P. Suggate & M. M. Cresswell, pp. 223–50. Royal Society of New Zealand, Wellington.

Meyen, S. V. (1970). On the origin and relationship of the main Carboniferous and Permian floras and their bearing on general paleogeography of this period of time. *Proc. Papers Second Gondwanan Symp.*, 551–5. CSIRO, South Africa.

Mikhazlov, Yu. A., Ustritshii, V. I., Chernyak, G. Ye & Yavshits, G. P. (1970). Upper Permian glaciomarine sediments of the northeastern USSR. *Akad. Nank. SSSR. Doklady,* **190,** 1184–7.

Milankovich, M. (1938). Neue Ergebnisse der astronomischen Theorie der Klinaschwankungen. *Bull. Acad. Sci. math. nat., Belgr., A,* **4,** 1–41.

Milner, A. R. & Panchen, A. L. (1973). Geographical variation in the tetrapod faunas of the Upper Carboniferous and Lower Permian. In *Implications*

of continental drift to the Earth sciences*, ed. D. H. Tarling & S. K. Runcorn, vol. 1, pp. 353–68. Academic Press, London.

Morris, N. (1975). The Rhacopteris flora of New South Wales. In *Gondwana geology*, ed. K. S. W. Campbell, pp. 99–108. ANU Press, Canberra.

Newell, N. D. (1957). Supposed Permian tillites in northern Mexico are submarine slide deposits. *Bull. geol. Soc. Am.*, **68,** 1569–76.

Newell, R. E., Herman, G. F., Gould-Stewart, S. & Tanaka, M. (1975). Decreased global rainfall during the past ice age. *Nature*, **253,** 33–4.

Nöel, M. & Tarling, D. H. (1975). The Laschamp geomagnetic event. *Nature*, **253,** 705–7.

Péwé, T. L. (Ed.) (1969). *The periglacial environment.* McGill-Queen's University Press, Montreal. 487 pp.

Plumstead, E. P. (1973). The enigmatic Glossopteris flora and uniformitarianism. In *Implications of continental drift to the Earth sciences*, ed. D. H. Tarling & S. K. Runcorn, vol. 1, pp. 413–24. Academic Press, London.

Ramsay, A. T. S. (1972). Aspects of the distribution of fossil species of calcareous nannoplankton in North Atlantic and Caribbean sediments. *Nature*, **236,** 67–70.

Rehmer, J. A. & Hepburn, J. C. (1974). Quartz sand surface textural evidence for a glacial origin of the Squantum 'Tillite', Boston Basin, Massachusetts. *Geology*, **2,** 413–15.

Robinson, P. L. (1973). Palaeoclimatology and continental drift. In *Implications of continental drift to the Earth sciences*, ed. D. H. Tarling & S. K. Runcorn, vol. 1, 451–76. Academic Press, London.

Rocha-Campos, A. C. (1967). The Tubaraco Group in the Brazilian portion of the Parana Basin. In *Problems of Brazilian Gondwana geology*, ed. J. J. Figarella *et al.*, pp. 27–102. Roesner, Curitiba.

Rognon, P., de Charpal, O., Biju-Duval, B. & Gariel, O. (1968). Les glaciations 'siluriennes' dans l'Ahnet et le Mouydir (Sahara central). *Service Geol. Algerie, Bull.* **38,** 53–82.

Runnegar, B. & Campbell, K. S. W. (1976). Late Palaeozoic faunas of Australia. *Earth Sci. Rev.*, **12,** 235–57.

Rust, I. C. (1975). Tectonic and sedimentary framework of Gondwana Basins in southern Africa. In *Gondwana Geology*, ed. K. S. W. Campbell, pp. 537–64. ANU Press, Canberra.

Saito, T. & Van Donk, J. (1974). Oxygen and carbon isotopes measurements of Late Cretaceous and Early Tertiary foraminifera. *Micropaleontology*, **20,** 152–77.

Savin, S. M., Douglas, R. G. & Stehli, F. G. (1975). Tertiary marine paleotemperatures. *Bull. geol. Soc. Am.*, **86,** 1499–510.

Schermerhorn, L. J. G. (1975). Tectonic framework of

Late Precambrian supposed glacials. In *Ice ages: ancient and modern*, ed. E. A. Wright & F. Moseley, pp. 241–74. Seel House Press, Liverpool.

Selwyn, A. R. C. (1859). *Geological notes on a journey in South Australia from Cape Jarvis to Mount Serle. Adelaide Parliamentary Papers*, South Australia, 20, 4 pp.

Shackleton, N. J. & Opdyke, N. D. (1973). Oxygen isotope and palaeomagnetic stratigraphy of equatorial Pacific core V28–238: oxygen isotope temperatures and ice volumes on a 10^5 year and 10^6 year scale. *J. Quaternary Res.*, **3**, 39–55.

Shah, S. C. & Sastry, M. V.A. (1975). Significance of Early Permian marine faunaus of peninsular India. In *Gondwana geology*, ed. K. S. W. Campbell, pp. 391–5. ANU Press, Canberra.

Spencer, A. M. (1971). *Late Pre-Cambrian glaciation in Scotland. Mem. geol. Soc.* London, **6**, 98 pp.

Spencer, A. M. (1975). Late Pre-Cambrian glaciation in the North Atlantic region. In *Ice Ages: Ancient and Modern*, ed. A. E. Wright & F. Moseley, pp. 217–40. Seel House Press, Liverpool.

Steiner, J. & Grillmair, E. (1973). Possible Galactic causes for Periodic and episodic glaciations. *Bull. geol. Soc. Am.*, **84**, 1003–18.

Suess, H. E. (1970a). Bristlecone-pine calibration of the radiocarbon time scale 5200 B.C. to the present. In *Radiocarbon variations and absolute chronology*, ed. I. U. Olsson, pp. 303–9. Wiley, New York.

Suess, H. E. (1970b). The three causes of the ^{14}C fluctuations, their amplitudes and time constant. In *Radiocarbon variations and absolute chronology*, ed. I. U. Olsson, pp. 595–604. Wiley, New York.

Sutherland, P. C. (1870). Notes on an ancient boulder-clay of Natal. *Q. Jl geol. Soc. Lond.*, **26**, 514–17.

Symons, D. T. A. (1975). Huronian glaciation and polar wander from the Gondwanda formation, Ontario. *Geology*, **3**, 303–6.

Taira, K. (1975). Temperature variation of the 'Kuroshio' and crustal movements in eastern and southeastern Asia, 7000 years B.P. *Palaeogeogr. Palaeoclimatol., Palaeoecol.*, **17**, 333–8.

Tarling, D. H. (1974). A palaeomagnetic study of Eocambrian tillites in Scotland. *J. geol. Soc. Lond.*, **130**, 163–77.

Tarling, D. H. & Mitchell, J. G. (1976). Revised Cenozoic polarity time scale. *Geology*, **4**, 133–6.

Theron, J. N. & Blignault, H. J. (1975). A model for the sedimentation of the Dwyka glacials in the southwestern Cape. In *Gondwana geology*, ed. K. S. W. Campbell, pp. 347–56. ANU Press, Canberra.

Theyer, F. (1972). Late Neogene paleomagnetic and planktonic zonation southeast Indian Ocean – Tasman Basin. Ph.D. thesis, Southern California University, 200 pp. (In Blank & Margolis, 1975.)

Turner, P. & Tarling, D. H. (1975). Implications of new palaeomagnetic results from the Carboniferous System of Britain. *J. geol. Soc. Lond.*, **131**, 469–88.

Washburn, A. L. (1973). *Periglacial processes and environments*. Arnold, London. 320 pp.

White, I. C. (1907). Permo-Carboniferous climatic changes in South America. *J. Geol.*, **15**, 615–33.

Williams, G. E. (1975a). Possible relation between periodic glaciation and the flexure of the galaxy. *Earth Planet. Sci. Lett.*, **26**, 361–9.

Williams, G. E. (1975b). Late Pre-Cambrian glacial climate and the Earth's obliquity. *Geol. Mag.*, **112**, 441–544.

Williams, M. A. J. (1975). Late Pleistocene tropical aridity synchronous in both hemispheres? *Nature*, **253**, 617–18.

Wollin, G., Kukla, G. J., Ericson, D. B., Ryan, W. B. F. & Wollin, J. (1973). Magnetic intensity and climatic changes 1925–1970. *Nature*, **242**(5392) 34–7.

Wright, A. E. & Moseley, F. (1975). (Eds) *Ice ages: ancient and modern*. Seel House Press, Liverpool. 320 pp.

2

Palaeobotany and climatic change

T. A. WIJMSTRA

One way of reconstructing former climates and therefore of recognizing climatic change is the reconstruction of former biota. This can be done by analysing those parts of these biota that fossilize in suitable environments. From the fact that it is not possible to cultivate the same crops everywhere we can conclude at a glance that a specific life form (plant or animal) makes certain demands on its environment. This reasoning may be reversed – the presence of a life form for a certain timespan means that the requirements of that plant or animal as to its environment were fulfilled. By this simple argument it is possible to demonstrate that a plant or animal species can be seen as a mapping from the abiotic world into the biotic world. By analysing the function or functions which define this mapping it is possible to reconstruct, from the presence of a certain combination of plants and animals, the abiotic world that accompanies this particular ecosystem.

Another way of obtaining climatic evidence from the past is the study of tree rings. Many kinds of trees grow only during a part of each year, and this is recorded by sheath-like layers of wood that appear as annual rings when the trunk is sectioned. The amount of wood formed in one year differs for each tree species according to the reaction of that species to climatic and edaphic conditions at the site. In some circumstances the chief factor controlling the annual growth is moisture and the rings vary mainly with precipitation. In other areas the controlling factor is temperature. By analysing the variation in tree-ring thickness in various trees a regional pattern can be constructed and this pattern can be translated by mathematical techniques into variation in temperature, precipitation and other climatic factors.

How can we analyse the connection between the biotic and abiotic world? We know that there are systems in nature which are in dynamic equilibrium with their surroundings. We call this entity composed of plants, animals and micro-organisms and accompanying factors, such as temperature, light, precipitation, evaporation, nutrients and so on, an ecosystem or biogeocoenose. This ecosystem can be defined as a subset of the biosphere in which the elements are in constant relation to each other. But the ecosystem is not a closed system; solar radiation and precipitation are external inputs, and an exchange of carbon dioxide, oxygen and water takes place between parts of the ecosystem and the atmosphere by convection and wind. At the same time, in humid areas, a part of the precipitation is delivered to the ground-water and a part leaves the system as run-off. Part of the incoming radiation is reflected and minerals are added by sedimentation of dust and so on, while exchange of organic material takes place through the movements of animals.

The effects of regional differences in climate on an ecosystem is best seen in the middle European forest system. In this system, beech, oak and hornbeam are in competition with each other. As the climate becomes warmer and drier, or more humid, the beech is replaced by different oak species. The oak, at increasing continentality, is in turn replaced by the hornbeam. In this system the growth conditions of plants belonging to the understory are dependent on the composition of the upperstory. This is not always the case. In a peat bog, for instance *Sphagnum* cover is the guiding element. Besides

Sphagnum, only those species that, even under poor nutrient conditions, produce so much matter that they are not overgrown by the *Sphagnum* will survive. So in a wet oceanic climate the growth of *Sphagnum* is so intensive that pine cannot maintain itself, while in more continental parts the growth of *Sphagnum* spp. is inhibited and the peat bog becomes more and more overgrown by pine. This example gives us an idea of how different climatic conditions (precipitation) lead to a different vegetation type which will also show up in a different way in the fossil record.

A close correlation exists between plant cover and climate; if we can reconstruct former vegetations, we have a concealed indication of former climates, as long as we can use plant species that are not extinct. We then make the assumption, of course, that the climatic requirements of a plant species do not change with time.

Table 2.1. *Production of pollen (number of pollen grains produced)*

(a) Pollen production of an anther from

Rumex acetosa	30 000
Secale cereale	19 000
Fraxinus excelsior	12 500
Calluna vulgaris	500
Plantago lanceolata	7 700
Sanguisorba officinalis	2 700

(b) Pollen production of a flower from

Pinus nigra	1 480 000
Picea excelsa	600 000
Rumex acetosa	43 500
Carpinus betulus	27 800
Sanguisorba officinalis	11 000
Tilia cordata	43 500

Pollen analysis has proved in the past 50 years to be an excellent tool for reconstructing former vegetations. The method relies on the fact that the outer walls of pollen grains consist of a very resistant organic material. The pollen grain has a few properties that makes it very useful for our purpose. It can, for instance, be preserved for many millions of years in sediments if protected against oxidation. Also, pollen grains are produced in enormous quantities by many species of plants (Table 2.1). The pollen rain produced during the flowering of the

plants will be distributed over a relatively large area and is to some extent related to the composition of the vegetation of that area. The outer wall of the pollen grain shows intricate specific patterns of ornamentation. Therefore, it is possible to identify the taxon from which the preserved remains of a pollen grain originated, sometimes even at the specific or lower level. Another advantage is the smallness of the pollen grain (10–150 μm), which makes it possible to extract even from small samples a sufficient number of individual grains to permit quantitative results. This enables narrow sampling intervals, sometimes one centimetre or less in suitable sediments, and makes it possible to distinguish changes in the pollen rain, and consequently changes in vegetation, within time intervals as small as one year.

Pollen analysis can contribute to the solution of various problems in different areas of the historical sciences. Palynological problems might involve the history of vegetation, the climatic history of a region, the correlation of sedimentary sequences, the subdivision of a taxon, the effect of Man on nature. The result of a palynological survey can give us some insight into principles and concepts pertaining to organization at community level and also into the development and evolution of former ecosystems. The results of pollen analyses are presented in pollen diagrams, a graphic presentation tion of the percentages of individual pollen types. In order to summarize the results in a way that better illustrates the observed changes in percentages, general diagrams are constructed. In these diagrams, groups of pollen types, belonging to plants of a certain biome, are taken together and so it is possible to observe the changes from a tundra to the northern coniferous biome, or from the desert biome to a chapparal, or a temperate deciduous forest biome to a temperate grassland biome.

The basic concept in palynological work is the pollen rain, a subset of all pollen that comes into circulation and only that part that is fossilized in the sedimentary sequence. The composition of the pollen rain is dependent on several factors: First, the frequency of the species in the region; secondly, on the absolute pollen production of the species. This production itself can vary specifically and individually according to the conditions under which the specimen grows – in the middle or at the boundary of its geographical area, in an open position or in a closed stand.

The frequency of flowering years is also important. There are forest trees which flower only at intervals longer than one year because sometimes a period of rest is needed before a species can flower profusely. The nearer one comes to the limit of the area of distribution, the longer the periods of rest between flowering. The dispersal mechanism of the pollen is also important; dry pollen is better dispersed than pollen that sticks together. And pollen production is not the same for every taxon (see Table 2.1), so the pollen rain must be corrected for the differences in pollen production. This problem is solved by comparison with the pollen rain of recent vegetation. For central Germany the relation between recent pollen spectra and the composition of the forest was as follows: *Pinus* 7.7, *Betula* 7.9, *Quercus* 7.8, *Picea* 1.1, *Fagus* 0.2, *Abies* 0.1 Tsukada (1966) established the following ratio between pollen percentages and area coverage of trees: *Pinus* 22.4, *Alnus* 5.8, *Betula* 1.5, *Pterocarya* 1.1, *Fagus* 0.9, *Tsuga* 0.7, *Quercus* 0.6, *Thuja* 0.6, *Tilia* 0.5, *Acer* 0.03. Using these ratios it is possible to interpret pollen rain records in terms of vegetation coverage. As long as the forest is composed of trees that produce comparable amounts of pollen, the actual composition of the forest is comparatively well reproduced. This is the case in northern Finland. The situation is more complicated if the main forest trees are seriously under-represented as is the case in the tropical rain forest.

As the scope of this book is climatic change I will now look into the different ways in which changes in pollen rain can be translated into changes in climate. The remains of different taxa are found in samples analysed for pollen grains: these include the pollen of higher plants, spores of mosses and lycopodes, and vegetative and generative remains of fungi, algae and so on. In general it can be stated that the last categories give us a different type of information from the first ones. From fluctuations in the first categories of remains, information is implied concerning the relation between regional vegetation and climate. The last categories give us more information on local edaphic circumstances. So we must be very critical in evaluating the information given by the various components of our record.

The problem is to find an algorithm that extracts climatic evidence from the autecological data which indicate the components of the vegetation present according to the palynological evidence. There are two general ways of approach: autecological, or synecological. In the autecological approach the climate is reconstructed by a detailed knowledge of the ecology of certain species. For instance, it is well known that water plants have several features which make them especially valuable for ascertaining Late Glacial temperatures, because they have a higher dispersal rate, partly because they are transported by migrating water birds over long distances, and also because they do not require any specific soil conditions before they can thrive. So a good knowledge of the autecology of aquatics will give us valuable information on temperatures. The ecology of land plants is, however, also useful.

Jasione montana and *Seseli libanotis*, for instance, indicate that during Allerød time the climate in southern Denmark was temperate with an average July temperature of 13–14 °C. The presence of *Pleurospermum* in the same period would seem to show that the winter temperature must have been at least 2 °C lower than today. On the other hand, the presence of *Armeria maritima* in the coldest phases of the Late Glacial suggests that the winters were not extremely cold, because this species today avoids areas with very cold winters, below 8 °C. Also, comparison of the recent geographical distribution of a certain taxon with isotherm and hypsitherm maps may give us information concerning temperature and rainfall. In this way it has proved possible to establish that July temperatures of at least 12–14 °C were present during some time in the period between 50 000 and 28 000 B.P. And it is along this line of thought that pollen analysis in Western Europe has long developed.

The second approach to the transfer of floristic data into climatological parameters has been chiefly developed in America. This method aims at the formulation of mathematical transfer functions; by means of these one can translate records of fossil pollen directly into quantitative estimates of climatic parameters. In this method a data base must first be constructed. In this data base all possible types of recent regional vegetations are incorporated with their pollen rain, together with their climatic data, such as temperature, rainfall, presence of air masses and so on. The assumption is made that a fossil pollen rain is comparable to recent vegetation, vegetation, present in our data bank. Then by comparing the recent vegetation with its climate, an impression of the fossil climate is obtained. This approach requires several assumptions: (*a*) the observed floristic composition of types, as assessed by the samples of (recent) pollen in a certain area, is in balance with the moderate climate; (*b*) the climate is the ultimate cause of changes of pollen records in the time; (*c*) the climatic response exhibited by each pollen type has remained constant during the time interval under discussion; (*d*) the relation between each climatic variable and a set of pollen types is linear.

Assumption (*a*) is fulfilled during the interglacials, but may not be fulfilled during the interglacial–glacial transitions and during the interstadials because the transition time or the duration of an interstadial was too short for the immigration, from their refuges, of more thermophilous elements and elements which produce copious amounts of pollen. For example, the general pollen rain recorded during the Bølling and Allerød times suggests that the Bølling interstadial developed under a different climate. However, analysis of the macroscopic remains suggests that the Bølling was at least not colder than the Allerød. During the interstadials only *minimum* values for the climatic response are found because the *optimum* values for the climatic response are not reflected in the pollen record.

A complication arises when the upheaval of a mountain chain, say, has created new areas for occupation. During such a process of local climatic changes and plant immigration new species are created and pre-existing taxa may respond to climatic variation in other ranges of their ecological amplitudes, so that new response functions become initiated and different transfer functions must be used. Assumption (*b*) is very often valid and is only not fulfilled during a transgression caused by isostatic readjustment and in newly originated habitats with a fast rate of evolution. Assumption (*c*) is correct in the middle range of the ecological amplitude of a taxon but when conditions are marginal, unusual responses may occur (an example is the

profuse flowering of plants when ecological conditions are extremely critical).

Recent reconstructions of the climate of our Earth
Japan
A palaeoclimatic study has recently been started on Lake Biwa (35° 15′ N, 136 ° 05′ E) in Japan. This lake has existed from Pliocene times onwards and shows a continuous record of deposition of material suitable for pollen analysis. A preliminary pollen diagram has been published, and gives the climatic record of the past 70 000 yr. In the two intervals 19–22 m and 12–13 m a strong influx of cool elements is notable; in time, these correspond to a first cold maximum of about 70 000 to 60 000 yr B.P. and a second at about 18 000 yr B.P., respectively. These intervals might well reflect the extreme cold periods of the most recent glacial and are also noticed in other places of the world (Colombia, Greece, Taiwan). There is also evidence in this diagram of the effects of the less extreme conditions of the Middle Pleniglacial (between 50 000 and 26 000 yr ago) reflected by a more temperate vegetation around Lake Biwa. Further information on this very important section is likely to be available after detailed analysis.

The next information comes from the island of Taiwan. A pollen section was taken in a lake at an altitude of 745 m (Fig. 2.1). This lake is situated at 23° 52′ N and 120° 55′ E in the centre of the island. The vegetation in that area showed the following zonation. Below 500 m subtropical rain forest is encountered. The most important trees in this vegetation are *Liquidambar*, *Trema orientalis*, *Mallotus paniculatis*, *Mallotus phillipensis*, *Mallotus japonica*, *Diospyros*, *Celtis formosana*, *Styrax suberifolia* and various bamboo species.

Between 500 and 1800 m a warm temperate forest is found with *Castanopsis hystrix*, *Castanopsis carlesii*, Lauraceae, *Camellia*, *Eurya* and *Schima*. Also found are *Podocarpus macrophyllum*, *Keteleeria davidiana*, *Caephalotaxus*, conifers and many tree ferns. The temperate forest (1800–2400 m) is composed of deciduous hardwood species, such as *Cyclobalanopsis*, *Quercus* species, *Ulmus uyematsuii*, *Zelkova serrata*, *Symplocos* spp., *Juglans cathyaensis*, *Carpinus kawakamie* and *Salix* spp., mixed with conifers like *Pinus morrisonicula*, *Cunnunghamia konishii*, *Taiwania cryptomeroides*, *Pseudotsuga wilsonia* and *Chamaecyparus formosen-*

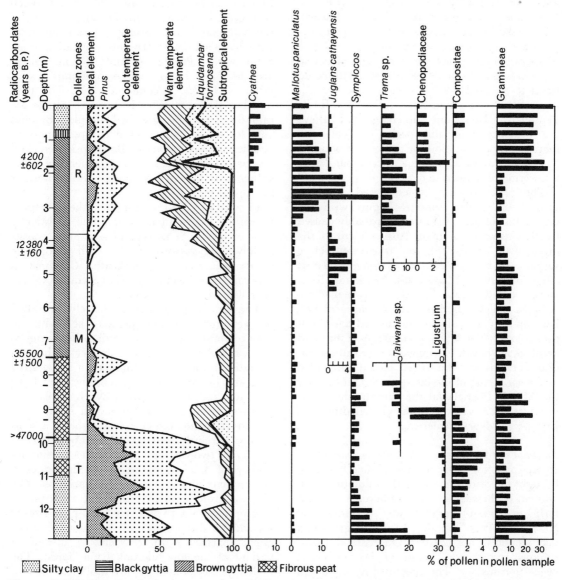

Fig. 2.1. Pollen diagram of a 12.79-m core from Jih Tan (745.5 m), central Taiwan (after Tsukada, 1966).

sis. In this forest no species of *Castanopsis* is to be found. Above 2400 m a boreal forest is found. In this forest trees of *Tsuga chinensis, Abies kawakamii* and *Picea morrisonicola* mixed with *Pinus* are found. In Fig. 2.1 the general diagram of this site is given, and the reconstruction of the vegetation sketched. From the ¹⁴C data it can be concluded that the greater part of the latest glacial is represented. During pollen zone *T* a

warm temperate forest was present around our site with a temperature somewhat lower than nowadays. This period might represent the end of the Brørup interstadial defined in Europe. The expansion in zone *M* of the boreal forest points to a temperature lowering of about 8–11 °C. At the boundary *M/T* a zone is encountered with an influx of subtropical forest elements. This zone has interstadial character and can probably be correlated with the Odderade interstadial of Western Europe. During the rest of pollen zone *M* a cool temperate forest was present at the site. In this cool temperate forest phase one expansion of boreal forest was found at about 35 000 B.P. This could point to the cool interval between the penultimate and last interstadial of the Middle Pleniglacial (Denekamp and Hengelo). I believe the evidence from this period indicates an increase in precipitation for the period between 35 000 and about 14 000 B.P., accompanied by a lowering of temperate of about 6 °C.

The *M/R* boundary is estimated at about 11 000 B.P. Pollen zone *R* begins with the destruction of primeval forests, probably caused by the influence of man. Also evident is the increase of subtropical and warm temperate species, such as *Mallotus paniculatis, Trema orientalis, Alnus* and *Castanopsis.* At the middle of zone *R*, dated as 4200–60 B.P., the steep increase of grass pollen together with *Liquidambar* and Chenopodiaceae indicates intensified agricultural activities. The higher percentages of pine and hemlock may be due to long distance transportation. It seems that the coolest climate prevailed in the early part of the latest glacial while after the Middle Pleniglacial the climatic deterioration was not as great as in the glaciated parts of Europe and North America.

Australia and the southern hemisphere in general
In the past few years more evidence has been brought forward concerning the climatic development of the southern hemisphere. A particularly important diagram was presented by Kershaw (1973) (Fig. 2.2). In this diagram of a lake section in northeastern Queensland a continuous record was found of 60 000 yr of vegetation history. The present-day vegetation of northeastern Australia consists of sclerophyl forests and woodlands with isolated patches of rain forest on sites with a high rainfall and/or high nutrient soils. In Queensland the rain forest dominates the eastern part. This rain forest decreases in structural complexity and floristic variety with increase in altitude and latitude or decrease in rainfall. We can recognize the following general sequence with increasing altitude.

In a tropical rain forest or mesophyl vine forest, Araliaceae, Eleocarpaceae, Lauraceae,

Meliaceae, Myrtaceae and Proteaceae reach their greatest development. The next member of the zonation is the subtropical rain forest or notophyl vine forest. In this vegetation type, Sapindaceae, Proteaceae and Monominiaceae are the most important. Also found in this forest is *Agathis*, a coniferous tree. The next member of the sequence is the warm temperate rain forest or the so-called simple notophyl vine forest. In this vegetation type important families are Saxifragaceae, Balanopsidaceae and especially Cunoniaceae. The last member of the sequence is the cool temperate rain forest or microphyl vine thicket.

In contrast with the rain forest the sclerophyl vegetation is structually simple and floristically poor in species. This type of vegetation consist mainly of *Eucalyptus* species with xerophytic shrubs or grasses underneath. *Casuarina* is generally found in this type of vegetation except on soils derived from basalt.

The climate of the region is largely controlled by the subtropical high-pressure belt with the derived southwest trades forming the dominating winds through much of the year. The precipitation is high in the summer. In Fig. 2.2 the diagrams of two sites are given, one from Lake Euramoo and the other from Lynch crater. The first site gives us a pollen record of the past 10 000 yr, while Lynch's crater shows a sedimentary record from 60 000 to about 10 000 B.P. So the two sites provide us with a continuous history of the vegetation and climate of a greater part of the most recent glaciation. The diagram from Lynch's crater illustrates pollen rain up to about 10 000 yr B.P. In pollen zones L5 and L6 the presence of *Araucaria* and *Podocarpus* together with *Dacrydium* points to a variant of the low microphyl vine forest. The pollen spectra of zone L5 are similar to modern spectra from moist araucarian forest, a forest type extensive nowadays in southeastern Queensland. In this forest type the annual rainfall lies between 900 and 1400 mm. The presence of *Dacrydium*, not found on the mainland of Australia today, suggests the existence of a fossil vegetation type.

In pollen zone L4 the increase in rain forest angiosperm pollen (such as *Elaeocarpus*) at the expense of rain forest gymnosperms (*Podocarpus* and *Araucaria*) and sclerophyl taxa, along with a slight rise in pteridophytic spore values suggest

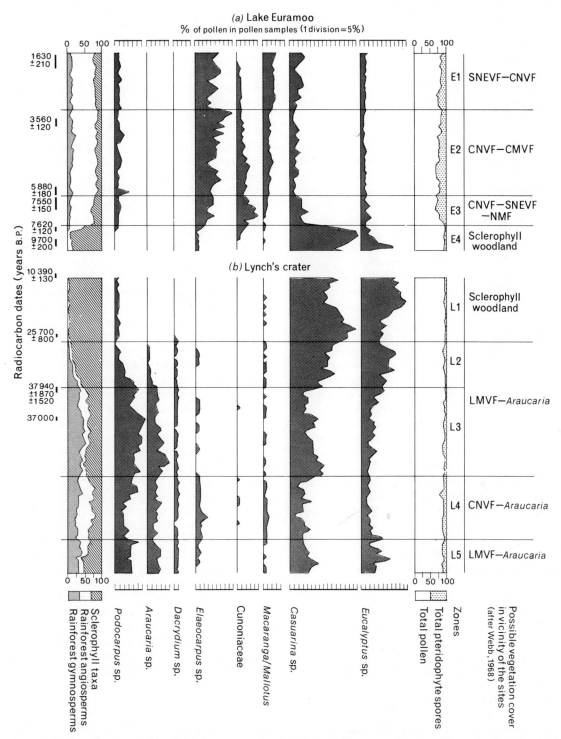

Fig. 2.2. Pollen diagrams of northeastern Queensland (after Kershaw, 1973). SNEVF = simple notophyll evergreen vine forest; CNVF = complex notophyll vine forest; CMVF = complex mesophyll vine forest; NMF = nanophyll mossy forest; LMVF = low microphyll vine forest.

the presence of the subtropical complex notophyl vine forest. This type of vegetation is found nowadays at sites with an increased precipitation as compared to the araucarian forest, from about 1200–1500 mm yr^{-1}.

In pollen zone L2 and L1 the pollen rain is dominated by sclerophyl woodland elements, such as *Casuarina* and *Eucalyptus*. This points to an expansion of the sclerophyl woodland at the cost of the rain forest indicating increasing drier conditions.

The pollen diagram from Lake Euramoo provides details of changes within the past 10 000 yr. Zone E4 with its dominance of sclerophyllous vegetation, as suggested by the pollen record, compares well with zone L4 of the Lynch's crater diagram. At about 7500 B.P. a change from a sclerophyl vegetation to rain forest takes place in the area. Throughout the period dominated by rain forest both sclerophyl and rain forest gymnosperm percentages were low in the pollen record, while generally speaking, the fern spore percentages show higher values. This indicates a rather moist type of climate. Zone E3 is characterized by high values of Cunoniaceae pollen, zone E2, on the contrary, by maxima in curves of *Elaeocarpus* while in zone E1 an increase of *Macaranga* and *Mallotus* percentages is worth mentioning, accompanied by higher values for sclerophyl pollen producers. From these pollen data an increase in precipitation and mean annual temperature for the period 6000–3000 yr B.P. is implied. Precipitation comparable to present-day levels is evident in the past 3000 yr of the pollen record, because this shows an increase of the drier rain forest and a partial immigration of sclerophyl vegetation. As yet, no such continuous record is available from other parts of Australia, but some of the evidence that is available can be summarized as follows.

In southern Australia, for the period before 40 000 B.P., evidence of well-developed soil profiles below dated lake deposits indicates a long and relatively dry period. Vegetation evidence from lake beds leads to similar consions. In the period 40 000–25 000 B.P. lake levels were rising in western New South Wales and the vegetation became denser around Lake Leak. Indicators in the Snowy Mountains suggest temperatures lower than today, while an increased aeolian activity in the Flinders range

might point to an absence of protecting vegetation cover on slopes and along rivers. The climate might be interpreted as cold and relatively dry by comparison with today; at the end of this period a lowering of lake levels took place and Lake Leak became dry.

From 25 000–15 000 B.P. Lake Leak continued dry, and sparse, open eucalypt vegetation occurred in the area, confirming conditions drier than at present. These conditions remained up to 11 000 B.P. Then lake levels were rising and vegetation became denser. Glaciation in Tasmania and the Snowy Mountains was at a maximum in this period, but from the presence of local vegetation on high levels around 16 500 on Mount Twynum a retreat of glaciers must be concluded.

In Holocene time, high lake levels with accompanying dense vegetation are recorded in Lake Leak between 800 and 3000 B.P. All this evidence might be summarized as follows.

Before 40 000 B.P., conditions were somewhat colder and drier than today. Between 40 000 and 26 000 B.P. the climate must have been wetter, after 26 000 B.P. an increase in dryness sets in, accompanied by a retreat of mainland glaciers. The synchronous nature of these events points to a regional change due either to a reduction in precipitation or to increased seasonality in the climate when rising summer temperatures were combined with colder winters, or a combination of both factors.

Africa

In the next part of the chapter we will give some attention to the development of the climatic research in Africa. An example of the tropical, high mountain areas of Africa is shown in the diagram of Sacred Lake on Mount Kenya (altitude 2400 m). In Fig. 2.3 the general pollen diagram is represented, the right hand side of which shows the vegetation types which were present at the lake site during the consecutive time intervals. The present distribution of the vegetational zones will be given, compiled from the work of Coetzee (1967) and Coe (1967).

The montane rain forest. This zone consists of two types of woody vegetation – a drier and a more humid kind of forest. On Mount Kenya the montane rain forest zone is represented on the

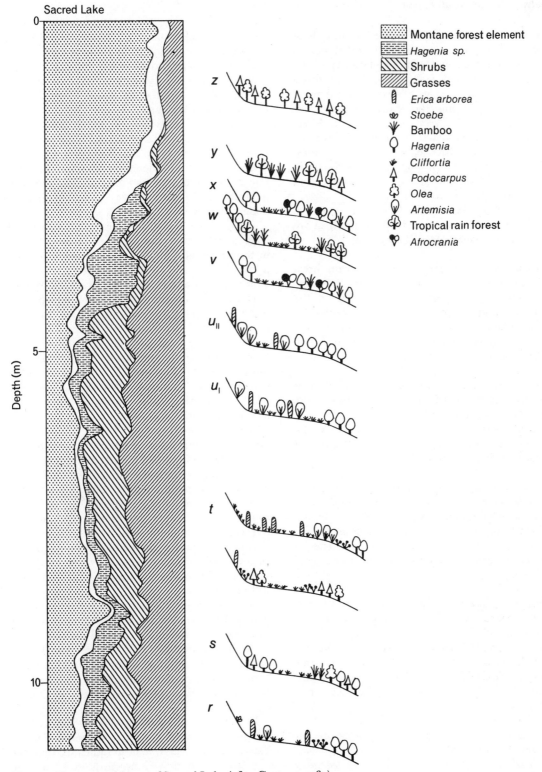

Fig. 2.3. Vegetation history of Sacred Lake (after Coetzee, 1967).

northeast and south slopes by the humid type of forest and on the west and northwest slopes by the dry montane rain forest. The latter type also occurs near Sacred Lake. The humid forest consists of many species of broad-leaved evergreen trees. Among the species present are *Afrocrania volkensii*, *Pygeum africanum* and *Macaranga kilimandscharica*. The lower part of this zone, up to 1670 m altitude, is called the semi-tropical rain forest. Among the trees found in this zone *Ficus*, *Sideroxylon*, *Celtis*, *Vitex* and *Trema* are conspicuous forest elements.

The dry montane rain forest has a closed canopy of fairly low trees. The principal taxa are *Juniperus*, *Podocarpus*, *Olea*, *Cassipourea*.

The bamboo zone. Above the montane rain forest zone lies the bamboo zone. This zone is associated with the greatest annual rainfall on the mountain and does not seem to occur above the dry montane rain forest. Its chief component is *Arundinaria alpina* and it contains scattered trees of the upper montane rain forest.

The Hagenia–Hypericum *zone.* This zone forms the tree line on Mount Kenya above the bamboo zone. The dominant trees are *Hagenia abyssinica* and *Hypericum lanceolatum.* Near the timber line and in the lower part of the belt of ericaceous forms *Cliffortia* is found.

The ericaceous belt. This zone consists of two parts, a lower one in which *Philippia*, *Stoebe* and *Cliffortia* are found and an upper zone in which *Erica arborea*, *Anthospermum* and *Artemisia afra* are more important. The last-mentioned species occurs in drier situations and is more fire-resistant.

The Afro-alpine belt. On Mount Kenya this belt is characterized by moist tussock grassland with *Senecio brassica*. Higher up, at 3800–4000 m, the tussock grassland becomes drier. *Senecio brassica* is replaced by *Senecio keniodendron* and on drier sites by *Alchemilla* scrub; *Lobelia telekii* is also found here. In *Carex* bogs *Lobelia keniensis* is found.

The nival zone. In this zone, plant life occurs on sheltered places where *Arabis alpina*, *Carduus platyphyllus*, *Lobelia telekii* and *Senecio keniodendron* occur. Above this zone *Helichrysum brownei* is

found up to 4800 m in scattered places which remain ice-free for longer periods of time.

Fig. 2.3 is a record of 33 350 ^{14}C-dated years of vegetational history. At present, Sacred Lake is situated within the montane forest in the humid part of this zone.

During the deposition of pollen zone *r* the surrounding vegetation of Sacred Lake belonged to the ericaceous belt. The high percentages of *Artemisia* point to drier habitats, and the tree line lay somewhat below the altitude of the lake, judging by the percentage (30%) of *Hagenia* pollen present.

Pollen zone *s* contains high percentages of *Hagenia*, indicating the presence of the tree line in the immediate surroundings of the lake; also present at the site were *Olea* and *Pygeum* trees. This indicates an amelioration of the climate.

During the deposition of zone *t* which took place between 14 050 and 26 000 B.P. the lake lay within the ericaceous belt. In the lower half of this interval the presence of *Cliffortia* and other elements of the lower ericaceous belt indicates that this lower belt then coincided with the actual position of the lake. *Podocarpus*, *Olea*, *Pygeum* and *Rapanea* were also present in the neighbourhood, although they grew at somewhat lower altitudes. These genera point to more arid conditions around Mount Kenya concomitant with a decline of temperature.

In the upper part of zone *t* the lake site was within the Afro-alpine zone. The high percentages of *Artemisia* point to dry conditions. Pollen zone U (14 050–10 560 B.P.): during this interval the lake site was still within the ericaceous belt, but the increasing percentages of *Hagenia* pollen are indicative of a gradually rising tree line. The conditions at the site were still rather arid.

In pollen zone *v* the tree line has migrated beyond the lake. Within the forest, the presence of *Hagenia*, *Afrocrania*, *Pygeum* and *Rapanea* points to a humid montane rain forest, the higher percentages of large, grass pollen pointing to a bamboo zone in the neighbourhood. The climate prevailing during this interval must have been very wet.

In the interval represented by pollen zone *w* the presence of pollen of *Macaranga*, *Myrica*, *Trema*, *Cordia* and *Galiniera* indicates that the tree line again moved upwards and the lake site was covered with a mixed forest consisting of bamboo and elements of the tropical rain forest. The climate was still very wet.

According to the higher percentages of *Hagenia* in pollen zone *x* the tree line was again depressed, suggesting cooler conditions than in the previous interval. From the presence of *Afrocrania* a continuously wet climate must be deduced.

In pollen zone *y* there was a maximum development of the bamboo forest, the presence of *Afrocrania* pointing to a wet type of climate.

In *pollen zone z* the predominant vegetation type was a mixed forest consisting of the dry montane rain forest and humid forest elements. This could imply a cooler and more misty climate.

In Africa, beside the results discussed above from palynological work on the slopes of Mount Kenya, evidence concerning changes of vegetation has been provided by Kendall (1969) using a core from Lake Victoria (altitude 1100 m). Kendall was able to show that the vegetation around Pilkington bay from at least 15 000–12 500 years ago was a savanna. After 12 500 B.P., forest trees became much more abundant, declining temporarily about 10 000 B.P. This initial forest seems to have been evergreen. At about 6000 B.P. this forest was replaced by a somewhat different one more adapted to drier conditions. During the past 3000 years sedimentation of all forest pollen types has declined. This change in the pollen rain is due to the conversion of forest land to agriculture and also because the indigenous crops produce too little to register clearly in the fossil record. The results from various studies in Africa have led to the same conclusion: prior to 12 000 years ago the climate of tropical Africa was dry. The beginning of this dry period is not yet known and seems to embrace the past 60 000 years. There is evidence, however, that this generally dry climate was repeatedly interrupted by transition periods of less severe drought. The past 12 000 years have been comparatively moist but the moistness was broken in at least tropical African areas by transitory dry phases at 10 000 and about 6000 years ago. The past 3000 years have been the driest of the past 12 millenia. The only place which has some indications of wetter conditions during the maximum extension of the glaciers in Europe is the area south of the high Atlas mountains, although the evidence is by no

means conclusive, because it does not come from continuous sections but only from isolated pollen-bearing horizons. These horizons could correspond, according to their ^{14}C dating, to interstadials as for example, found in the northern part of Greece. They therefore represent slightly better climatic conditions in the Atlas area.

The European records

One site in Greece is chosen here as an example of the vegetation history in the northern part of the Mediterranean area (Fig. 2.4). In an intermontane basin in the eastern part of Macedonia a continuous peat section is present which represents the Holocene and the whole of the most recent glacial period. The present-day vegetation in that area may be described as follows. Up to 500 m altitude a scrub vegetation is present in which evergreen oaks and maquis elements are conspicuous; between 600 m and 800 m a forest with deciduous oaks, *Castanea*, *Fraxinus*, *Acer* and *Ulmus* is found, followed by a belt of coniferous forest with *Abies* or pine trees. In this region no beech forests are present. The pollen diagram of the first 35 m of deposits is represented in Fig. 2.4. The left hand side of this figure shows the general or composite diagram and the right side the corresponding vegetation types. During the latest interglacial Pangaion (=Eemien), at first deciduous oak forest with *Carpinus* was present (Fig. 2.4*a*). The uppermost belt at that time consisted of a coniferous (*Abies*) forest. As time passed this sequence of forest zones along the slopes was succeeded by a more Mediterranean-like zonation consisting of evergreen oak forest in the lowermost zone; the tree line being formed by pine forest (Fig. 2.4*b*). After the end of the Pangaion interstadial this area was covered with an *Artemisia*-grass steppe containing some Chenopodiaceae.

During the Doxaton interstadial the predominant vegetation type was dry, open oak forest with Cistaceae and other maquis elements and in the higher parts pine forest had developed. This indicates that during this interval the climate was rather warm and of a 'Mediterranean' type with winter rains corresponding with that now found in the Crimean area.

During the period between the Doxaton and Drama interstadials an *Artemisia*-grass steppe

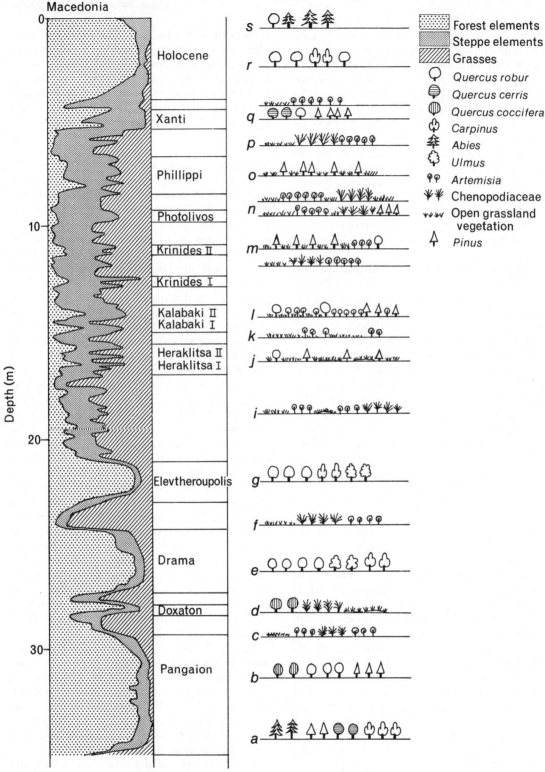

Fig. 2.4. Vegetation history of Macedonia.

was present. During the Drama (=Brørup) interstadial deciduous oak forest with *Carpinus* and *Fagus* was found (Fig. 2.4*e*). There is no palynological evidence of the occurrence of evergreen mediterranean forest during that interval. Temperatures must have been a little lower than they are today, but the precipitation must have been more evenly distributed because the predominant vegetation type belongs to that of the present sub-Mediterranean zone. After this interstadial the first very dry and cold steppe phase set in (Fig. 2.4*f*). The extreme conditions during this time are indicated by the high percentages of Chenopodiaceae, more particularly by the presence of *Eurotia* pollen in this zone. The average summer temperature must have been at least 5–8 °C lower than it is today and was during the Eemian–Pangaion.

In Fig. 2.4*g* the type of forest vegetation of the Elevtheroupolis interstadial is indicated. It is also of the sub-Mediterranean type and not greatly different from the Brørup–Drama forest type, so that the prevailing climatic conditions must also have been similar. The subsequent zone, that of the lower Pleniglacial, is represented in Fig. 2.4*j*. It consisted of an *Artemisia*–Chenopodiaceae steppe, at first with stands of pine in it or, more probably, with a pine forest zone in the higher parts of the mountains. Towards the end of this steppe zone, even this pine belt must have disappeared and the conditions must have resembled those now occurring in central Asia, where a low winter temperature and a low amount of precipitation prevail.

The next period, the Inter-Pleniglacial, is examplified in the vegetation sketches *i, j, k, l* and *m*. This period is characterized by an alternation of phases with steppe vegetation (*j, k*) and with steppes intermingled with stands of pine or oak during the interstadials. From the pollen diagram it can be concluded that the Kalabaki interstadial represents the climatic optimum of this period. The other interstadials (those of Heraklitsa and Krinides) also had a warmer climate; they are represented by steppe phases relatively rich in Chenopodiaceae, indicative of drier conditions. In the Upper Pleniglacial the predominant vegetation type was a steppe. In the middle of this steppe phase the pine-rich interstadials of Photolivos and Philippi are found (in the period between 16 000 and 22 640 B.P.).

Especially during the Philippi (= Lascaux) interstadial extensive stands of pine were present in the steppe vegetation and also a background of oak forest pollen was present in this interval. The climatic conditions of this interval are not very much different from those in the Drama and Elevtheroupolis interstadials, but perhaps a little drier.

The last part of the Upper Pleniglacial was very dry and cold (Fig. 2.4*p*) indicated by the presence of *Eurotia* and the dominance of *Artemisia* and Chenopodiaceae in this interval.

In the Xanthi interstadial, sub-Mediterranean conditions returned in the area, indicated by the presence of *Quercus cerris* pollen type suggesting a rise in temperature and a considerable increase in precipitation. The frequent occurrence of *Pistacia* suggests a rather warm climate and an open type of forest. Conditions must have been very much like those now found in the Crimea, as also evidenced by the presence of *Juniperus* pollen. In this area about 30% of the precipitation now falls during the winter; the average winter temperature is about 5 °C.

After a predominance of steppe vegetation during the Younger Dryas (Fig. 2.4*r*). the Holocene forest phase sets in. At first the stands of woody forms are of a sub-Mediterranean type (Fig. 2.4*s*), changing into a Mediterranean type with evergreen oaks, maquis elements and an uppermost *Abies* zone.

As a representative case for northwest Europe the vegetational development in the Netherlands is chosen (see Fig. 2.5). The recent vegetation type without human interference must have been oak forest. In more nutrient-rich sites *Carpinus* and *Fagus* are found and on the poorest soils a *Betula*-rich variant of the oak forest is the predominant vegetation type.

During the Eemian, after an initial pine–birch forest phase, oak forest established itself in this area. At first elm was an important constituent of this forest, but later hazel was more common. A change in the composition of the forest took place again with *Carpinus* becoming an important element. At the end of the Eemian the oak forest was replaced by a pine–spruce forest as a result of the deterioration of the soil, a situation also indicated by the expansion of heather.

After the Eemian the forest opened at first and a sub-arctic park landscape prevailed. The

Netherlands

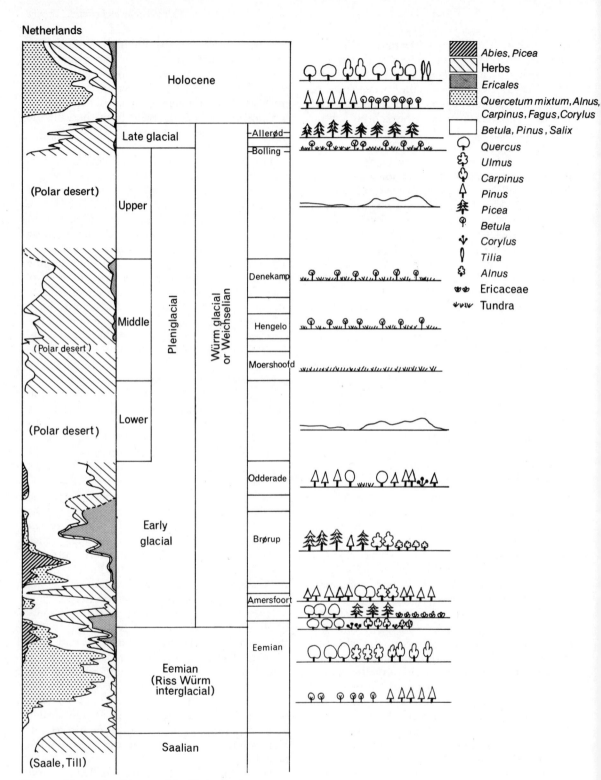

Fig. 2.5. Vegetation history of the Netherlands.

amount of precipitation was rather high, as shown by the development of heathland vegetation. During the Amersfoort interstadial the area became reafforested. The forest was principally a type of pine wood although the higher values of *Quercus, Fraxinus* and *Ulmus* pollen in spectra of the upper half of this interstadial point to a re-immigration of these thermophilous trees and the establishment of a *Quercetum mixtum* type of high forest at more favourable sites. The temperature conditions must have been comparable to those prevailing during the Allerød. After this Amersfoort interstadial an opening up of the forest indicates a deterioration of the climate. Some arctic–alpine plants were found in the area, but genuine tundra conditions most probably never occurred.

The next interstadial, that of the Brørup, was characterized by a spruce forest, with some scattered thermophilous elements. These thermophilous elements (oak, elm) did not contribute much to the overall composition of the forest. After the Brørup interstadial an open arctic vegetation was found. During the Odderade interstadial a pine forest with thermophilous elements became established.

During the Lower Pleniglacial no higher plant life could maintain itself. In the Middle Pleniglacial, during the interstadials of Moershoofd, Hengelo and Denekamp, vegetation types comparable to those found in arctic and polar areas were found. The Moershoofd interstadial shows a vegetation dominated by cyperaceous and other herbs; there are no remnants of trees and shrubs. Apart from the plants already mentioned, other plants now occurring in tundra vegetation were found, such as *Saxifraga oppositifolia*, and many heliophilous elements, such as *Plantago maritima* and *Helianthemum*, as well as plants such as *Salix polaris, S. reticulata, S. herbacea* and the dwarf birch *Betula nana*. During the Hengelo interstadial small stands of birch could perhaps maintain themselves at favourable sites. *Dryas octopetala*, *Artemisia* and Chenopodiaceae were also present. Some of these plants point to a high fall of snow. In several, shorter periods the occurrence of *Artemisia* and Chenopodiaceae, on the other hand, points to drier conditions. According to Zagwijn (1974) these steppe vegetation-bearing beds are the equivalents of the shrub tundra beds in the Hengelo interstadial and can be explained by a difference in habitat.

During the Upper Pleniglacial the conditions were too severe for the maintenance of plant life.

In the Late Glacial, during the Bølling and Allerød interstadials, a park-like birchwood landscape and pine forest were found in the Netherlands. It is questionable whether these differences in vegetation cover were the result of a real difference in climate or if they are only apparent and attributable to the appreciable distance the more thermophilous trees had to bridge during their migrations. The aquatic elements of the flora suggest that the Bølling interstadial was not much colder than and might even be as warm as the Allerød. In the intervening period before and after these interstadials, herbaceous forms were more important elements of the vegetation cover. The presence of *Empetrum* in the Younger Dryas may point to a wetter type of climate (judging by the present distribution of *Empetrum*), but may also be related to soil and soil development.

In Fig. 2.5 the three stages of the Holocene vegetation development are given: first a pioneer birch–pine forest developed, followed by the oak–elm-rich forest of the hypsithermal and culminating in the natural oak–beech and oak–birch forest types of the present. The distributional ranges of these vegetation types are governed by edaphic conditions.

In northeast France there is in a peat bog called La Grande Pile a continuous record of Weichselian vegetation changes. The bog lies in a basin formed in an earlier glacial period. The first pollen zone from the diagram made by Woillard (1974) is dominated by pollen of Cyperaceae, grasses, *Artemisia, Thalictrum, Helianthemum, Sanguisorba* and *Selaginella*. This indicates an open vegetation developed under cold conditions. Next in this section three forest phases are to be seen (Eemian, St. Germain 1 and St. Germain 2). All three phases show a zone with higher percentages of pollen produced by thermophilous trees, such as *Quercus, Ulmus, Tilia, Acer, Fagus, Abies*. In my opinion these three forest phases can be correlated with the Pangaion, Drama and Elevtheroupolis sequence found in the early Weichselian (see above). These forest phases are separated by intervals in which open vegetation prevailed. In this vegetation *Helianthemum, Sanguisorba* and so on were found frequently. After these forest phases a long period of open vegetation is represented in the pollen diagram. In this open vegetation *Artemisia,* Chenopodiaceae, *Helianthemum,* Caryophyllaceae are frequently found, indicating open, steppe-like vegetations. In this predominating open vegetation period several oscillations occur, caused by increasing percentages of pine and birch pollen. Two of these oscillations were dated by radiocarbon analysis as 34 600 and 30 500 B.P. This result made it highly probable that these oscillations, which according to their pollen record must represent an amelioration of climate, can be correlated with the Denekamp and Hengelo interstadials. From the aquatic flora present during the extreme phase it can be deduced that the July temperature was about 13–14 °C, while the presence of *Armeria* might point to a January temperature less than −8° C.

Vegetation changes in northern Eurasia

In general, one can say that the tundra vegetations were restricted to areas which at that time had a relatively oceanic climate, such as the southern part of England and the Netherlands. But beginning with central and eastern Belgium and extending to the Mediterranean area, Hungary and beyond eastwards, steppe-like communities seem to have been more characteristic. These open vegetation types were dominated by *Artemisia* and Chenopodiaceae and several herbs and grasses, together wth appreciable numbers of halophytic plants. The presence of halophytic plants led Russian palaeobotanists to suggest that the soils must have been salty in eastern Europe and Siberia during the height of glacial times. Scanning electron microscopic analysis of chenopods has shown that this assumption may be correct. In the area under discussion no forest belt existed during the height of the two most recent glaciations. The forest trees were confined, in general, to several small mountainous areas and to the northern border of the Caspian Sea. The forest area around the Caspian was much wider during the early glacial phases than it was later. In most areas of northern Eurasia the soil was frozen throughout the year and the glacial periods were characterized by maximum distribution of various steppe communities as we have seen. The interglacial

periods were, on the other hand, characterized by forest which covered the continent from the Atlantic coast to the Japanese sea and Bering Strait. The wave-like succession of cold and warmer climates during the Pleistocene had important consequences for the species composition of the forest vegetation in the successive interglacials. Due to the east–west trend of the important mountains (Alps, Himalayas) barriers against plant immigration were created. This eventually caused an impoverishment in floristic composition of the interglacial forest. So, for instance, the colder climates at the Plio–Pleistocene border caused the annihilation of *Sequoia, Taxodium, Glyptostrobus, Nyssa, Liquidambar, Fagus, Zelkova* and *Liriodendron*. At the same time *Tsuga* disappeared from Siberia and vegetation types comparable to the recent taiga were established already in the beginning of the Pleistocene. Since that time this forest, adapted to extreme cold climates, has changed very little in species composition though the taiga survived only within very small areas in a predominant open vegetation during the glacials.

The new world

The next example comes from Colombia (South America) where on a high plain in the eastern Cordillera (altitude 2580 m) a section was drilled in the Lake of Fuquene and palynologically analysed. These sections cover probably the whole of the latest glacial and part of the preceding interglacial. The general diagram and its tentative vegetational history is presented in Fig. 2.6. The Andean vegetation belts in this area are today as follows.

The *Subandean forest* is found between 1000 and about 2300 m altitude. In this zone such genera as *Acalypha, Alchornea* and *Cecropia* are the richest pollen producers. Between roughly 2300–3400 m the *Andean forest* belt is present. In this belt two groups of associations can be distinguished, one rich in *Quercus* and the other rich in *Weinmannia*. The difference in ecological conditions seems to be that the oak forest prefers sites with a higher annual precipitation and more continuous misty conditions than some species of *Weinmannia*. Edaphic differences seem to play a minor role, since both types of forest occur mainly on sandstones. In wetter places in the Sabana de Bogotá *Alnetum* is found. The next higher belt is formed by the *high-Andean*

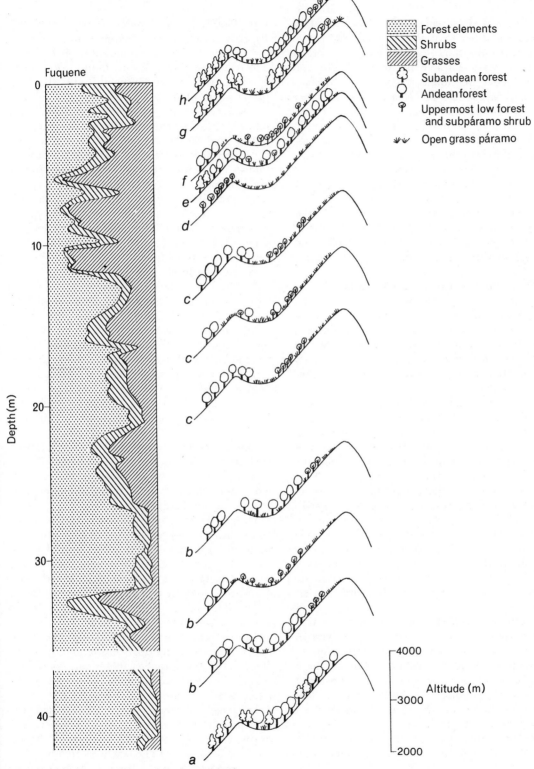

Fuquene

Depth (m)

Forest elements
Shrubs
Grasses
Subandean forest
Andean forest
Uppermost low forest
and subpáramo shrub
Open grass páramo

4000

Altitude (m)

3000

2000

Fig. 2.6. Vegetation history of the Bogotá basin.

41

dwarf forest and *subpáramo scrub formations*. This irregular belt begins at 3000–3200 m and peters out at 3500 m locally even reaching 3800 m. In the lower part of this belt various shrubs and small trees are found, such as *Vaccinium, Miconia, Gaultheria* and *Hesperomeles*. Another type of 'subpáramo' shrub is composed predominantly of Compositae and may occur in patches as high as 3800 m. A rather special type of dwarf forest, the *Polylepis* forest, occurs in patches above the actual forest limit.

The páramo belt proper extends from 3300 or 3500 m to 4000–4200 m. Characteristic taxa are *Espeletia, Gentiana, Valeriana, Geranium, Plantago* and *Paepalanthus*. Locally, an association is found in which *Acaena* is the dominant species and extends to altitudes between 4000 m and 4300 m.

The altitudinal forest limit nowadays lies at some 3500 m in places with a more humid climate and in drier sites at about 3300 m. In wetter, higher habitats oaks extend to the tree line. In drier places, associations in which *Weinmannia tomentosa* is common are frequently encountered. The present situation is represented in Fig. 2.6*h*.

The vegetational history of the latest interglacial (the equivalent of the Eemian–Pangaion interglacial in northern Europe, see Fig. 2.6*a*) is as yet insufficiently known. What we do know is that during certain intervals in that period the Sabana de Bogotá was occupied by a Subandean forest, as indicated by the presence of *Acalypha, Alchornea* and *Cecropia* pollen in sediments of this time sequence. At wetter sites in the Sabana, *Myrica* and alder carr must have been present. It follows that the Andean forest was found at higher altitudes than at present. For the time being it is not possible to indicate the exact position of the tree line during that time.

A tentative reconstruction of the vegetation type during the early interstadials is given in Fig. 2.6*b*. A distinct lowering of the tree line to below 3000 m can be observed. On the high plain of Bogotá, remnants of the Andean forest with alder and gale carr were present. The precise position of the Subandean–Andean forest limit is not well known.

The vegetational aspect during the Middle Pleniglacial interstadials is reconstructed in Fig. 2.6*c*. On the high plains low forest or shrub was found locally (alder, gale, *Symplocos*). *Polylepis–Acaena* is very frequent and there are indications for local abundance of *Dodonaea*. The oak forest proper is only found below about 2400 m, that is, at lower altitudes than the present high plain.

During the most extreme conditions of the latest glacial the vegetational conditions were as represented in Fig. 2.6*d*. The tree line was depressed to below 2000 m; at this altitude the high-Andean dwarf forest was found. In the high plain, páramo and super-páramo conditions prevailed at that time, judging by the presence of *Draba*, today restricted in its occurrence (in the area under discussion) to super-páramo vegetation. The climate was, moreover, extremely dry, as is shown by the very low lake level.

In the Late Glacial the tree line started to rise. During the Guantativa interstadial (the equivalent of the Allerød and Bølling) the forest again invaded the high plain. The Andean forest, mainly consisting of oak, was found on the slopes above the plain up to 3000 m. Locally, pioneer vegetation of *Myrica* and *Dodonaea* became established. At the beginning of the Late Glacial the lake levels rose suddenly, indicating an increase of effective rainfall.

During the El Abra stadial (Younger Dryas) the forest limit was depressed by about 800 m and lay below the high plain level (see Fig. 2.6*f*). Some pioneer forest may have been present in sheltered valleys, however.

During the hypsithermal of the Holocene the conditions were as represented in Fig. 2.6*g*. The upper limit of the Andean forest moved above 3400 m. On the high plain a Subandean forest with *Acalypha* and *Cecropia* was found, and *Weinmannia* became more frequent on the higher slopes. This all points to a rise in temperature and to rather moist conditions.

At the beginning of the Subatlanticum the forest limit was depressed to about 3200 m (Fig. 2.6*h*) and the vegetational conditions suggest a deterioration of the climate as compared with the conditions of Fig. 2.6*g*.

In general, we may say that the tropical montane Andean climatic belts were created during the upheaval of these mountains during the Late Pleistocene. The newly created belts became populated by processes of adaptation of elements from the local neotropical flora and by the arrival of elements which immigrated from the Holarctic and Antarctic floral areas. This process continued during the entire Pleistocene. For instance, *Weinmannia* originally came from the south, while *Myrica*, *Alnus* and *Quercus* came from the north. In the transitional and sub-páramo forest there is an abundance of local genera, while in the páramo vegetations, elements from remote areas are more frequent. From the Holarctic came elements like *Bartschia*, *Draba*, *Hypericum* and *Berberis*. While the Antarctic provided this vegetation with elements like *Acaena*, *Azorella* and *Muehlenbeckia*, the Pleistocene shows a sequence of several glacials and interglacials comparable with those of the northern hemisphere. The contemporaneity of the changes of temperature could be sustained for the past 50 000 yr by ^{14}C dating. The depression of the Andean forest limit was of the order 1200–1500 m. We must accept a lowering of temperature of 6–7 °C to explain the total depression of the altitudinal forest limit. During the coldest period of the most recent glacial the climate on the high plains was much drier and annual precipitation may have been about 3 °C lower than today. This means that the temperature gradient in the northern Andes was much steeper than today. During glacial time the surface area occupied by the páramo vegetation was much larger than its present extent. Many now isolated 'islands' were linked together during colder periods.

In the coastal lowlands of northern South America today the following vegetation sequence occurs. Along the estuaries and the muddy shores of the seas mangrove forest is found. Often the outermost zone is formed by *Rhizophora*, with *Avicennia* forest following behind on drier but periodically inundated soil. Behind the mangrove belt there is an alternation of more elevated former beach ridges and swampy areas. The ridges are covered with a type of forest characteristic of drier soils, while in the swamps open, herbaceous vegetation or swamp forest is present. Further inland the relatively dry Wallaba forest or more humid tropical forest are found. Further still inland the savannas are found.

From a great number of bore holes in the coastal lowland the vegetational succession during an eustatic climatic cycle can be indicated as follows. In the period before 45 000 B.P. in many sites in the coastal area a reflection of the mangrove belt is found. There follows an extension of swamp forest elements as the sea seems to have retreated from the area. In the next phase mangrove elements disappear completely, while open grass savanna elements become dominant in the coastal zone. This zonation lasts until the beginning of the Holocene when the area gradually becomes invaded by the sea. First, the *Avicennia* belt passed through the zone and subsequently *Rhizophora* become dominant. So all the available data point to the erstwhile dominance of real savanna vegetation in the present coastal area of Guyana and Surinam during glacial times. This indicates that during that time a savanna climate was present in that area, with a lower precipitation and/or more pronounced dry and wet seasons than at present.

From several pollen diagrams from lakes in the savannas of the Llanos orientales of Colombia and the Rupununi savannas in Guyana the following vegetation history may be derived. During the Upper Pleniglacial the area was covered by a grass savanna of a very dry character. In the diagram there are some indications that during the Middle Pleniglacial a savanna woodland existed, indicating a slightly wetter climate. Between 75 000 B.P. and 13 000 B.P. these areas were covered by *Byrsonima* woodland, indicating wetter conditions than today. There followed a major period of open savanna, apparently drier than today to about 3800 B.P. After 3800 B.P. a savanna woodland is reflected in our diagram. This period of drier forest existed to about 2000 B.P. After this period the present vegetation of grass savanna becomes evident in our diagrams.

In the Amazon basin, an area at present covered with rain forest, some palynological data are available that strongly favour the idea of the former existence of savannas in areas now covered with tropical rain forest.

In the last part of this chapter some remarks will be made concerning the vegetational history of North America. The most recent interglacial was, according to the pollen analysis of sediments, characterized in the central and southeast US by a vegetation similar to that of today. During the early part of the latest glacial the vegetation changed in that area to principally a boreal forest. At first this forest was dominated by *Picea* in the northeast part of Kansas. In the eastern part this boreal forest consisted of *Picea* and *Pinus*. In the southern Appalachians, a forest with *Pinus banksiana* was found. At the maximum of the most recent glacial, pine was excluded from the forest. Tundra vegetation was present in the higher parts of the Appalachian mountains and in a narrow belt in New England and northern Minnesota. At the retreat of the glaciers this boreal forest changed into communities of *Pinus* and *Betula* in the more northerly areas, *Ulmus* and *Quercus* south of the great lakes, open grassland in the west and oak and pine in the south Appalachians. Meanwhile in the western Cordillera the tree line was depressed 800–1000 m and alpine vegetation was vastly expanded in areal coverage. The various forest belts were lowered as well and many of the now desert basins of the southwest were covered by shrub steppe. At the end of the Glacial the climatic change caused the various vegetation belts to ascend the mountains, changing their composition in the process.

From the evidence brought forward in this chapter and the literature cited below it must be concluded that, in general, the great fluctuations in climate were contemporaneous in the southern and northern hemispheres. Secondly, the advances of the glaciers, coinciding with the extreme phases of the glacials, were contemporaneous with the generally drier climate of the Earth. The only place where an indication of a wetter climate than now was encountered, is the area of the southwest of the US and possibly an area south of the Atlas mountains. Within glacial periods, ameliorations of climate occurred, called interstadials. During these phases a rise in temperature, accompanied by increasing precipitation, took place. The wetter phases, formerly called pluvials, do not coincide with the extreme phases of the glacials, but occur, in general, at the transitions from glacial to interglacial conditions. Within the period from roughly 50 000–22 000 B.P., a general rise in temperature took place, accompanied by an increase in precipitation.

Bibliography

Since each piece of climatic evidence from palaeobotanical evidence rests on so many sources, it seems appropriate to give the background material for this chapter in groups relating to the geographical distribution of that evidence.

General

Faegri, K. & Iversen, J. (1975). *Textbook of pollen analysis*. Munkgaard.

Fritts, H. C. (1966). Growth-rings of trees, their correlation with climate. *Science,* **114,** 973–9.

Iversen, J. (1973). The development of Denmark's nature since the last glacial. *Danm. geol. Unders. Raekke V,* No. 7C.

Kershaw, A. P. (1973). Late Quaternary vegetation and climate in northeastern Australia. In *Quarternary studies,* ed. R. P. Suggate & M. M. Cresswell, pp. 181–7. The Royal Society of New Zealand, Wellington.

Odum, E. P. (1971). *Fundamentals of ecology*. Saunders, Philadelphia.

Walter, H. (1968). *Die Vegetation der Erde in öko-Physiologischer Betrachtung,* vol. I. Gustav Fischer Verlag, Jena.

Africa

Coe, M. J. (1967). *The ecology of the alpine zone of Mount Kenya*. Junk Publishers, The Hague. 136 pp.

Coetzee, J. A. (1967). Pollen analytical studies in East and southern Africa. *Palaeoecology of Africa* III, ed. E. M. van Zinderen Bakker. Balkema, Cape Town.

Kendall, R. L. (1969). An ecological history of the Lake Victoria Basin. *Ecol. Monogr.,* **39**(2), 121–76.

Livingstone, D. A. (1975). Late Quaternary climatic change in Africa. *A. Rev. Ecol. System.,* **6,** 332–41.

South America

Van der Hammen, T. (1974). The Pleistocene changes of vegetation and climate in tropical South America. *J. Biogeogr.,* **1,** 3–26.

North America

Wright, H. E. (1971). Late Quaternary vegetation history of North America. In *The Late Cenozoic glacial ages,* ed. K. K. Turekian, pp. 425–65. Yale University Press, New Haven.

Asia

Frenzel, B. (1968). The Pleistocene vegetation of northern Eurasia. *Science,* **161,** 637–49.

Fuji, N. (1973). Changes of climate during Wisconsin stage based on palynological study of a 200 m core sample of Lake Biwa in Japan. *Proc. Japan Acad. Sci.,* **49,** 737–48.

Tsukada, M. (1966). Late Pleistocene vegetation and climate in Taiwan (Formosa). *Anthropology,* **55,** 543–9.

Europe

Farrand, W. R. (1971). Eastern Mediterranean Late

Quaternary palaeoclimates. In *The Late Cenozoic glacial ages*, ed. K. K. Turekian, pp. 493–529. Yale University Press, New Haven.

Frenzel, B. (1968). *Grundzüge der pleistozänen Vegetationsgeschichte Nord-Eurasiens*. Franz Steiner Verlag, Wiesbaden.

Van der Hammen, T., Wijmstra, T. A. & Zagwijn, W. H. (1971). The floral record of the Late Cenozoic of Europe. In *The Late Cenozoic glacial ages*, ed. K. K. Turekian, pp. 38–91. Yale University Press, New Haven.

Wijmstra, T. A. (1969). Palynology of the first 30 m of a 120 m deep section in northern Greece. *Acta bot. neerl.*, **18**(4), 511–27.

Woillard, G. (1974). Exposé de recherches palynologiques sur le Pleistocène dans l'est de la Belgique et dans les Vosges Lorraines. Thèse doctorale, Universite catholique de Louvain, Louvain.

Zagwijn, W. H. (1974). Vegetation, climate and radiocarbon datings in the Late Pleistocene of the Netherlands. Part III: Middle Weichselian. *Meded. Rijks geol. Dienst* N. S., **25**(3), 101–10.

3

Isotope studies

JEAN-CLAUDE DUPLESSY

Most elements occur in nature as a mixture of stable isotopes, such as hydrogen (^{1}H, ^{2}H) carbon (^{12}C and ^{13}C) and oxygen (^{16}O, ^{17}O, ^{18}O). Urey & Greiff[1] established the existence of comparatively large differences in the chemical properties of hydrogen and deuterium compounds and of smaller differences in the chemical properties of the isotopic compounds of other elements of low atomic weights. Urey's theoretical study of the thermodynamic properties of isotopic substances[2] showed that carbonates precipitated from a single aqueous solution at different temperatures should have a different ^{18}O content. McCrea[3] investigated the isotopic composition of calcium carbonate precipitated inorganically from water, and determined the abundance of ^{18}O in calcium carbonate relative to water as a function of temperature. Epstein, Buchsbaum, Lowenstam & Urey[4] established an empirical temperature scale based on isotopic analysis of calcium carbonate deposited by marine invertebrates grown in natural environments of known temperature. This empirical relationship, while in accordance with McCrea's data, is more precise than the latter. Emiliani[5] presented the first oxygen isotopic analyses of pelagic and benthic foraminifera in deep-sea cores, interpreting the isotopic data in terms of the Pleistocene climatic variations.

Subsequent work on numerous cores from the Atlantic,[5-7] the Caribbean Sea,[5, 8-10] the Pacific,[11, 12] the Indian Ocean[13, 14] and the Mediterranean Sea[15, 16] has provided highly correlated isotopic records. The four curves in Fig. 3.1, which show oxygen isotopic variations in the Late Cenozoic pelagic foraminifera in wide-spaced deep-sea cores, give similar general trends but vary in detail. I shall concentrate here on the significance of these curves and their variations. Particular attention will be given to the factors controlling the isotopic composition of a given foraminifera specimen and to the assessment of which of these factors are likely to have changed in the past.

The procedure of oxygen isotope analyses of carbonate can be divided into three steps. First, the carbonate to be analyzed is cleaned so as to ensure that a pure carbon dioxide sample is obtained. Next the carbon dioxide is extracted and purified for mass spectrometric analysis, and finally, the oxygen isotopic composition of this carbon dioxide is compared in the mass spectrometer with that of a laboratory standard. The isotopic composition is expressed as the per mil difference δ^{18}O between the ^{18}O/^{16}O ratio in the sample and that in the laboratory standard

$$\delta^{18}O = \left\{ \frac{(^{18}O/^{16}O) \text{ sample}}{(^{18}O/^{16}O) \text{ standard}} - 1 \right\} \times 1000$$

At present, mass spectrometers that are commercially available are capable of determining δ^{18}O to a precision of $\pm 0.05\%_0$

Cleaning techniques

The ^{18}O/^{16}O ratio in a sample is obtained by measuring the abundance ratio of mass 46 to the sum of the masses 43, 44, 45 (and 47 in some cases) in the mass spectrometer. So any organic or inorganic molecules or radicals formed in the ionization chamber of the mass spectrometer which have masses in the range 43–47 may bias the isotopic ratio.

Organic matter is one of the major sources of contamination. The purification process devel-

46

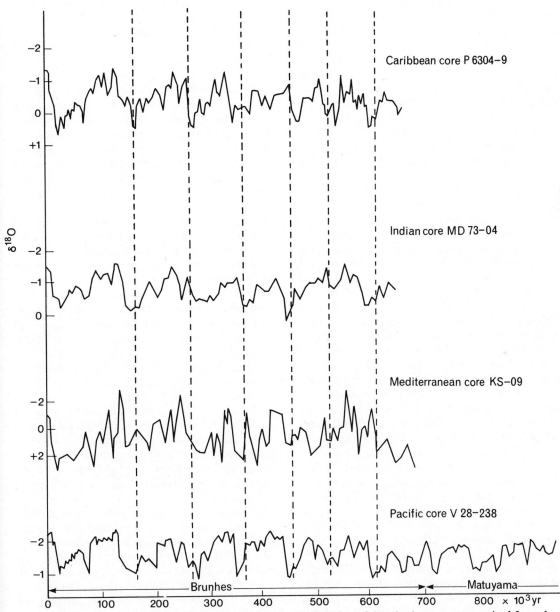

Fig. 3.1. Oxygen-18 variations of planktonic foraminifera as function of time in deep-sea cores raised from the Caribbean sea,[9] the Indian Ocean,[14] the Mediterranean Sea (C. Grazzini, unpublished) and the Pacific Ocean.[11] Odd numbers represent warm isotopic stages defined by Emiliani.[5]

oped by Epstein *et al.*[4] involves grinding the carbonate to fine powder and roasting it in a stream of helium at 475 °C. A second process, developed by Naydin, Teys & Chupakhin,[17] involves roasting the powdered sample at 475 °C in a vacuum. A third process involves digesting the powdered carbonate in chlorox.[9]

Emiliani[9] reports that the different preparation procedures may give somewhat different results, depending on the technique adopted, as well as the material used. Samples that had been roasted in helium gave $\delta^{18}O$ values systematically 0.55‰ smaller than those which were not roasted. Shackleton[18] pointed out that the roasted samples were analyzed in Chicago while PDB standard was still in use and unroasted samples were analyzed in Miami when PDB was used up. He thus suspects that the discrepancy reflects mainly a calibration problem.

At Gif, samples are roasted in a vacuum at a temperature of 400 ± 25 °C. We have always observed small variations due to this purification pre-treatment. The magnitude of the variations depends on the material used. For example: PARIS–SIRAP is a fine powder (less than $22 \mu m$) ground from a pure white marble (supplied by Dr R. Létolle) and Marble V is a crushed, clean white marble (supplied by Dr N. J. Shackleton) with a grain size of 125–180 μm. For both standards, there is no difference (± 0.01 ‰) between the roasted and unroasted treatments. Fletton (supplied by Shackleton) is a belemnite with the same grain size as Marble V; roasted runs give results that are 0.22 ‰ lighter than the unroasted runs. The same difference is also observed in Shackleton's laboratory at Cambridge. On the other hand, roasted NBS 20 standard is 0.22 ‰ heavier than unroasted NBS 20. PARIS–SIRAP and Marble V are pure calcite but the NBS 20 and Fletton standards have some noticeable impurities. Thus it is believed that the presence of the impurities has caused the observed differences between the roasted and unroasted results. Our study shows that the most important factor affecting the observed variations is the difference of isotopic ratio between the carbonate sample and its impurities. For this reason, a roasting pre-treatment is necessary in order to compare samples from different locations which do not have the same impurities.

Another source of possible contaminants is fine sediments and remains of coccoliths trapped inside the chambers of foraminifera. To eliminate such a contamination, Emiliani[9] crushes the foraminiferal samples, washes them in distilled water in an agitator for 15 min and dries them at 90 °C. At Gif, we use Shackleton's method,[11] that is, the foraminifera tests are washed in an ultrasonic bath for a few seconds and then rinsed three times in methanol.

Extraction of carbon dioxide

In the acid extraction of carbon dioxide from the carbonate only two-thirds of the carbonate oxygen goes into the carbon dioxide. This partition of oxygen atoms must occur reproducibly. In McCrea's classical method[3] the samples and the 100 % phosphoric acid are first degassed under vacuum for at least 24 h and then reacted under agitation at 25.2 °C in a thermostat bath. After a reaction time of at least 4 h, the carbon dioxide is condensed with liquid nitrogen. It is then released by warming up with a mixture of acetone and dry ice (which retains the water) and transferred to the mass spectrometer.

Shackleton[18, 19] has developed a method which allows accurate analyses of carbonate samples weighing as little as 0.1 mg. The carbonate sample is placed in a small thimble and crushed. After roasting under vacuum and cooling, the thimble is placed in a reaction vessel (Fig. 3.2). Before acidification of the sample, the reaction vessel, with 100 % phosphoric acid in the side arm, is outgassed for a few hours at 50 °C to a pressure of less than 10^{-5} mm Hg. The acidification takes only a few minutes. The gas which evolves is a mixture of water and carbon dioxide. Water vapor is condensed out by passing the mixture through a trap cooled with dry ice and acetone (-80 °C) and the carbon dioxide is then frozen in a liquid nitrogen trap.

It is very important that the reaction takes place at low pressure and that the carbon dioxide and water are not in contact with each other. (We have previously[7] described the problems with reaction occurring at other than low pressure.) Following the reaction, any traces of non-condensible gas are pumped out and the sample is transferred to the mass spectrometer system. This procedure, routinely used at Gif, takes only about 20 min.

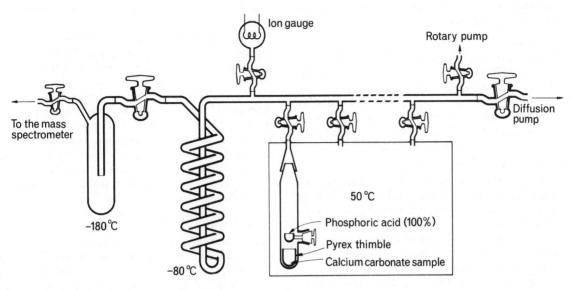

Fig. 3.2. Carbonate preparation using Shackleton's method.[18,19]

The problem of grain size

Emiliani,[9] using McCrea's preparation method, showed that the grinding of the laboratory standard Marble V to fine powder tends to enrich the lighter isotope. (^{16}O) by 0.2 ‰. He also stated that this effect is not detected if the grinding is performed under chloroform.

Fritz & Fontes[20] analyzed limestone with grain sizes ranging from 250–315 μm to less than 40 μm. Their results show an enrichment of 0.7 % in the finest fraction. They concluded that precise paleotemperature measurements can be obtained with McCrea's method only if samples of the same grain size are analyzed. Walters, Claypool & Choquette[21] reacted pure calcite at 25 °C with phosphoric acid and showed that the $^{18}O/^{16}O$ ratio of the evolved carbon dioxide increases by as much as 3 ‰ during the course of the reaction. They also found that the speed of the reaction increased with decreasing grain size and so attributed the change in isotopic composition of the evolved carbon dioxide in part to the particle size of the reacting carbonate (as did Fritz & Fontes), and also suggested the possible importance of particle surface strain caused by crushing and grinding.

As it is not practical to reduce small samples of foraminifera to grains of uniform size, Shackle-ton[18] studied his reaction method for possible grain size effects. He found that at 50 °C the very fast reaction of phosphoric acid with carbonate gave no systematic variation of $\delta^{18}O$ with grain size over the size intervals 125–180 μm to less than 45 μm, which is the size range commonly used for isotopic analyses.

From all these results, it may be concluded that the grain size effect is probably strongly related to the speed of the reaction and this has to be taken into account by all users of McCrea's method. In Shackleton's method at 50 °C, the reaction rate is so fast that all the carbonate sample, whatever the grain size, should be completely reacted in a few minutes, and the evolved carbon dioxide should have an isotopic composition independent of grain size.

One point remains to be considered: why does the grinding of Marble V in carbon tetrachloride or carbon trichloride suppress the grain size effect? Possibly, in a humid ambient environment (such as Emiliani's laboratory at Miami), isotopic exchange with water vapor may have occurred during grinding.

The paleotemperature scale

Establishing the relationship between the $^{18}O/^{16}O$ ratio in calcium carbonate precipitated by

49

marine shells and the temperature of this precipitation is the first necessary step in paleoclimatological studies.

The Epstein et al. *paleotemperature scale*

Urey and his colleagues[4] grew calcareous marine animals at known temperatures, measuring simultaneously the $^{18}O/^{16}O$ ratio of the carbonate shells (δ^{18}) and the $^{18}O/^{16}O$ ratio of the water (δ^{18}_w) in which the shell grew. A least-squares fit of their data gives the paleotemperature equation (where t is expressed in °C):

$$t = 16.5 - 4.3 \ (\delta^{18} - \delta^{18}_w) + 0.14 \ (\delta^{18} - \delta^{18}_w)^2$$

Craig[22] pointed out that the δ values used by Epstein *et al.*[4] to generate this equation were raw δ values uncorrected for any systematic errors in the measurements of the mass spectrometers. As these systematic errors will be different on different instruments, corrections must be applied in order to give well-defined isotopic values independent of laboratory measurements. The various correction factors involved have been listed by Craig,[23, 24] who suggested a modified equation based on a least-squares fit to the corrected data:

$$t = 16.9 - 4.2 \ (\delta^{18} - \delta^{18}_w) + 0.13 \ (\delta^{18} - \delta^{18}_w)^2$$

Calcium carbonates are said to be deposited in isotopic equilibrium if they follow the Epstein paleotemperature scale. The work of Keith & Weber[25] demonstrated that many organogenous carbonates do not deposit their shells in isotopic equilibrium with their growth medium. It is thus necessary to check whether or not this also applies to the foraminifera.

Isotopic composition of benthic foraminifera

It is relatively easy to check the equilibrium precipitation of the benthic foraminifera test carbonate as this occurs in a known place where the temperature and $^{18}O/^{16}O$ ratio of the water can be measured and then compared with the $^{18}O/^{16}O$ ratio of the foraminiferal tests living in that water.

Smith & Emiliani[26] analyzed modern samples of benthic foraminifera, collected at different depths on the continental shelf and slope off western-central America. The isotopic temperatures obtained agree closely with the temperatures measured in the field. But the four samples in the same core (C9) raised from a water depth of 885 m indicated a scattering of 0.26 ‰ in $\delta^{18}O$. The corresponding temperature scatter

is a little more than 1 °C, which is much greater than the annual temperature variations at this depth.

Vinot-Bertouille & Duplessy[27] have shown that individuals of the same species of large, heavy (5–10 mg), recent benthic foraminifera living in shallow waters differ by up to 3 ‰ in their carbon and oxygen isotopic compositions and yield calculated paleotemperatures higher than the actual ambient values. This effect is not due to the variation of magnesium carbonate in foraminiferal tests but depends on the species. If a large number of individuals was analyzed, a mean value could be obtained which is different for different species even if the respective populations have developed in the same conditions. Some of these species are in isotopic equilibrium but most of them are not.

In studying the fossil benthic foraminifera, Duplessy, Lalou & Vinot[28] observed that different species of benthic foraminifera from the same level in an Atlantic core yielded different oxygen isotopic values. Because the core was raised from the flat top of a large seamount, the sediment is not reworked and these discrepancies confirm the disequilibrium observed for most of the benthic foraminiferal species. The two benthic species *Planulina wuellestorfii* and *Pyrgo* sp. gave the same isotopic variations all along the core, indicating that the variations have a paleo-oceanographical meaning and that the difference between the isotopic composition of a benthic species and the equilibrium value has remained constant in the past.

Further studies[11, 18, 29] have confirmed this phenomenon and indicate that very precise paleoclimatological information cannot be obtained by analyzing a mixture of benthic species. Since the deviation between any two species is essentially constant, however, a generalized benthic curve can still be obtained by taking one species as a reference and correcting the $\delta^{18}O$ values of the other species by some constant whose value will depend on the particular species involved.

The large scattering observed by Vinot-Bertouille & Duplessy[27] has not been evident in small, deep benthic forms. Replicate analyses of samples (about 0.3 to 0.5 mg of calcium carbonate) of *Planulina wuellestorfii*, *Cibicides pseudo ungerianus* or *Nonion* sp. present the same standard deviation as pure carbonate standards; for

Pyrgo sp. the standard deviation is generally higher. At present, we do not know whether this greater scattering for *Pyrgo* sp. was due to faulty cleaning of the samples or to metabolic processes. In the case of large, shallow water benthic foraminifera, part of the scattering could have resulted from seasonal temperature fluctuations, but these variations are not large enough to account for all the isotopic variations observed.[27]

The final step necessary in order to use the $^{18}O/^{16}O$ ratio of benthic foraminifera as an index of the temperature variations of deep water in the past is the establishment of an isotopic temperature calibration for the particular organisms in question. This is necessary for two reasons. First, most of the foraminifera are in pure calcite[30-32]. The Epstein paleotemperature scale is an empirical best fit to data obtained on mollusc carbonate. Some of the molluscs deposit their shell in calcite, others in aragonite; Tarutani, Clayton & Mayeda[33] report a small but significant fractionation between aragonite and calcite.

Secondly, the temperature range in which Epstein *et al.* grew their molluscs is 7–29 °C. Most of the deep benthic foraminifera live today at temperatures between −1 and +6 °C. This is outside the temperature range studied experimentally.

Shackleton[18] has presented a relationship of calcite–water fractionation in the temperature range 0–25 °C by expanding the results of O'Neil, Clayton & Mayeda[34] (applicable to 0–500 °C) at about 16.9 °C. The relationship obtained is similar to the Epstein paleotemperature scale in the temperature range of 15 to 25 °C, but at lower temperatures the two scales diverge significantly. With the present state of knowledge, it is therefore not practical to use a paleotemperature scale more precise than the following form for temperatures below 16.9 °C:

$$t = 16.9 - 4.0\,(\delta^{18} - \delta^{18}{}_w)$$

In his analysis of cores from the Pacific Ocean, where the temperature range is 0.8 °C to 7 °C, Shackleton noted that the benthic foraminifera *Uvigerina* and *Pyrgo murrhina* have $^{18}O/^{16}O$ ratios which fit the preceding paleotemperature scale and thus concluded that these forms deposited their test in isotopic equilibrium. Other species, like *Planulina wuellestorfii*, *Nonion* sp., *Cibicides* sp., show deviations from

this paleotemperature scale by a constant value. However, more precise measurements are needed to obtain a precise relationship as *wuellestorfii* has been reported to be isotopically lighter than the equilibrium value by 0.64 ‰ in the Pacific,[18] by about 1 ‰ in the North Atlantic Ocean,[28] and by 1.15 ‰ ($\pm$0.03) in the Norwegian Sea.[29] We do not know whether these different results are real or are a consequence of mixing and disturbances in the analyzed cores.

Clearly, then, the isotopic composition of the benthic foraminifera depends not only on the temperature and $^{18}O/^{16}O$ ratio of the sea water in which the animal lives, but also on the species. Isotopic variations of the same benthic species in a deep-sea core record the resultant of two effects: the variations of the temperature and the $^{18}O/^{16}O$ of sea water.

Oxygen isotopic composition of planktonic foraminifera
The problems with planktonic foraminifera are somewhat different. Most of the species live in the temperature range where the Epstein *et al.* paleotemperature scale applies. However, living specimens have been found not only near the surface but also as deep as 1500–2000 m.[35,36]

Thus when we analyze specimens of planktonic foraminifera from sediments or plankton tows, we cannot be certain of the real depth at which the test was secreted.

Emiliani[37] reported that different species of pelagic foraminifera from the same core levels have different oxygen isotopic composition. Assuming that the calcitic tests were deposited in isotopic equilibrium, he interpreted differences as an indication that different species of pelagic foraminifera deposit their shell material at different depths: the deeper the colder. Moreover, a comparison of samples from the Gulf of Mexico, the equatorial Atlantic and the eastern equatorial Pacific indicated that the same species may vary considerably in its depth habitat in adjusting itself to proper temperature and water density. This interpretation has received strong support from collections made using nets that could be opened and closed at depth. The stratification of pelagic foraminifera collected *in situ* by such devices is in good agreement with that determined by oxygen analysis.[38-40] From the above studies, later workers have tacitly assumed that planktonic foraminifera deposit their tests in isotopic equilibrium, and have employed the oxygen isotopes as a tool for ecological and micropaleontological studies.

Emiliani[41] has compared the isotopic composition of three size groups of recent pelagic foraminifera from the Caribbean Sea. (The smaller samples are interpreted as the younger specimens.) There was no detectable difference in the $^{18}O/^{16}O$ ratio between size groups for *Globigerinoides ruber*, *G. sacculifer*, *Globoquadrina eggeri*, *Pullienatina obliquiloculata* and *Sphaeroidinella dehiscens*, indicating that these species deposit their shell material at constant depths. A small decrease in the $^{18}O/^{16}O$ ratio with specimen age was observed for *Globigerinoides conglobata* and *Orbulina universa* from the Mediterranean which was interpreted as a slightly decreasing depth habitat with age. On the other hand, a marked increase in the $^{18}O/^{16}O$ ratio with specimen age for *Globorotalia menardii* and *G. truncatulinoides* from the Caribbean suggests that these species continue to grow their shells while sinking through the water column, a conclusion in agreement with the observations of Bé & Lott.[42] In marked contrast with the Caribbean specimens, *G. menardii* does not seem to change its depth habitat in the Pacific and Indian Oceans appreciably.

Longinelli & Tongiorgi[43] have found marked differences in oxygen isotopic composition between right- and left-coiling foraminifera, both living and fossil. These differences were explained as indicating that the two forms deposited their shell material during different periods of the year; the right-coiling forms during the summer, and the left-coiling ones during the cold seasons. This interpretation was supported by field observations in the Sargasso Sea off Bermuda.[44] More generally, Bé[44] observed a seasonal alternation of temperate and subtropical species living mainly in the euphotic zone of the oceanic waters around Bermuda, and remarked that this sequence of planktonic species with unlike ecologic tolerances in the same geographic locality during different seasons is reflected as a mixing assemblage in the sediment. In this case, the variations of isotopic temperature values are both necessarily due to different depth levels but could also be dependent upon seasonal temperature variations within the euphotic zone in which test secretion took place.

Hecht & Savin[45, 46] have gone further and attempted to use isotopic determinations to find whether morphological variations within a population of foraminifera are due to varying ecological stress. They have observed, for example, that individuals of *Globigerinoides ruber* with a diminutive final chamber have a higher $^{18}O/^{16}O$ ratio than normal specimens and have deduced that the presence of diminutive final chambers indicates growth in a stressed environment, such as in deeper, colder water. Other implications of such studies have been discussed by Bé & Van Donk.[47]

In a reassessment of the above studies Shackleton, Wiseman & Buckley[48] reported that living specimens of *G. ruber*, *G. dutertrei* and *Pullienatina obliquiloculata* collected in plankton tows within the isothermal layer of the Indian Ocean have significantly different $^{18}O/^{16}O$ ratios. With the exception of *P. obliquiloculata* the isotopic data give warmer temperatures than actual ones. Isotopic disequilibrium for planktonic foraminifera is inferred, assuming that the analyzed individuals deposited their shell material at the depth at which they were collected. However, this assumption is not warranted since test secretion is a discontinuous process[49,50] and vertical migration in the water column of planktonic individuals has been described.[39] If the reported disequilibrium should be confirmed, it will be important to establish whether the metabolic effect remains constant for each species, as is the case for the benthic foraminifera. Only then will it be possible to use oxygen isotopes for ecological studies. From the data available to date, it seems that this constancy has been established for *Globigerinoides ruber* and *G. sacculifer*, as Emiliani[5] has shown that these two species having the same depth habitat record exactly the same variation during the last four climatic cycles in core A 179–4. Further discussion of the implications of the isotopic disequilibrium of planktonic foraminifera is given by Shackleton *et al.*[48]

Carbon isotopic composition of planktonic foraminifera
The carbon isotopes in foraminifera are not as well studied as the oxygen isotopes. But results obtained so far do indicate that at least in some foraminiferal species isotopic disequilibrium occurs between the carbonate carbon and the carbon dioxide dissolved in sea water.

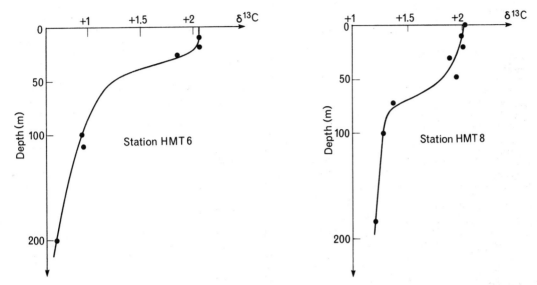

Fig. 3.3. Carbon isotopic composition of total dissolved carbon dioxide as function of depth in equatorial Atlantic waters: Station HMT 6 (4° 32′ N 19° 38′ W); Station HMT 8 (0° 08′ S 18° 32′ W).

Fig. 3.3 shows two typical profiles of the $^{13}C/^{12}C$ ratio of total dissolved carbon dioxide in the equatorial Atlantic Ocean. The high $\delta^{13}C$ value in surface waters reflects isotopic exchange with the atmospheric carbon dioxide. The decrease of $\delta^{13}C$ with depth is related to the carbon dioxide produced from oxidation of organic matter which is generally depleted in ^{13}C.[51] Similar profiles have been observed at all latitudes in the Atlantic[51] and Pacific[52] Oceans.

If we assume carbon isotopic equilibrium with sea water during the foraminiferal shell precipitation, we can predict that shallow depth habitat species (e.g. *Globigerinoides ruber* and *G. sacculifer*) should have similar $^{13}C/^{12}C$ ratios that are heavier than those of the deeper habitat species (e.g. *Globoratalia menardii*).

Table 3.1 presents results of carbon and oxygen isotopic analyses on pelagic foraminifera from the tops of trigger-weight cores raised from the northern Indian Ocean. The oxygen isotopic composition and their differences between species from the same core are very similar to those observed by Emiliani.[37] On the other hand, carbon isotopic compositions are not in agreement with the range predicted by assuming

isotopic equilibrium between the carbonate shell and the carbon dioxide dissolved in sea water: the shallow depth habitat species *Globigerinoides ruber* and *G. sacculifer* do not have the same $^{13}C/^{12}C$ ratios, the former being 0.64 ± 0.20 ‰ lighter than the latter. Furthermore, *Pullienatina obliquilaculata* which lives in shallower depths than *Globoratalia menardii* is not isotopically heavier but, on the contrary, is on the average 0.98 ‰ lighter than *G. menardii*.

Moreover, none of these species is in carbon isotopic equilibrium with their environment: calcite has $^{13}C/^{12}C$ ratios that are about 1 ‰ heavier than the dissolved bicarbonate from which the mineral is experimentally precipitated under equilibrium condition.[53] While no $\delta^{13}C$ profiles have been available from the Indian Ocean, if we assume that the isotopic composition of the dissolved carbon dioxide in the Indian Ocean surface waters is similar to those measured in the Atlantic[51] and Pacific,[52] we can see from Table 3.1 that *Globigerinoides ruber*, *G. sacculifer* and *P. obliquiloculata* are not in carbon isotopic equilibrium with the dissolved bicarbonate; i.e. these species are depleted in ^{13}C. Vinot-Bertouille & Duplessy[27] have reported

53

Table 3.1. *Carbon and oxygen isotopic composition (VSPDB)[a] of pelagic foraminifera from the tops of trigger-weight cores from the northern Indian Ocean*

Core	Latitude	Longitude	Globigerinoides ruber		G. sacculifer		G. menardii		Pullienatina obliquiloculata	
			$\delta^{18}O$	$\delta^{13}C$	$\delta^{18}O$	$\delta^{13}C$	$\delta^{18}O$	$\delta^{13}C$	$\delta^{18}O$	$\delta^{13}C$
V 14101 B	8° 39′ N	58° 34′ E	−1.54	+1.05	−1.37	+1.77	−0.37	+1.18	—	—
V 14104	13° 26′ N	53° 27′ E	−1.98	+1.12	−1.69	+2.08	—	—	—	—
DAL 7 60	23° 15′ N	61° 28′ E	—	—	−1.66	+1.30	−0.51	+1.38	−0.74	+0.55
DAL 1	22° 44′ N	64° 34′ E	—	—	—	—	−0.64	+1.40	−0.60	+0.41
MAH 22	10° 56′ N	52° 20′ E	—	—	—	—	−0.47	+1.45	−0.58	+0.78
MAH 93	14° 23′ N	66° 12′ E	—	—	—	—	−0.34	+1.71	−0.12	+0.59
V 14107	11° 51′ N	46° 45′ E	−2.23	+1.24	−1.57	+1.70	—	—	—	—
RC 9 155	07° 24′ N	72° 48′ E	−2.57	+0.76	−2.20	+1.57	—	—	—	—
RC 12328	03° 57′ N	60° 36′ E	−2.33	+1.19	−2.02	+1.83	—	—	—	—
RC 12331	02° 30′ N	69° 52′ E	−1.44	+0.39	−1.04	+1.46	—	—	—	—
RC 17126	7°——N	72°——E	−2.47	+0.89	−1.75	+1.78	—	—	—	—
V 19176	07° 07′ N	76° 33′ E	−2.04	+1.20	−2.34	+1.50	—	—	—	—
V 19177	07° 35′ N	74° 13′ E	−2.37	+0.99	−2.13	+1.47	—	—	—	—
V 19178	08° 07′ N	73° 15′ E	−2.84	+0.89	−1.95	+1.79	—	—	—	—
V 19183	08° 07′ N	62° 47′ E	−2.89	+1.16	−1.94	+1.55	—	—	—	—
V 19185	06° 42′ N	59° 20′ E	−2.31	+0.97	−2.03	+1.67	—	—	—	—
V 19188	06° 52′ N	60° 40′ E	−1.82	+1.04	−1.74	+1.70	—	—	—	—
RC 12339	09° 08′ N	90° 02′ E	−2.65	+1.12	−2.31	+1.53	−1.03	+1.31	—	—
RC 12340	12° 42′ N	90° 01′ E	−2.93	+0.95	−2.18	+1.55	—	—	—	—
RC 12341	13° 03′ N	80° 35′ E	−2.77	+1.11	−2.68	+1.65	—	—	—	—
RC 12344	12° 46′ N	96° 04′ E	−3.36	+0.92	−3.26	+1.59	—	—	—	—
RC 12 347	09° 20′ N	93° 20′ E	−2.78	+0.90	−2.27	+1.51	—	—	—	—
RC 14 36	00° 28′ S	90° 00′ E	−2.66	+0.92	−2.24	+1.60	—	—	—	—
RC 14 39	05° 51′ N	90° 31′ E	−2.42	+1.08	−2.45	+1.52	—	—	—	—
RC 14 44	05° 58′ S	102° 27′ E	−2.55	+1.35	−2.26	+1.87	—	—	—	—

[a] PDB is the international standard.

depletions in both ^{18}O and ^{13}C in the benthic foraminiferal tests. This depletion assumes a linear relationship and may be explained either by introduction in the calcite test of metabolic carbon dioxide (which is very much depleted in both heavy isotopes) or by a fractionation whereby the lighter isotopes pass through the cellular membranes more readily than the heavy ones. If the metabolic activity is the cause of isotopic disequilibrium, conceivably it could also explain the carbon disequilibrium mentioned above and the non-equilibrium oxygen isotope fractionations between sea water and planktonic foraminifera tests, as noted by Shackleton *et al.*[48]

The long-term isotopic record

In agreement with Emiliani's pioneering work,[5] oxygen isotope analyses of planktonic and benthic foraminifera from various deep-sea cores have generally revealed the same trends of quasi-periodic fluctuations. However a precise interpretation of these fluctuations is still a matter of controversy.

Mixing, reworking and dissolution of the sediments

Isotopic curves are obtained by analyzing monospecific foraminiferal samples from a series of core levels. It is then tacitly assumed that all individuals taken at the same core level lived at the same time. This is an unwarranted assumption because the deep-sea floor is continuously reworked.

Turbidity currents are generated by continental shelf slumps, often triggered by earthquakes.[54,55] Moving downslope, these currents have sufficient erosive power to uncover and mobilize previously deposited sediments, which settle out in a region where decrease in the bottom slope occurs. Thus, in the erosion of slumping areas, hiatuses are generated in the sediment stratigraphic record. On the other hand, in the deposition area, widespread, thick layers of mixed sediments are deposited in a short time interval.[56] Such cores are not suitable for paleoclimatological studies.

Sediment reworking also occurs in protected areas where skeletal remains of planktonic micro-organisms settle on the deep-sea floor,

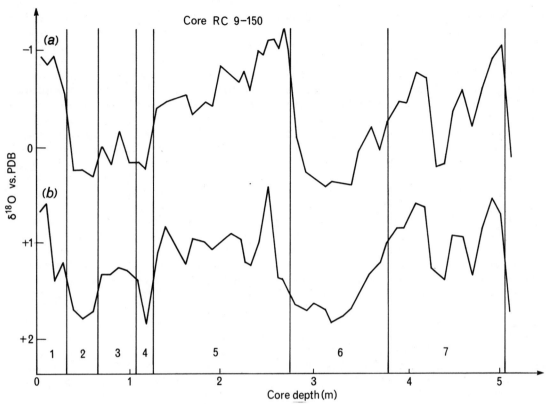

Fig. 3.4. Oxygen isotopic record of the planktonic species (*a*) *Globigerinoides sacculifer* and (*b*) *G. inflata* in the Indian core RC 9–150 (31° 17′ S, 114° 36′ E, 2703 m). The numbers 1 to 7 are Emiliani's stages.[5]

from the overlying water column. As deep water is generally aerobic, benthic macro-organisms live on or in the bottom of the sea floor[57], many of which are mud-dwellers and mud-eaters. As a result of their activity, the sediment is disturbed, reducing the time resolution of the deep-sea cores and smoothing the sedimentological variations; for example, Shackleton & Opdyke[12] compared the peak-to-peak amplitude of isotopic changes measured in cores V 28–238 and V 28–239, both of which were raised from the Salomon Plateau. This amplitude is systematically higher in core V 28–238 which had approximately double the sedimentation rate, so that its time resolution was less affected by the mixing.

The paleoclimatological record will be particularly affected by mixing when an abrupt climatic change occurs. Fig. 3.4 compares the isotopic records of two planktonic foraminiferal species in the Indian core RC 9–150 from the

Australian margin coast. Two abrupt climatic changes are observed, the first one at 10–30 cm (limit between Emiliani's stages 1 and 2) and the second one at 255–275 cm (limit between Emiliani's stages 5 and 6). However, the effect of each climatic change is not reflected at the same core depth by the two foraminiferal species with *Globigerinoides sacculifer* becoming systematically isotopically lighter before *G. inflata*. For core RC 9–150, the sedimentation rate is about 2×10^{-3} cm yr^{-1}.[58] Thus the lag between *G. sacculifer* and *G. inflata* represents about 6×10^3 yr. This lag cannot be explained by any oceanographic pattern as the mixing time of the ocean is about 10^3 yr.[59] The best explanation is given by a mixing model,[60–62] with the warm species (*G. sacculifer*) becoming more abundant as the climate warms up. Upon reaching the bottom, mixing in the sediment introduces warm individuals of the same species deposited in these lower layers during the cold period. Therefore

55

an isotopic analysis performed on a mixture of about 10–20 individuals of *G. sacculifer* will give $\delta^{18}O$ values characteristic of a warm climate before the warming really occurred. The cold temperature species *G. inflata* tends to become relatively less abundant as the climate warms up. The same mixing effect introduces in the upper (warm) layers of the sediment a large number of individuals deposited during the cold period, so that measured $\delta^{18}O$ values of this species indicate a climatic warming after it has really occurred.

Dissolution tends selectively to remove the more easy soluble species[63, 64]; as these species often lived in shallower water, the resultant fossil faunal assemblages look cooler than the living population. Moreover, in a single species, dissolution selectively removes from the population those members that live closer to the surface. Therefore, the fossil assemblages which have suffered more dissolution register a lower isotopic temperature as a consequence of that dissolution. This is evident in core V 28–239; the temperatures indicated by *G. sacculifer* from the top of this core are about 1 °C cooler than those indicated by the same species from the top of core V 28–238.[12]

Interpretation of isotopic records
I shall now focus on the interpretation of the isotopic records of deep-sea cores raised from depths above the compensation depth. While this will minimize the dissolution effects, the mixing effect, which strongly reduces the time-resolution of cores with sedimentation rates less than about 2 cm 10^{-3} yr, must not be forgotten.

As discussed above, the $\delta^{18}O$ value of a monospecific foraminiferal sample will reflect the temperature and $^{18}O/^{16}O$ ratio of the water in which these foraminifera lived. This is true even for species which are not in isotopic equilibrium, if the deviation from equilibrium remained constant. This constancy has been virtually proved in the pelagic and benthic foraminifera by the identity of the isotopic variations recorded during the Late Cenozoic by species from the same depth habitat.[5,28]

As originally stated by Emiliani,[5] both the temperature and sea water $^{18}O/^{16}O$ ratio vary with the climate. As the vapor pressure of $H_2^{16}O$ is greater than that of $H_2^{18}O$, atmospheric water vapor is depleted in heavy isotopes

relative to the sea water from which it evaporates.[65–67] Conversely, condensation from a limited amount of vapor gives an enriched condensate and a residual vapor depleted in the heavy isotopes relative to the initial vapor. This has been confirmed by many investigations which demonstrated that there is a correlation between the $^{18}O/^{16}O$ ratio of precipitations and the temperature of the air; the cooler the air, the lighter the rain.[67] In Greenland and the Antarctic area, precipitations are very depleted in heavy isotopes relative to sea water: 29 ‰ in Camp Century–Greenland;[68] 34.5 ‰ in the crest of the Greenland ice sheet;[69] and from 20 ‰ in coastal stations to 58 ‰ at the pole of relative inaccessibility in Antarctica.[70]

Therefore, as the amount of ice present on the continent increases (glaciation), the water extracted from the ocean will be more and more depleted in heavy isotopes while the remaining ocean becomes isotopically heavier. This ^{18}O enrichment is entirely reflected in the calcite test of the foraminifera. On the other hand, if the temperature of precipitation of the foraminiferal test decreases, the $\delta^{18}O$ of the calcite increases. Thus, heavy oxygen isotope ratios indicate cold climate. Conversely, temperature increases and ice-cap melting will be reflected by light oxygen isotope ratios.

In order to evaluate the two unknown variables of temperature and salinity, Emiliani,[5] using Flint's data,[71] estimated the amount of ice stored on the continents and suggested an isotopic composition for Scandinavian and Laurentide ice sheets not far from the average snow now falling on these areas, i.e. $\delta^{18}O = -15$ ‰. With these assumptions, the glacial/interglacial amplitude of ocean $^{18}O/^{16}O$ variations is 0.5 ‰ and isotopic curves are interpreted as showing that the surface temperature of the Caribbean Sea and the equatorial Atlantic fluctuated by about 6 °C during the most recent climatic cycles.[5,6,8,9] This result is in good agreement with that inferred by micropaleontologists investigating faunal changes represented in the same cores.[76–78] In spite of this striking agreement, many workers have doubted the validity of Emiliani's assumptions for the changing isotopic composition of the ocean during the waxing and waning of the continental ice sheets. Therefore, many attempts have been made to measure separately, either the amplitude of temperature variations or the amplitude of sea water isotopic variations occurring during a Pleistocene climatic cycle.

To check his estimates, Emiliani[72,73] deduced the amplitude of glacial/interglacial temperature variations at low latitude, from methods other than oxygen isotope analysis, obtaining mean values of 7 to 8 °C for the Caribbean and 5 to 6 °C for the equatorial Atlantic, in agreement with previous estimates. Using the recent data of Dansgaard & Tauber,[74] Emiliani then calculated a value of −9 ‰ for the average oxygen isotopic composition of the North American and European ice-caps during the Wisconsin period. This value seems very heavy, compared with that of the present ice-caps; however, Emiliani[72,73] pointed out that the North American and European ice-caps extended into the middle latitudes and thus were near the oceanic areas supplying moisture to the atmosphere, so that precipitation could have fallen on the ice-cap directly as snow, without having undergone the extensive process of fractional precipitation so apparent today in Greenland and Antarctica.

Imbrie, Van Donk & Kipp[75] used the same approach in studying Caribbean core V 12–122, but came to an opposite conclusion. They estimated the amplitude of glacial/interglacial temperature variations (derived from the transfer function method applied to the total fauna[75]) to be only 2.2 °C. The magnitude of pelagic foraminiferal $\delta^{18}O$ change during the shift from late glacial to post-glacial times, is 2.2 ‰. Using a back-calculation in the Epstein paleotemperature equation, these authors estimate the associated change in isotopic water composition to be 1.8 ‰; at least 0.4 ‰ of this change is attributed to a local evaporation precipitation effect, and the balance ($\leqslant 1.4$ ‰) to the global ice volume effect.

Emiliani[5] reported that the amplitude of isotopic variation between glacial and interglacial levels of Caribbean core A 179–4 was greater in the case of shallow water species (*Globigerinoides ruber* and *G. sacculifer*) than in the case of those species living at greater depths (*Globorotalia menardii* and *G. dubia*), which he attributed to smaller temperature variations experienced by the deeper water masses. Lidz, Kehm & Miller[40] extended this work through additional glacial and interglacial intervals using the Caribbean core P 6304–8, previously

studied by Emiliani.[9] Fourteen planktonic foraminiferal species were isotopically analyzed in five warm core levels, four cold core levels and in one level showing intermediate temperature. To evaluate the isotopic change of the water in glacial conditions, they plotted the amplitude of the glacial/interglacial isotopic change against the mean interglacial isotopic composition for each species studied, obtaining a gross relationship in which the deeper species (*Globorotalia tumida, G. truncatulinoides, G. crassaformis*) exhibited the smallest amplitude, which was about 50 % of that of the shallower species. They concluded that variations of the isotopic composition of the sea water cannot account for more than half of the isotopic variations measured in the foraminiferal shells, that is 0.9 ‰.

Lidz *et al.*[40] tacitly assumed that the depth distribution of each species remained constant as the climate varied. Shackleton[79] suggested an alternative interpretation: that the isotopic variations of the shallower specise *Globigerinoides ruber* and *G. sacculifer* are entirely due to change in the isotopic composition of sea water, and that the deeper species must have lived in warmer water during the glacial periods than during the interglacials. One simple explanation is that those species living at some depth migrated upwards to compensate for the density increase concomitant with glaciation. This hypothesis is supported by Emiliania's conclusion that water density is an important factor controlling the depth habitat of pelagic foraminifera.[37] So it seems that this approach is of use in establishing the depth habitat of pelagic foraminifera during the glacial periods but does not allow determination of the amplitude of the glacial/interglacial sea water $^{18}O/^{16}O$ variation.

Evidence from benthic foraminifera
Ocean bottom water is formed in the Antarctic region during the winter, when sea water freezes around Antarctica. As the ice forms, it rejects most of the salt from the water, so that the salinity and density of the water under the ice increase. As a result, this denser water sinks, producing Antarctic bottom water which flows towards the north in all the oceans. In the northern Atlantic ocean, another deep-water mass, the North Atlantic deep water, originates mainly in the Norwegian Sea[80] where an inflow of light warm surface water balances the out-flow

of cold dense water formed by the action of the atmosphere at the sea surface within the basin. In both cases, these water masses, which fill the deeper part of all the oceans, sink at a temperature close to freezing. Thus, even during glacial conditions, the bottom water of the oceans could not have been significantly cooler than today.

from the Pacific core DWBG 114. More precise analyses in the North Atlantic,[28] the equatorial Atlantic,[7] the equatorial Indian[14] and the equatorial Pacific[11] Oceans showed that benthic foraminifera always show variations higher than 1.4 ‰, the precise amplitude depending on the sedimentation rate in the core and thus on the sediment mixing.

Table 3.2. *Oxygen isotope composition of the benthic genus* Nonion *in core MD 73025* (u) *indicates* Uvigerina *analyses to which 0.34%* *has been subtracted*

Depth (cm)	$\delta^{18}O$ PDB	Depth (cm)	$\delta^{18}O$ PDB	Depth (cm)	$\delta^{18}O$ PDB	Depth (cm)	$\delta^{18}O$ PDB
0–3	+3.13	390	+4.15	815	+3.88	1230	+3.66
8	+3.16	400	+4.28	840	+4.36	1240	+3.06
23–30	+2.87	410	+3.99	850	+3.85	1260	+3.51
30–33	+2.82	420	+4.21	860	+4.33	1280	+2.89
40–45	+3.07	430	+4.14	865	+4.30	1310	+3.02
52	+3.02	440	+4.10	885	+4.14	1320	+2.99
60–65	+3.06	450	+4.42	890	+3.96	1330	+2.99
72	+3.14	460	+4.26	910	+3.82	1350	+2.80
80–88	+3.09	490	+4.19	920	+3.92	1360	+2.74
102	+3.28	500	+4.21	930	+3.80	1370	+3.75
112	+3.32	525	+4.27	940	+3.76	1380	+3.56
122	+3.25	540	+4.04	950	+3.86	1390	+4.30
134	+3.28	550	+4.14	960	+3.57	1400	+4.60
145	+3.35	570	+4.24	970	+3.87	1410	+4.25
167	+3.48	580	+4.29	980	+3.72	1425	+3.55(u)
182	+3.55	590	+4.25	990	+3.57	1450	+3.16(u)
202	+3.41	605	+4.28	1020	+3.63		
212	+3.66	625	+4.18	1040	+4.32		
222	+3.82	640	+4.24	1060	+4.14		
232	+3.84	650	+4.07	1070	+4.09		
242	+3.90	660	+4.16	1080	+3.67		
252	+3.92	670	+4.19	1110	+3.81		
262	+3.94	680	+4.39	1120	+3.84		
272	+4.29	690	+4.41	1130	+3.78		
282	+4.64	710	+4.11	1140	+3.62		
292	+4.62	720	+4.21	1150	+3.88		
300	+4.72	730	+4.10	1160	+4.07		
310	+4.65	740	+4.13	1170	+4.26		
330	+4.49	760	+4.09	1180	+4.01		
340	+4.31	770	+4.03	1190	+3.90		
350	+4.47	780	+4.18	1200	+3.69		
370	+4.05	790	+4.04	1210	+3.51		
380	+3.78	800	+4.23	1220	+3.65		

As a consequence, $^{18}O/^{16}O$ ratios of monospecific benthic foraminifera in deep-sea cores will directly reflect the variations of ocean water isotopic composition. Shackleton, who suggested the method,[81] applied it to Caribbean core A 179–4, and demonstrated that the benthic foraminifera record variations of the same magnitude as the pelagic species previously analyzed by Emiliani in the same core.[5] He also measured a difference of about 1.4 ‰ between post-glacial and glacial benthic foraminifera

In order to obtain a precise measurement of the variation of benthic foraminifera isotopic composition during a climatic cycle, we (at Gif) have analyzed the high sedimentation rate core MD 73025 (43° 49' S, 51° 19' E), raised from a depth of 3284 m during the cruise OSIRIS 1 of the French M/S *Marion Dufresne*. The mean sedimentation rate of this southern Indian core was about 10^{-2} cm yr^{-1} until 18000 ago and about 16×10^{-3} cm yr^{-1} between that time and the present. The benthic genus *Nonion* sp. was

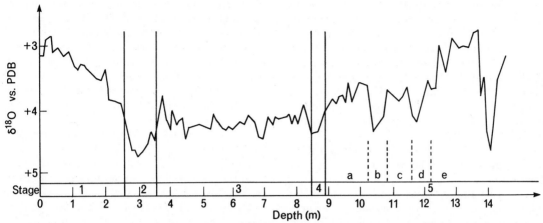

Fig. 3.5. Oxygen isotopic record of the benthic genus *Nonion* in core MD 73025 (43° 49′ S, 51° 19′ E, 3284 m). For an explanation of stages 1 to 5, see Emiliani.[5] The letters *a–e* define substages within stage 5.

analyzed in the upper 14 m of this core. Below this level, *Nonion* specimens were scarce and additional samples of *Uvigerina* sp. were measured. A difference of 0.34 ‰ (±0.11) was found between *Nonion* and *Uvigerina*. In obtaining mean values for each level, we have subtracted 0.34 ‰ from the *Uvigerina* values (Table 3.2). The oxygen isotopic record (Fig. 3.5) is continuous and extends back to stage 5 (using the nomenclature introduced by Emiliani[5]). This stage 5 is cut by what seems to be a sharp glacial episode of short duration. It is not clear whether this episode is real or results from a disturbance in the sediment record. The maximum $\delta^{18}O$ value during stage 2 (mean $\delta^{18}O$ of the levels 280–310 cm) is +4.66 ‰. The minimum $\delta^{18}O$ value during stage 1 (mean $\delta^{18}O$ of levels 0–45 cm) is +3.01 ‰. Therefore, from stage 2 (most recent glacial) to stage 1 (present interglacial) the $\delta^{18}O$ values of the benthic foraminifera decrease by 1.65 ‰. Lower values, measured in cores with lower sedimentation rates, are a consequence of sediment mixing which smooths the isotopic record.

The bottom temperature at the coring site is estimated to be 0.6 °C (±0.1) from temperature measurements obtained during the same cruise of the M/S *Marion Dufresne*.[82] The freezing point of surface sea water is −1.9 °C, and the adiabatic warming for water sinking from the surface to 3000 m is 0.2 °C[83], so that the lower limit of temperature for any sea water at 3000 m is thus −1.7 °C. Therefore at the location of

core MD 73025 the largest possible magnitude of bottom water cooling would be 2.3 °C. Such a cooling would induce a $\delta^{18}O$ increase of 0.55 ‰, so that the minimum value for sea water $\delta^{18}O$ change between stage 2 and stage 1 may be taken as 1.1 ‰.

From these measurements, we conclude that the amplitude of glacial/interglacial $\delta^{18}O$ sea water variations is greater than 1.1 ‰ and can be estimated as 1.6 ‰ if we assume no bottom-water temperature variations in the past. Thus any model which suggests a value smaller than 1.1 ‰ for this amplitude is physically impossible and must be rejected.

Ice-cap evidence

The ^{18}O content of sea water is directly related to the extent of the continental ice volume; the larger the ice sheets formed on the continent, the heavier the isotopic composition of the sea water. The present isotopic composition of the Greenland and Antarctic ices was used by many authors to evaluate the sea water $^{18}O/^{16}O$ change during a glacial period.

The factors governing the isotopic composition of high-latitude precipitations are well known.[67,84–86] Using Dansgaard's results,[66] Olausson[87] calculated the average isotopic composition in the Greenland and Antarctic surface ice sheets. He assumed that a two-dimensional average is a good approximation of the three-dimensional mean isotopic composition of the whole ice sheet. He then extended the calculations to the most recent glacial ice sheet using various air temperature estimates, and obtained a value of $\delta^{18}O = -33$ ‰ for the isotopic composition of continental excess ice. This value is much lighter than Emiliani's estimate ($\delta^{18}O = -15$ ‰). Olausson suggested that the entire planktonic foraminifera isotopic change may be ascribed to change in the isotopic composition of the ocean water. This suggestion has been criticized by Emiliani[9] who objects to the arbitrary nature of the air temperature values assumed by Olausson for the glacial Pleistocene.

New calculations have been presented by Shackleton[81] who used Nye's dynamic ice sheet model[88] in which a glacier is assumed to be in a state of dynamic equilibrium; snow accumulates above the firn line, and snow and ice melt in the ablation zone. Ice formed in the accumulation zone moves towards the ablation zone along flow lines that are deeper in the ice sheet as the site of formation approaches the ice divide.[88–90] Thus, ice formed in the periphery of the accumulation zone follows superficial flow lines and is melted in the ablation zone within a short time. On the other hand, the bulk of the ice sheet originates from precipitation far inland on the glacier. As a consequence, ice in the ice sheet periphery has a much lower $\delta(^{18}O)$ value than precipitation in the source area of the upper layers,[86,91] and the residence time for the most isotopically light ice (formed in the center) is considerably greater than that for the ice formed on the periphery.

Using this flow model and without assuming any decrease in air temperature during glacial stages, Shackleton[81] calculated for the continental excess ice a value close to that obtained by Olausson[87] and concluded (in good agreement with the benthic foraminifera evidence) that oxygen isotopic variations of foraminifera in deep-sea cores reflect only the sea water oxygen isotope variations due to waxing and waning of the continental ice-caps.

These evaluations have received strong support from drillings through the Greenland and Antarctic ice sheets. Stable isotope variations of the ice provide paleoclimatic information and enable more precise estimation of the isotopic composition of the ice during glacial periods.

Two long and complete ice cores have been

recovered from ice sheets and analyzed: one at Camp Century, northwest Greenland,[68] and the other at Byrd Station, West Antarctica.[92] Analyses have also been made on samples from the Vostok bore-hole, East Antarctica, which reach a depth of 950 m.[96]

The isotopic composition of polar glacier ice ($^{18}O/^{16}O$ or D/H) depends chiefly on the temperature of formation of the snow at the sheet. Decreasing temperatures of formation lead to decreasing $\delta^{18}O$ in snow.

The age of an ice at various levels is estimated from the measured accumulation rates and from calculations of thinning through flow using Nye's model.[88] Layers near the bottom of an ice sheet, because of their extreme thinning by flow, are often difficult to age precisely.

As an example, calculations applied to the Camp Century ice core gave an age of 157 000 yr B.P. for the bottom layer. Oxygen-18 analyses of the ice indicated large variations which are correlated with the well-known continental and oceanic paleoclimatic record.[95] To establish the time scale, Dansgaard, *et al.*[95] have calculated the Fourier power spectrum of the $\delta^{18}O$ data obtained on the top 283 m of the ice core. Two dominant cycles with respective periods of 78 and 181 yr appear. The fact that both periods are close to those noticed previously for sunspot cycles has given the authors confidence in the time-scale they obtained for that part of the ice core, that is, about the last eight centuries.

Between 296 m and 1150 m depth in the core, the same power spectrum analysis indicates a 350-yr period which the authors correlate with the 405-yr period of ^{14}C concentration fluctuations in the atmosphere (previously detected in tree rings). Therefore the time-scale for this part of the ice core is adjusted in order to obtain the 405-yr period. This corrected time-scale is considered correct to within 10 %, as well-known interstadials (Bølling, Allerød, Dryas, etc.) appear in the ice record with ages identical to those determined by ^{14}C or by the varve chronology. For the whole length of the core (1373 m) power spectrum analysis indicates a 2000-yr period back to 45 000 yr B.P. Beyond this age, the periodicity seems to vary from 2000 to 4000 yr. The authors suggested correlation of these calculated periods with the 2400-yr period noted in the tree-ring^{14}C concentration variations. By applying this 2400-yr period to the isotope record of the whole ice core,[95] the age of the bottom of layer becomes 125 000 yr B.P. instead of 157 000 yr B.P. as obtained by Nye's model. These results clearly show the difficulty and some ambiguities involved in the dating of deep ice cores.

Nye's model has also been applied to the Byrd[92] and Vostok[96] ice cores. Both cores record the most recent glacial period but their time-scales are much less precise than that of the Camp Century ice core. Johnsen *et al.*[97] have attempted to apply power spectrum analysis to the Byrd isotopic results. They concluded that this method does not give a reasonable time-scale, and suggested several tentative time-scales which have to rely on the Camp Century record. At present, it seems that pole-to-pole correlations have remained uncertain, except for the more pronounced features and general trends.

Precise temperature determinations deduced from ice isotopic record have been obtained only for the past century.[69] An empirical equation relating the mean $\delta^{18}O$ of precipitation with mean temperature t (°C) of surface air in northern latitudes is[67]

$$\delta = 0.7t - 13.6$$

Direct comparisons between $\delta^{18}O$ of the snow deposited during the last century at Camp Century and the measured air temperature at Upernavik on the West Greenland coast suggest that the equation is also valid for temporal variations of short duration in the precipitation δ and temperature at a given locality.[68] These results have been recently extended to the past several centuries.[69]

It is, however, difficult to deduce from a long, isotopic ice record the past air temperature variations at the coring site, for the following reasons.[68]

(1) The deeper strata originated further inland where more severe climatic conditions could have prevailed.

(2) The isotopic composition of sea water which provides the moisture for precipitation has changed; but this effect is smaller than 2 ‰ and theoretically could be taken into account with reasonable accuracy using data on the benthic foraminifera (see above).

(3) The ratio of summer to winter precipitation

and the main meteorological wind pattern[99] could have changed.

(4) The flow pattern and thickness of the ice sheet might have changed. Raynaud & Lorius[93,94] showed that the total gas content in ice at Camp Century 1200–1400 m down (that is, the Wisconsin–Würm ice) is on average 12 % lower than the Holocene ice. They concluded that the site formation altitude of the Wisconsin ice was 1200–1400 m higher than that of the present Camp Century. The temperature decrease related to this altitude change (without any global temperature variation) accounts for more than half of the 11 ‰ isotopic variation between the Holocene and last Wisconsin ice.

For the Byrd core, similar measurements of the total gas content indicate that the site formation altitude of Wisconsin and Holocene ice were similar.[94] The amplitude of the oxygen isotope variation is only 7 ‰[92], very similar to that measured at the Vostok Station whose formation site is also assumed to have experienced only small altitude changes.[96] These isotopic data allow estimation of the mean isotopic composition of ice sheets during the most recent glacial period. At present, the most precise budget has been presented by Dansgaard & Tauber[74] who obtained the following results.

(1) For the Laurentide ice sheet, the ice divide ran very close to the present isoline, $\delta^{18}O = -20$ ‰ for the mean annual precipitation. The surface of the ice sheet was about 2 km above the present ground level. Due to this altitude effect, the mean surface temperature was at least 13 °C lower than the present annual mean at ground level. It corresponds to a lowering of 9 ‰ for the mean $\delta^{18}O$ values of the precipitation in this area. The general temperature decrease during the most recent glaciation produced a 4 ‰ lowering for the mean isotopic composition of the precipitation.[94] $\delta^{18}O = -30$ ‰ is thus considered as an upper limit for the mean isotopic composition of the North American ice sheet during the Wisconsin.

(2) For the European ice sheet, $\delta^{18}O$ of the precipitation should have been as low as that of the same latitude in the North American continent, since the warm Gulf

Stream was completely cut off from the Norwegian Sea and displaced 15° of latitude to the south in the eastern Atlantic[98] (see also below) during glacial times.

(3) For the Antarctic ice sheet a 5 ‰ $\delta^{18}O$ lowering, compared to the present isotopic composition, has been estimated for the Wisconsin ice. (Results from the Byrd and Vostok drillings indicate that a 7 ‰ lowering seems more realistic[92,96]).

In summary, an upper limit of $\delta = -30$ ‰ can be placed for the mean isotopic composition of continental excess ice during the latest glacial maximum. The isotopic balance between sea water and continental ice demonstrates that the increase of the mean $^{18}O/^{16}O$ of ocean water during the last glacial period was at least 1.2 ‰, in line with Olausson's estimates and with the benthic foraminifera evidence.

The overall pattern

The amplitude of glacial/interglacial sea water isotopic change is greater than 1.2 ‰, based on the benthic foraminifera and ice-cap evidence. A probable value for this amplitude in the Pacific and Indian Oceans (which have similar water masses) would be 1.6 ‰, by assuming no temperature change for the Indian Ocean bottom water. This value can be measured in ideal deep-sea cores free of sediment mixing, such as core MD 73025 and other cores with high sedimentation rates.[100]

Micropaleontological results indicate that the amplitude of glacial/interglacial sea surface temperature variation in the tropical and equatorial areas is about 2 °C in most of the oceans.[101] The inferred isotopic change is thus 0.5 ‰, corresponding to less than 30 % of the sea water $\delta^{18}O$ changes.

Thermodynamics requires that both the temperature and sea water effects be recorded in pelagic foraminifera. In near-perfect cores (e.g. core P 6304–9) isotopic variations of more than 2 ‰ have been observed.[9] In such cores, the temperature effect can be calculated by subtracting the benthic record from the pelagic record. This approach has been used by Duplessy, Chenouard & Reyss[7] in the equatorial Atlantic and by Shackleton & Opdyke[11] in their re-interpretation of Emiliani's Caribbean results. The glacial temperature decrease thus estimated is between 2 and 3 °C, in agreement

with the CLIMAP micropaleontological estimates.[101]

In most cases, however, sediment mixing has smoothed the isotopic record, causing its amplitudes of variation to reduce. It becomes then difficult, if not impossible, to calculate the temperature effect which is small compared to the sea water effect. In spite of this smoothing effect, $^{18}O/^{16}O$ variations remain detectable in cores whose sedimentation rate is higher than 5–10 mm per 1000 yr. These variations reflect mainly the sea water effect. As the time constant for oceanic mixing is about 1000 yr, we expect sea water isotopic changes to be spacially synchronous. Foraminiferal isotopic analysis thus enables correlations between wide-spaced deep-sea cores to be made (see Fig. 3.1). It should therefore be emphasized that the time sequence which Emiliani[9] was able to obtain from analyses of many deep-sea cores is the best stratigraphic reference for the whole Brunhes epoch.

Paleo-oceanographic applications

Besides providing a stratigraphic tool as discussed, oxygen isotope studies of pelagic and benthic foraminifera can also reveal past changes in the oceanic circulation patterns.

Agulhas current and subtropical convergence fluctuations in the Indian Ocean

Bé & Duplessy[58] have analyzed two mid-latitude sediment cores from opposite margins of the Indian Ocean. Core RC 1769 (31° 30′ S, 32° 36′ E, 3308 m deep) is located 500 km southeast of Durban, Africa, and core RC 9–150 (31° 17′ S, 114° 36′ E, 2703 m deep) comes from the continental slope 125 km northwest of Perth, Australia. Although the latitude of both cores is the same, the surface water over the African site is on average 2.5 °C warmer than the water at the Australian location. This temperature contrast is related to the strong Agulhas current which transports warm waters southwards along the African coasts, whereas near southwest Australia an eastern boundary current is weakly developed.[102] Surface waters are 35.5 ‰ at the location of the African core and 35.6 ‰ at the location of the Australian core. Their oxygen isotopic composition is therefore the same within 0.1 ‰.[103]

Fig. 3.6 compares the $\delta^{18}O$ time records of the same species *Globigerinoides sacculifer* in both

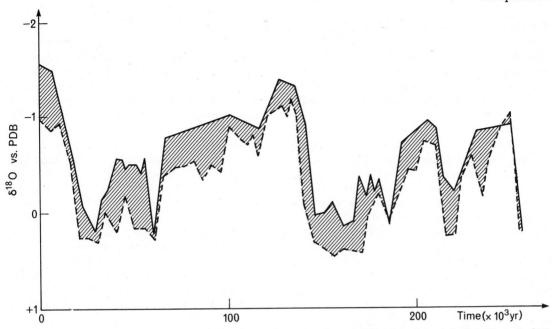

Fig. 3.6. Oxygen-18 variations of *Globigerinoides sacculifer* vs. time in core RC 9–150 (——) (31° 17′ S, 114° 36′ E, southwest Australia) and in core RC 1769 (– – – –) (31° 30′ S, 32° 36′ E, southeast Africa).

cores. In the core tops, pelagic foraminifera from the African site are isotopically 0.6 ‰ lighter (that is, 2.5 °C warmer) than those from the Australian site, in agreement with the present oceanographic data.

During glacial conditions, a stronger latitudinal dependence of the isohalines has been inferred.[101] We may thus assume that during the whole Late Quaternary, the difference between the $\delta^{18}O$ values of sea water at the two sites has remained zero. Therefore, the past surface water temperature difference between Africa and Australia can be computed from the $\delta^{18}O$ difference of the pelagic foraminifera from the same species in the two cores.

At present, the strong Agulhas current deflects the isotherms southward along Africa, whereas off Australia the eastern boundary current is so weak that there are no strong gradients of the isotherms.[104] The measurement of the temperature difference between Africa and Australia is thus directly related to the strength of the Agulhas current: Africa was warmer when the Agulhas current was stronger.

Fig. 3.6 is an example of temperature measurement using ^{18}O analysis and shows that during glacial stages 2, 4 and 6, African and Australian surface waters had almost the same temperature, an indication that the Agulhas current was much weaker than today. Moreover, comparison of the two isotope records (Fig. 3.6) clearly shows that the temperature difference between Africa and Australia was never as high as today. These results have been confirmed by micropaleontological studies and extended over the past 540 000 yr (Bé & Duplessy, unpublished). During this period, African surface waters were always cooler than today. The general weakness of the Agulhas current during glacial stages and most parts of interstadial stages has been related to a more northern position of the subtropical convergence in the Indian Ocean.

Laurentide ice sheet meltwater in Gulf of Mexico
The Laurentide ice sheet advanced to its maximum 18 000 yr ago, and covered 13×10^6 km² of the northern North American continent. After that time, this ice sheet began to shrink; its melt discharged via the Mississipi river.

Oxygen isotope measurements of *Globigerinoides sacculifer*[105] and *G. ruber*[106] in Late Quaternary deep-sea cores from the Gulf of Mexico show anomalously light values superimposed on a characteristic oceanic oxygen isotopic curve.

This peak was interpreted as a result of an influx of isotopically light glacial meltwater via the Mississippi River. It produced in the Gulf of Mexico a low salinity (31–33 ‰) surface layer which persisted until the time when melts of the Laurentide ice sheet discharged via the Hudson River.

Radiocarbon ages on bulk core material have been obtained.[106] Taking into account some possible bias by older carbonate, the age of the isotopic peak was estimated at 11 000–12 200 B.P. corresponding to the Valders ice re-advance.[107] The peak has been interpreted as a surge, a catastrophic event that could be the explanation for the deluge stories common to many Eurasian, Australian and American traditions.

Paleo-oceanography of the Norwegian Sea

The Norwegian Sea is today a typical Mediterranean basin[108] in that inflow of light, surface water balances the outflow of deep dense water formed by cooling and evaporation at the sea surface within the basin. The outflowing, deep water contributes to the North Atlantic deep water and has been followed by temperature and salinity measurements from the North Atlantic to the Pacific Ocean.[109]

Micropaleontological and sedimentological studies[110] indicate that during the most recent glacial period the Norwegian Sea was ice covered, probably all year round. Oxygen-18 analyses made on *Globigerina pachyderma* from a Norwegian Sea core K 11 (71° 47′ N, 01° 36′ E, 2900 m deep) have indicated the presence of a very cold water mass during isotopic stages 2, 3, 4, and last part of stage 5.[[111]]

Oxygen-18 analyses of benthic foraminifera from the same core have shown that during the latest glaciation, the Norwegian Sea deep water was warmer than its modern equivalent[29] and it had the same temperature, oxygen isotope ratio and salinity as the North Atlantic deep water. This suggests that the glacial Norwegian deep water was not produced by the sinking of surface waters, as it is now, but rather was a part of the North Atlantic deep water invaded into the Norwegian Basin. If this was the case, it implies from the water budget that during glacial times surface currents flowed from the Norwegian Sea to the Atlantic Ocean, which is the reverse of the present situation.

Other developments

I have been primarily concerned here with the application of isotopic analysis of oxygen in benthic and pelagic foraminifera in order to increase our knowledge of the Quaternary paleoclimate. However other isotopic studies are also in progress which both complement the work on foraminifera and extend the time periods involved into the Prequaternary.

Paleoclimatic information on the continent has been obtained from the $^{18}O/^{16}O$ variations of cave concretions which record the isotopic composition of precipitations,[112, 113] and from the hydrogen, carbon and oxygen isotope analysis of organic matter. Isotopic analyses of wood from central Germany were correlated with the existing weather records for the years 1712–1954 and experimental relationships between air temperature and the $^{13}C/^{12}C$, D/H and $^{18}O/^{16}O$ ratios from organic matter have been determined.[114] The proportion of plants following the Calvin photosynthetic pathway relative to that following the Hatch Slack photosynthetic pathway has also been suggested as a climatic index.[115]

In order to follow oceanic surface paleotemperatures, Labeyrie[116] determined the isotopic fractionation between water and silica for both diatoms and sponges, as a function of the temperature of formation. Because the slope of Labeyrie's experimental regression line is close to that of the carbonate paleotemperature scale, accurate values of $\delta^{18}O$ for sea water cannot be extracted by simultaneous solution of the two linear equations; however, since diatoms live in the euphotic zone, the comparison between $\delta^{18}O$ in fustules of diatoms and in associated tests of benthic foraminifera may be used to follow the variations of surface water temperature during the Pleistocene.

Margolis *et al.*[117] demonstrated that calcareous nannofossils showed isotopic variations similar to that presented by pelagic foraminifera from the same DSDP core and concluded that coccoliths can also be used as a paleotemperature indicator.

The pioneer work of Emiliani[118] showing the temperature decrease of Pacific bottom waters during the Tertiary has been extended with the availability of Tertiary sea-floor sediment samples through the DSDP $\delta^{18}O$ analyses of pelagic and benthic foraminifera indicate that prior to the middle Miocene, high- and low-latitude temperatures changed in a parallel fashion.[119, 120] The Tertiary–Cretaceous boundary is only marked by a small temperature drop with the main temperature drop occurring in the earliest Oligocene. This rapid 5 °C reduction in bottom-water temperature may be attributed to the onset of Antarctic bottom-water formation at a temperature close to the freezing point.[121] A temperature rise through the Early Miocene was followed in the Middle Miocene by a sudden divergence of high- and low-latitude temperatures, which reflects the establishment of the Antarctic ice sheet.[119, 120] The first northern ice sheet accumulated rather rapidly 2.6 Myr ago[122] and since that time climatic conditions have been modulated by the waxing and waning of both polar ice sheets.

I am greatly indebted to J. Labeyrie for continuous support and encouragements; to T. L. Ku and M. Kritz for reviewing and considerably improving the manuscript; to B. Lecoat, J. Antignac and A. Beaugrand for performing the analyses.

Cores MD 73–04 and MD 73025 were raised during cruise OSIRIS 1 of M/S *Marion Dufresne* and the support of our Indian Ocean studies by Les Terres Australes et Antarctiques Françaises is gratefully acknowledged.

Cores RC 9–150 and RC 17–69 and the trigger-weight core tops from the northern Indian Ocean come from the Lamont-Doherty Geological Observatory collection. Samples have been provided by A.W.H.Bé and their study is an outgrowth of the CLIMAP project.

References

1. Urey, H. C. & Greiff, L. J. (1935). *J. Am. chem. Soc.*, **57**, 321–7.
2. Urey, H. C. (1947). *J. chem. Soc.*, 562–81.
3. McCrea, J. M. (1950). *J. chem. Phys.*, **18**, 849–57.
4. Epstein, S., Buchsbaum, R., Lowenstam, H. A. & Urey, H. C. (1953). *Bull. geol. Soc. Am.*, **64**, 1315–28.
5. Emiliani, C. (1955). *J. Geol.*, **63**, 538–78.
6. Emiliani, C. (1958). *J. Geol.*, **66**, 264–75.
7. Duplessy, J-C., Chenouard, L. & Reyss, J-L. (1974). In *Variation du climat au cours du Pleistocène*, ed. J. Labeyrie, pp. 251–8. CNRS, Paris.
8. Emiliani, C. (1964). *Bull. geol. Soc. Am.*, **75**, 129–44.
9. Emiliani, C. (1966). *J. Geol.*, **74**, 109–26.
10. Emiliani, C. (1972). *Science*, **178**, 398–401.
11. Shackleton, N. J. & Opdyke, N. D. (1973). *Quaternary Res.*, **3**, 39–35.
12. Shackleton, N. J. & Opdyke, N. D. (1976). In *Investigation of Late Quaternary paleooceanography and paleoclimatology*, ed. R. M. Cline & J. Hays, pp. 449–64.
13. Oba, T. (1969). *Sci. Rep. Tôhoku Univ., Sendai*, 2nd Ser. (Geology), **41**, 129–95.
14. Vila, F. & Duplessy, J-C. (1975). *3e reunion des sciences de la terre*, Montpellier, 381.
15. Emiliani, C. (1955). *Quarternaria*, **2**, 87–98.
16. Vergnaud Grazzini, C. (1975). *Science*, **190**, 272–4.
17. Naydin, D. P., Teys, R. V. & Chupakhin, M. S. (1960). *Geochemistry* **1956**.
18. Shackleton, N. J. (1974). In *Variation du climat au cours du Pleistocène*, ed. J. Labeyrie, pp. 203–9. CNRS, Paris.
19. Shackleton, N. J. (1965). In *Stable isotopes in oceanographic studies and paleotemperatures, Spoleto* **1965**, ed. E. Tongiorgi, pp. 155–9. Consiglio Nazionale delle Ricerche, Laboratorio di Geologia Nucleare, Pisa.
20. Fritz, P. & Fontes, J. C. (1966). *C. R. Acad. Sci.* (Paris), **263**, 1345–8.
21. Walters, L. J., Claypool, G. E. & Choquette, P. W. (1972). *Geochim. cosmochim. Acta*, **36**, 129–40.
22. Craig, H. (1965). In *Stable isotopes in oceanographic studies and paleotemperatures, Spoleto* **1965**, ed. E. Tongiorgi, pp. 161–82. Consiglio Nazionale delle Ricerche, Laboratorio di Geologica Nucleare, Pisa.
23. Craig, H. (1957). *Geochim. cosmochim. Acta*, **12**, 133–49.
24. Craig, H. (1961). *Science*, **133**, 1833–4.
25. Keith, M. L. & Weber, J. N. (1964). *Geochim. cosmochim. Acta*, **28**, 1787–816.
26. Smith, P. B. & Emiliani, C. (1968). *Science*, **160**, 1335–6.
27. Vinot-Bertouille, A.-C. & Duplessy, J-C. (1973). *Earth Planet. Sci. Lett.*, **18**, 247–57.
28. Duplessy, J-C., Lalou, C. & Vinot, A-C. (1970). *Science*, **168**, 250–1.
29. Duplessy, J-C., Chenouard, L. & Vila, F. (1975). *Science*, **188**, 1208–9.
30. Le Calvez, J. (1953). In *Traité de zoologie*, vol. 1, ed. P. P. Grassé, pp. 149–265. Masson et Cie, Paris.
31. Emiliani, C. (1955). *J. Micropaleont.*, **1**, 377–80.
32. Blackmon, P. D. & Todd, R. (1959). *J. Paleont.*, **33**, 1–15.
33. Tarutani, T., Clayton, R. N. & Mayeda, T. K. (1969). *Geochim. cosmochim. Acta*, **33**, 987–96.
34. O'Neil, J. R., Clayton, R. N. & Mayeda, T. K. (1969). *J. chem. Phys.*, **51**, 5547–58.

35. Phleger, F. B. (1960). *Ecology and distribution of recent foraminifera.* John Hopkins Press, Baltimore.

36. Vergnaud Grazzini, C. (1973). Etude écologique et isotopique de foraminifères actuels et fossiles de Mediterranée. Thesis, University of Paris.

37. Emiliani, C. (1954). *Am. J. Sci.*, **252**, 149–58.

38. Jones, J. L. (1967). *Micropaleontology*, **13**, 489–501.

39. Berger, W. H. (1969). *Deep Sea Res.*, **16**, 1–24.

40. Lidz, B., Kehm, A. & Miller, H. (1968). *Nature*, **217**, 245–7.

41. Emiliani, C. (1971). *Science*, **173**, 1122–4.

42. Bé, A. W. H. & Lott, L. (1964). *Science*, **145**, 823–4.

43. Longinelli, A. & Tongiorgi, E. (1964). *Science*, **144**, 1004–5.

44. Bé, A. W. H. (1960). *Micropaleontology*, **6**, 373–92.

45. Hecht, A. D. & Savin, S. M. (1970). *Science*, **170**, 69–71.

46. Hecht, A. D. & Savin, S. M. (1972). *J. Foramin. Res.*, **2**, 55–67.

47. Bé, A. W. H. & Van Donk, J. (1971). *Science*, **173**, 167–8.

48. Shackleton, N. J., Wiseman, J. H. D. & Buckley, H. A. (1973). *Nature*, **242**, 177–9.

49. Le Calvez, J. (1938). *Archs Zool. exp. gén.*, **80**, 163–333.

50. Arnold, Z. M. (1956). *Contr. Cushman Fdn foramin. Res.*, **7**(1), 1–14.

51. Duplessy, J-C. (1972). La géochimie des isotopes stables du carbone dans la mer. Thesis, University of Paris.

52. Kroopnick, P., Deuser, W. G. & Craig, H. (1970). *J. geophys. Res.*, **75**, 7668–71.

53. Rubinson, M. & Clayton, R. N. (1969). *Geochim. cosmochim. Acta.*, **33**, 997–1002.

54. Ericson, D. B., Ewing, M. & Heezen, B. C. (1952). *Bull. Am. Ass. Petrol. Geol.*, **36**, 489–511.

55. Heezen, B. C. & Ewing, M. (1952). *Am. J. Sci.*, **250**, 849–73.

56. Belderson, R. H. & Laughton, A. S. (1966). *Sediment.* **7**, 103–16.

57. Owen, D. M., Sanders, H. L. & Hessler, R. M. (1967). In *Deep-Sea photography*, ed. J. B. Hersey, pp. 229–34. John Hopkins Press, Baltimore.

58. Bé, A. W. H. & Duplessy, J-C. (1976). *Science*, **194**, 419–22.

59. Broecker, W. (1963). In *The sea*, vol. 2, ed. M. N. Hill, pp. 88–108. Interscience Publications, London.

60. Berger, W. H. & Health, G. R. (1968). *J. mar. Res.*, **26**, 135–43.

61. Ruddiman, W. F. & Glover, L. K. (1972). *Bull. geol. Soc. Am.*, **83**, 2817–36.

62. Broecker, W. S. (In Press.)

63. Berger, W. H. (1971). *Mar. Geol.*, **11**, 325–58.

64. Savin, S. M. & Douglas, R. G. (1973). *Bull. geol. Soc. Am.*, **84**, 2327–47.

65. Epstein, S. & Mayeda, T. (1953). *Geochim. cosmochim. Acta*, **4**, 213–24.

66. Dansgaard, W. (1961). *Meddr Grønland*, **165**, 1–120.

67. Dansgaard, W. (1964). *Tellus*, **16**, 436–68.

68. Dansgaard, W., Johnsen, S. J., Moller, J. & Langway, C. C., Jr (1969). *Science*, **166**, 377–81.

69. Dansgaard, W., Johnsen, S. J., Reeh, N., Gundestrup, N., Clausen, H. B. & Hammer, C. U. (1975). *Nature*, **255**, 24–8.

70. Lorius, C. (1974). *Antarctica : survey of near surface mean isotope values.* Cambridge Workshop Monograph. MIT Press, Cambridge, Massachusetts.

71. Flint, R. F. (1947). *Glacial geology and the Pleistocene epoch.* Wiley, New York.

72. Emiliani, C. (1971). In *Late Cenozoic glacial ages*, ed. K. K. Turekian, pp. 183–97. Yale University Press, New Haven.

73. Emiliani, C. (1970). *Science*, **168**, 822–5.

74. Dansgaard, W. & Tauber, H. (1969). *Science*, **166**, 499–502.

75. Imbrie, J. Van Donk, J. & Kipp, N. G. (1973). *Quaternary Res.*, **3**, 10–38.

76. Ericson, D. B. & Wollin, G. (1956). *Deep Sea Res.*, **3**, 104–25.

77. Ericson, D. B. & Wollin, G. (1956). *Micropaleontology*, **2**, 257–70.

78. Lidz, L. (1966). *Science*, **154**, 1448–51.

79. Shackleton, N. (1968). *Nature*, **218**, 79–80.

80. Sverdrup, H. U., Johnson, M. V. & Fleming, R. H. (1942). *The oceans.* Prentice Hall, Englewood Cliffs.

81. Shackleton, N. (1967). *Nature*, **215**, 15–17.

82. Le Floch, J. & Tanguy, A. (In Press.)

83. Defant, A. (1961). *Physical oceanography*, vol. 1. Pergamon Press, Oxford.

84. Dansgaard, W. (1954). *Geochim. cosmochim. Acta*, **6**, 241–60.

85. Aldaz, L. & Deutsch, S. (1967). *Earth Planet. Sci. Lett.*, **3**, 267–74.

86. Lorius, C. & Merlivat, L. (1975). In *Symposium on isotopes and impurities in snow and ice*, p. 133. International Union of Geodesy and Geophysics, Grenoble.

87. Olausson, E. (1965). In *Progress in oceanography*, ed. M. Sears, vol. 3, pp. 221–52. Pergamon Press, Oxford.

88. Nye, J. F. (1963). *J. Glaciol.*, **4**, 785–8.

89. Nye, J. F. (1951). *Proc. R. Soc.*, **207**, 554–72.

90. Nye, J. F. (1959). *J. Glaciol.*, **3**, 493–507.

91. Merlivat, L., Lorius, C., Majzoub, M., Nief, G. & Roth, E. (1967). In *Isotopes in hydrology*,

pp. 671–81. International Atomic Energy Agency, Vienna.

92. Epstein, S., Sharp, R. P. & Gow, A. J. (1970). *Science,* **168,** 1570–2.

93. Raynaud, D. & Lorius, C. (1973). *Nature,* **243,** 283–4.

94. Raynaud, D. & Lorius, C. (1975). In *Symposium on isotopes and impurities in snow and ice,* p. 135. International Union of Geodesy and Geophysics, Grenoble.

95. Dangsaard, W., Johnsen, S. J., Clausen, H. B. & Langway, C. C., Jr (1971). In *Late Cenozoic glacial ages,* ed. K. K. Turekian, pp. 37–56. Yale University Press, New Haven.

96. Barkov, N. I., Gordienko, F. G., Korotkevich, E. S. & Kotlyakov, V. M. (1975). In *Symposium on isotopes and impurities in snow and ice,* p. 123. International Union of Geodesy and Geophysics, Grenoble.

97. Johnsen, S. J., Dangsaard, W., Clausen, H. B. & Langway, C. C., Jr (1972). *Nature,* **235,** 429–34.

98. McIntyre, A. (1967). *Science,* **158,** 1315–17.

99. Gates, W. L. (1976). *Science,* **191,** 1138–44.

100. Ninkovitch, D. & Shackleton, N. J. (1975). *Earth Planet. Sci. Lett.,* **27,** 20–34.

101. CLIMAP Project Members (1976). *Science,* **191,** 1131–7.

102. Wyrtki, K. (1973). In *The biology of the Indian Ocean,* ed. B. Zeitschel, pp. 18–36. Springer-Verlag, Berlin.

103. Duplessy, J-C. (1970). *C. R. Acad. Sci.* (Paris), **271,** 1075–8.

104. Wyrtki, K. (1971). *Oceanographic atlas of the international Indian Ocean expedition.* National Science Foundation, Washington D.C.

105. Kennett, J. P. & Shackleton, N. J. (1975). *Science,* **188,** 147–50.

106. Emiliani, C., Gartner, S., Lidz, B., Eldridge,

K., Elvey, D. K., Huang, T. C., Stipp, J. J. & Swanson, M. F. (1975). *Science,* **189,** 1083–8.

107. Bloom, A. L. (1971). In *Late Cenozoic glacial ages,* ed. K. K. Turekian, pp. 355–79. Yale University Press, New Haven.

108. Worthington, L. K. (1970). *Deep Sea Res.,* **17,** 77–84.

109. Reid, J. L. & Lynn, R. L. (1971). *Deep Sea Res.,* **18,** 1063–88.

110. Kellogg, T. B. (1976). In *Investigation of Late Quaternary paleooceanography and paleoclimatology,* ed. R. M. Cline & J. Hays, pp. 77–110.

111. Duplessy, J-C., Chenouard, L. & Vila, F. (1975). In *Proceedings of the 9th international congress on sedimentology,* Nice 1975, pp. 57–63.

112. Hendy, C. H. & Wilson, A. T. (1968). *Nature,* **219,** 48–51.

113. Duplessy, J-C., Labeyrie, J., Lalou, C. & Nguyen, H. V. (1971). *Quaternary Res.,* **1,** 162–74.

114. Libby, L. M. & Pandolfi, L. J. (1974). In *Variation du climat au cours du Pleistocène,* ed. J. Labeyrie, pp. 299–310. CNRS, Paris.

115. Lerman, J. C. (1974). In *Variation du climat au cours du Pleistocene,* ed. J. Labeyrie, pp. 163–81. CNRS, Paris.

116. Labeyrie, L. (1974). *Nature,* **248,** 40–2.

117. Margolis, S. V., Kroopnick, P. M., Goodney, D. E., Dudley, W. C. & Mahoney, M. E. (1975). *Science,* **189,** 555–7.

118. Emiliani, C. (1954). *Science,* **119,** 853–5.

119. Shackleton, N. J. & Kennett, J. P. (1975). *Initial Report of DSDP,* **29,** 743–55.

120. Savin, S. M., Douglas, R. G. & Stehli, F. G. (1975). *Bull. geol. Soc. Am.,* **86,** 1499–510.

121. Kennett, J. P. & Shackleton, N. J. (1976). *Nature,* **269,** 513–15.

122. Shackleton, N. J. & Kennett, J. P. (1975). *Initial Report of DSDP,* **29,** 801–7.

4

Climatic change in historical times

JOHN GRIBBIN & H. H. LAMB

The principal features

Interest in the subject of climatic change was aroused during this century by the warming trend which persisted from the late 1890s through to the 1940s; although this trend was recognised in discussions of the Royal Meteorological Society as early as 1911, widespread public discussion of the implications of continued warming only occurred after the trend had already reversed, in the late 1940s and early 1950s. In recent years, there has been some public concern that the cooling trend which followed the warming of the first half of this century might now persist to produce severe problems. The lesson of the 1950s, when there was some speculation that by the end of this century the Arctic might be ice-free but in fact the climatic trend had already reversed, warns of the dangers in such naive extrapolation, and of the necessity to examine the patterns of climatic change over as long an historical period as possible in order to obtain a more reliable guide to the likely limits of climatic change in the years up to the end of this century and beyond.

The first and most important lesson of the historical record is that there is no such thing as a climatic 'normal' in the sense that the word was used 50 or more years ago, implying that if a long enough series of observations could be amassed it would produce an average value of each element describing the climate, an average to which the climate at the observing site would always tend to return. Far from being 'average weather', climate is always changing on a variety of different time-scales; the problem facing us is to determine the extent of these changes and to unravel the different time-scales of change as far

as possible from the historical record, in the hope that this will provide information about the underlying mechanisms of climatic change which interact to produce the complex pattern. Only with some understanding of which mechanisms dominate the changes occurring at present can there be any hope of reliable prediction of climatic change.

Unfortunately, the need for long historical records in studying time variations of climate conflicts with the need in studying global climate for records covering a broad area. The best historical records come from only two parts of the world, China and Japan and northwest Europe, and analysis of such geographically limited data cannot fail to be less than perfect in providing an overall picture of how the climate of our planet has varied. Nevertheless, a detailed study of either region is likely to indicate the broad outlines of the picture, and this expectation is reinforced by the discovery that the outline looks very similar whether it is derived from the European record or the Chinese record, which has been discussed in some detail by Chu Ko-chen.[1] There is evidence that some part of the pattern of climatic change drifts westward around the globe at an average rate of 0.6° of longitude per year, so that a warming recorded in the Far East in the sixteenth century A.D. reached Europe in the eighteenth century.[2,3] This may explain the non-synchroneity of the European and Chinese records, and also offers an intriguing potential insight into the global processes of climate, discussed elsewhere in this volume, although this westward 'wave' accounts for less than half of the variance of mean winter temperatures (taken as decadal averages) for the past 1000 yr and seems distinct from global trends of warming and cooling.

Historical records from the southern hemisphere are almost non-existent, and this has been a particularly worrying aspect of the attempt to describe global changes in climate from the historical record. Recently, however, a major effort to investigate the patterns of climatic change in Australia and New Guinea has been carried out by groups using the techniques of pollen and isotope analysis, changes in lake levels, glaciers and tree lines, and geological evidence. These data cover a far longer time scale than the historical records available for some parts of the northern hemisphere, but they show that major changes in climate, at least, occur with near synchroneity in the two hemispheres.[4] However, this may not be the complete story since there is some evidence from a study of temperature changes near New Zealand and elsewhere during this century that much of the Southern Ocean zone cooled between 1900 and 1945, and has since experienced a warming trend[5] – the opposite of the pattern deduced from data from the zones north of 30° S. Again, this pattern, if real, must indicate some fundamental feature of the global situation; for our present purpose, however, the lesson seems to be that while historical records from northwest Europe probably do provide a good guide to the broad features of recent climatic change in the northern hemisphere, we cannot with confidence claim that the smaller and shorter-term changes in particular can be interpreted as global trends. With that in mind, we must make the best use of the data we do have, which require extensive interpretation if the historical record of climate is to be extended back before the 1650s to the period when the barometer and thermometer were not available and no direct, quantitative records of temperature and pressure could be made.

Four climatic epochs since the end of the most recent phase of glaciation some 10 000 yr B.P. are of particular interest and merit further study. These are: the post-glacial climatic optimum (or warmest times; the old term 'climatic optimum' is now out of favour – optimum where and for what?) culminating between about 7000 and 5000 yr B.P.; the colder climatic epoch of the Iron Age, culminating between about 2900 yr B.P. and 2300 yr B.P.;

the secondary 'climatic optimum' of the early Middle Ages, roughly from 1000 to 800 yr B.P.; and the Little Ice Age, a cold period particularly marked between about 550 yr B.P. and 125 yr B.P., peaking in severity in Britain in the late seventeenth century. These outstanding phases of climate are of very different durations, and of course much less information is available about the older phases; it is unlikely that the complex of causes was the same for each, but they represent the extremes of recent conditions. In some cases extraneous factors – such as volcanic dust in the atmosphere – may have been particularly important.

The post-glacial warmest times
Between about 5000 and 3000 B.C. the eustatic sea-level was rising rapidly as the last remnants of the former great ice sheets melted after the ice age, and around 2000 B.C. it may have been a little higher than today, by about 3 m. This implies roughly 10^{15} m^3 less glacial ice than today on the mountains of that time, but with a distribution of the remaining ice not unlike that of the present climatic regime. The minimum extent of the ice cover on land was reached in the very last stages of the main warm period; although the rise in sea-level from the end of the previous full ice age was primarily due to the reduction of the extent of ice on land, care is needed in interpreting the records because of isostatic effects associated with the loading and unloading of ice on the land which altered the geography near the ice sheets (cf. references 7 and 8). During this optimum, the extent of sea-ice on the Arctic Ocean was also reduced, with open water extending beyond the channels of the Canadian Archipelago (reference 6, p. 143). A variety of data (see reference 9 and references cited therein), including fossil marine fauna from Spitsbergen and evidence of a wetter regime than today in the Sahara and the deserts of the Near East, all help to fill in the overall picture, with vegetation belts displaced northward and temperatures in Europe in summer averaging 2–3 °C above those of today. For the period up to 4000 B.C. in particular, the data are consistent with a displacement of the subpolar depressions and the axis of the main anticyclone belt northward, perhaps placing the high-pressure belt as far north as 40–45° N.

With the high-pressure belt well north of the

Mediterranean, Africa and the Near East would have experienced a significant shift of the trade wind and equatorial zones, with more widespread monsoon rains in the summer, and winter rainfall as well in the Mediterranean region. This seems to be consistent with other northern hemisphere data, and during the climatic optimum period Hawaii also experienced increasing rainfall, presumably associated with a northward shift of the trade wind zone; in addition, southern hemisphere data show that this was a global climax of warmth. The southern data come primarily from studies of the extent and distribution of forest species in southernmost South America and New Zealand and from the recent work in Australia and New Guinea mentioned earlier.[4] The Antarctic also was warmer at about this time, with temperatures 2–3 °C higher than now in the Antarctic continent, Tierra del Fuego and the Himalayan Mountains.[10]

Iron Age cold epoch

Although the period between 900 and 300 B.C. or somewhat later was a cold epoch, the most striking feature from the European record is the associated increase in wetness, with evidence of a widespread re-growth of bogs after a much drier period.[11] This change, producing a so-called 'recurrence surface', is a conspicuous feature of peat bog sections from all parts of northern Europe, including Ireland, Germany and Scandinavia; adaptation and abandonment of ancient tracks across the increasingly marshy lowlands in England and elsewhere provides confirmatory evidence.[12] The forests of Russia spread further south at this time than during the preceding warm epoch, and a change in the dominant species indicates some lowering of the summer temperature.

Further south, in the Mediterranean region and in North Africa, the climate during the Iron Age cold epoch seems to have become drier, but not so dry as it is today. Roman records are of value here; agricultural writers, such as Saserna, noted around 100 B.C. that the vine and olive were spreading north in Italy to regions where the weather had previously been too severe, suggesting that the centuries before 100 B.C. had indeed been cooler in the Mediterranean. Southern hemisphere data also indicate a cooler period roughly around 500–300 B.C.

Secondary optimum

For the climatic optimum which occurred around 1000–1200 A.D. and later epochs, more comprehensive data are available and historical records assume greater importance; both this epoch and the later Little Ice Age will be dealt with at greater length below, but first we provide a quick overview to complete our reconstruction of the principal features of climate since the end of the latest glaciation.

The secondary optimum shows many of the same features as the post-glacial optimum, but to a lesser degree and for a shorter duration. Melting of the Arctic pack-ice and restriction of the southward spread of drift ice played an important part in the founding and development of Norse colonies across the northern North Atlantic to America; in western and central Europe vineyards extended 3–5° further north in latitude and 100–200 m higher above sea-level (for a discussion of the vine as a climatic indicator, see references 13 and 21) and this evidence, together with estimates of the limits of tree lines, implies mean summer temperatures a little more than 1 °C higher than those now considered normal.

Archaeological and other evidence from North America suggests that the little optimum lasted until about 1300 A.D. before there was a change to cooler, wetter conditions;[14,15] to the south there is evidence of a wetter period contemporary with the little optimum, but perhaps longer lasting, in Central America, Cambodia, the Mediterranean and the Near East. In Antarctica, there is evidence that the penguin rookery at Cape Hallett was first colonised just before this period, presumably during a phase of improving climate, and that it has since remained occupied permanently.[16]

The Little Ice Age

Although the climate of the past 1000 yr was at its most severe in Britain during the second half of the seventeenth century, the Little Ice Age is generally taken as the longer period, from 1430 to 1850, during which there was a predominance of harsher conditions than those of this century but also some periods of relatively equable climate. During this epoch the Arctic pack-ice expanded considerably, with important effects for Greenland and Iceland which will be discussed below, and by 1780–1820 temperatures across the North Atlantic everywhere above 50° N seem to have been 1–3 °C less than those of the present day;[17] indirect evidence suggests that these temperatures had in fact been reached by the 1600s.

Forests at high levels in central Europe declined catastrophically after 1500, and eyewitness reports from Scotland describe widespread dying off of woods in exposed localities.[18] But although winters were very severe, it seems that in Europe around 50° N there were times when the prevailing summer temperatures were mostly around their present level, and even slightly above in the 1700s. Some evidence suggests a greater year-to-year variability of summer (and perhaps of winter) temperatures than in the present century. Severe winters also affected the Mediterranean, and glaciers advanced generally in Europe, Asia Minor and North America, while snow was recorded lying for months on the high mountains of Ethiopia where it is now unknown. The Caspian Sea rose during the Little Ice Age and maintained a high level until 1800, but the White Nile, fed from the rains of the equatorial belt, was low and these rains seem to have been weakened or shifted south. The overall picture is one in which there was an equatorward shift of the prevailing depression tracks in the northern hemisphere, with more prominent polar anticyclones.

This pattern is not fully reflected in the southern hemisphere, however, which seems to have partly escaped the cold epoch until the nineteenth century. Between 1760 and 1830 the fringe of Antarctic sea-ice seems to have been a little south of its present position, with the southern temperate rainbelt also displaced southward – that is, *away* from the equator. After about 1830, until 1900, the rain zone and depression tracks moved north with advances of the glaciers in the Andes and South Georgia at about the time that the northern hemisphere was recovering from the Little Ice Age.

Summary
While the first three epochs seem to have been associated with global shifts of climatic zones towards or away from the equator roughly synchronously in both hemispheres, the Little Ice Age data suggest a different pattern, with an equatorward shift in the northern hemisphere accompanying a synchronous (but smaller)

poleward shift in the southern hemisphere, followed by return movements in both hemispheres, though the *temperature* trends over most of the two hemispheres may have been more nearly parallel. The amplitude of surface temperature fluctuations has been greatest at high latitudes so that the meridional gradient at the surface, at least between 40° and 60° N, must have been markedly less in the warm epochs than in the cold ones; but temperatures on mountains even in low latitudes were 2–3 °C higher than today in warm epochs and 1–2 °C cooler than today in cool epochs.[10] Possibly the meridional equator–pole gradient of upper air temperatures was actually less during the cold epochs, with surface temperatures at high latitudes particularly unrepresentative at such times because of frequent strong inversions.[17]

There is at least a hint here that the climatic changes of the Little Ice Age had a somewhat different pattern of causes from those of the three other epochs which, with the Little Ice Age, represent the greatest departures from present-day conditions since the end of the latest full glaciation; this is particularly interesting since the Little Ice Age is the most recent of these epochs, and it is even possible that the warming of the late 1800s and early 1900s represents a temporary fluctuation from Little Ice Age conditions, to which we may shortly return. One possible mechanism (rather, a pair of associated possible mechanisms) for the Little Ice Age fluctuations in particular is discussed on pp. 146–7; here, we continue our survey of recent climatic changes without offering interpretation of their causes.

Fifteen hundred years around the North Atlantic
A variety of data from lands bordering the North Atlantic provides a more detailed picture of climatic changes over the past 1500 yr, and their effects on the developing cultures of mankind in the region. The best single, continuous record of climate over that period comes from analysis of a single 404-m long ice core drilled from the glacier at Crête in central Greenland.[19] The data from this core are particularly interesting in the context of the impact of climatic changes on Man's cultural activities, since they tell us about a region of time and space in which the Norse culture developed, flourished and suffered setbacks.

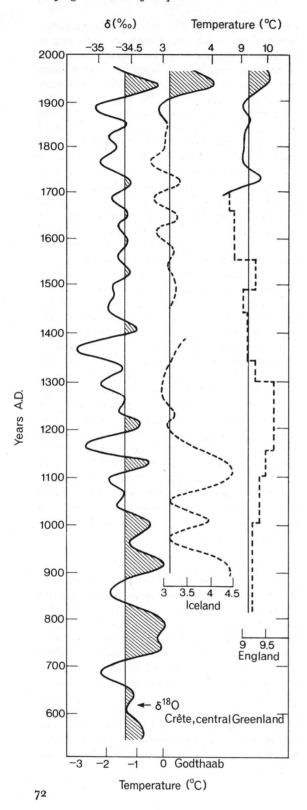

Changes in the isotopic composition of the oxygen in the ice provide a guide to temperature changes, as discussed in Chapter 3; but climatic interpretations have generally been hampered, as Dansgaard *et al.*[19] point out, by a poorly determined relationship between δ, the per mil deviation of ^{18}O from that in standard mean ocean water, and the actual temperature changes. A long-term change in δ over a long ice core may be partly caused by a change in surface altitude, not directly a measure of temperature variations associated with any climatic change. This effect is negligible for a relatively short core from a stable part of the ice sheet, as in the case for the Crête core; Dansgaard *et al.* also remove low frequency noise from the 'spectrum' of climatic fluctuations with low pass digital filter techniques removing all oscillations shorter than 10, 30 or 60 yr for different aspects of their analysis. Finally, it is necessary to identify layers in the core with calendar years, by the laborious process of counting successive δ summer maxima from the top; as yet, this has not been done for the 404-m core, but a reliable empirical rule relating depth and age has been established from such counts on two other cores from the region, going back to 1232 A.D. and 1176 A.D., respectively, and this rule has been used to date approximately the layers in the Crête core.

The results of this analysis are shown in Fig. 4.1 (from Dansgaard *et al.*) where they are compared with other observations from Iceland and Greenland. The 404-m ice core covers a span of 1420 yr, much longer than the span of the direct records shown by solid lines in the other two curves (see also Fig. 4.2); both England and Iceland records can be extended to cover a comparable period from indirect evidence.[20,21] The agreement between the (smoothed) curves over the past 150 yr or so

Fig. 4.1. Comparison between the ^{18}O concentration (left) in snow fallen at Crête, central Greenland (δ scale at top) and temperatures for Iceland and England. The curves are smoothed by a 60-yr low pass digital filter, except for England 800–1700 A.D. The solid curves are based on systematic, direct observations. The dashed curves rely on indirect evidence. After Dansgaard *et al.*,[19] reproduced by permission of the authors and *Nature*.

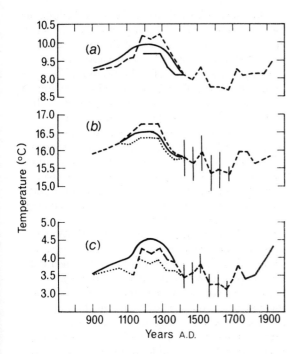

Fig. 4.2. Prevailing temperatures over England (°C) as 50-yr averages over the past millenium. Although there are some differences between the values determined by different interpretations of the data, these are less significant than the general trends common to all. After Lamb.[21] (*a*) Full year; (*b*) Summer (July and August); (*c*) Winter (December–February).

suggests that the Crête core does provide a good guide to temperature changes, and these have been calibrated by comparing the 0.9 per mil shift in δ between 1890 and 1935 with the 2.1 °C shift in temperature recorded at Godthaab between 1876 and 1935.

Although the δ curve seems to be a reliable indicator of climatic changes on time-scales of 60 to 200 yr, the important lower frequency changes which are obvious in the curve must be interpreted with caution. Systematic records of temperature even in England have been available only since 1659;[22–24] for earlier periods such climatic indicators as notes in Icelandic annals about sea-ice, mild or severe winters, records of good and bad harvests and so on assume a greater importance. The literary tradition in the early Icelandic society which produced the sagas of the period 1100–1300 has

proved particularly useful in establishing agreement between some features of the δ curve and historical changes in climate. The most striking result of such interpretations of historical and isotope evidence is that both the little optimum and the Little Ice Age seem to have had counterparts at Crête some 250 yr *before* the changes occurred in Europe, although data from both Camp Century[25] (further west in Greenland) and North America[26] suggest that such long-term changes were in phase in Europe and North America. This may imply a general cooling trend in Europe and North America over the next century, but Dansgaard *et al.* are cautious about any such prediction, especially in view of the probable disturbing influences of Man (see Section 5).

It may be possible to interpret this delay by considering the longest quasi-stationary Rossby waves in the general circulation, more particularly wave-number one, which concerns eccentricity of the circumpolar vortex as a whole. The concurrence of a cold regime in China with the warmest phase of the little optimum climate around the tenth to eleventh centuries in Greenland (on the opposite side of the hemisphere) may be linked with such eccentricity, its most remarkable feature being the cold regime in the Far East in the middle of a period of some centuries duration in which most of the globe was experiencing a climax of warmth. The westward movement of the major anomaly feature suggested by King[3] (see pp. 68–9) might bring an early onset of cooling in Greenland (observed as early as the 1190s) towards the Little Ice Age regime, while strengthening of the upper westerlies tended to increase wavelength in wave-numbers two, three and four, making Europe for another century more subject than before to warm ridge conditions. The changing wind pattern, affecting the course and intensity of the Gulf Stream, would be particularly important in influencing climate in Greenland, Iceland and England.

Whatever the exact cause of these changes it is instructive to consider their impact on Norse society. The first recorded attempt to settle in Iceland was made in 865 A.D. by a farmer, Floke Vilgerdson, in the last of an earlier series of cold fluctuations which was nearing its end. He lost his cattle in a severe winter, and returned to Norway with tales of 'a fjord filled up by sea-ice ... Therefore he called the country Iceland' (Landnam Saga, about 1200 A.D., quoted in reference 19). By 874, successful settlers reached Iceland, and the colony flourished in the rapid climatic warming which shows clearly in the δ curve. By 985, the colony was in a position to send its own colonists to Greenland, so named (according to the Greenlander Saga) by Eirik the Red in an attempt to persuade people to join his settlement there; but as Dansgaard *et al.* point out, this may do Eirik an injustice, since the δ curve suggests that Eirik reached Greenland towards the end of a warm period longer than any that has occurred there since. 'So, the drastic climatic change late in the ninth century may be part of the reason why Iceland and Greenland did not get the opposite names, which would have been more natural had they been discovered simultaneously.'[19] Drastic climatic change was certainly a major contributory factor in the collapse of the Greenland colony during the cold fourteenth century, and even the Iceland colony survived by only the barest margin through the worst ravages of the Little Ice Age.

Bryson[15] has summarised a great deal of information about climatic change since the most recent glaciation, emphasising the integral behaviour of the atmosphere which brings the varying patterns of related changes over the whole globe. The Arctic expansion of around 1900 B.C. caused a retreat of northern forests from regions which had been suitable for them for at least 1000 yr before that time, and this change may have adversely affected civilizations as far south as the Indus; the climatic changes which affected the Norse colonies after about 1200 A.D. were also related to a cooling of the Arctic and expansion of sea-ice and climatic zones southward. In the North American continent, associated changes would have shifted the summer westerlies southward from what is now southern Canada into what is now the northern US, with reduced rainfall in the northern plains, the present corn belt, west Texas and much of the intermontane west, but with increased rainfall in the Pacific northwest, the southeast and much of the east coast.[27] Just as the Atlantic changes of the period affected the Norse culture, so the continental changes affected the cultures of the American Indians.

One particular culture, that of the Mill Creek people, seems to have been profoundly affected, judging from the evidence of the archaeological record. These people occupied a region near the present American corn belt, living in the centuries just before 1200 A.D. in a region of tall-grass prairie uplands with wooded valley terraces and valley floors, hunting deer and growing corn. The Arctic expansion and associated rainfall changes produced a long drought – probably because of westerly winds, more persistent than before, crossing the Rocky Mountains and putting the region into the 'rain shadow'. The tall-grass prairies were replaced by short grass and most of the forest disappeared; in the west, farming villages disappeared entirely, and everywhere in the region the staple source of meat became the bison as the deer died out. The particular significance of these changes, as Bryson[15] stresses, is that the drought persisted for 200 yr, so that the archaeological and pollen records show unambiguously that minor climatic changes of the kind which have definitely occurred in the past millennium are sufficient to bring 200 yr of drought in the corn belt of North America, the region which today produces the surplus of grain on which many parts of the world depend for survival.

The pattern of climatic changes in the northern hemisphere since the middle of this century has shown shifts in the same direction as those associated with the Arctic expansion of 1200 A.D., although the magnitude of the deterioration is not yet as great. But similar coolings in the past millenium have never lasted for less than 40 yr,[15] which suggests that at the very least the present trend may return us to a situation more typical of the past few hundred years than of the past few decades. It is therefore likely to be of immediate practical value if we can obtain some insight not only into the broad features of climate in historical times but also into the detail of synoptic wind-flow patterns for specific recent epochs. Although the meteorological instrument record covers at best only the period since the middle of the seventeenth century, and only the area of northwest Europe for even that length of time, it has proved possible to produce a 'first (iteration) approximation' to many of the climatic features of the past three centuries. As well as being important in itself, such an approximation provides a comparison and check for the inferences deduced from other records, such as long, tree-ring records (see, for example, reference 26). It is also possible that new developments in analysing tree rings, which involve measurement of the deuterium/hydrogen ratio within each ring, may eventually provide a means of deducing actual temperatures from year to year in temperate regions before 1650, in much the same way as changes in the oxygen isotope ratio in ice provide a guide at high latitudes. This new technique has already been applied to trees in central Europe and western North America, and relatively young trees of a particularly broad-ringed species in New Zealand;[28–30] as the technique is used increasingly it will be of great interest to compare the tree-ring 'thermometer' with the historical record wherever possible.

The earliest years of the meteorological instrument record

The barometer and thermometer were both invented somewhat before the middle of the seventeenth century and were used in connection with daily weather records at an increasing number of sites in Europe from 1650–1700 onwards; during the 1650s, the first standardised instruments were distributed by the *Academia del Cimento* in Florence. The longest available records of rainfall go back to about 1700 in Europe, and although there are reports of gauges being used in Korea as early as the 1440s (and in Palestine nearly 2000 yr earlier) those records have not survived. Manley[24] has published details of monthly mean temperatures in central England since 1659, gleaned from a variety of historical sources; these records begin with the coldest decades of the Little Ice Age in England, so that fortuitously we do have some reliable quantitative indications of the severity of the winter of 1683–84, the coldest (taking the three-month mean) in the English instrumental record, and in all probability the coldest winter in England since quite early post-glacial times. This was the winter of the greatest of all the 'frost fairs' on the Thames, and also provided the raw material for the description of severe winter weather in *Lorna Doone*.

According to Manley,[31] reviewing the winter of 1683–84 as a whole, over much of the country the mean temperature in October was fully 4 °F below the average for the century, with November temperatures 2 °F below the average, December 7 °F below, January 10–11, February 8 and March 4 °F below the century's average. Severe cold prevailed over much of west Europe, with many great rivers and Lake Constance being frozen, but the winter of 1683–84 was not the coldest of the past 300 yr across the entire European region, and in France, for example, the winter of 1709 was regarded as more severe. An analogy has been drawn between the conditions of 1683–84 and those of the recent severe winter 1962–63 (see reference 31), with high pressure towards Iceland developing early in the winter into a 'blocking high' with associated clear skies and northerly winds over England for a critical fortnight in late December-early January, when snow spread over a wide area from Scotland to the east and south of England, and perhaps further. The North Sea and English Channel seem to have been cooler in the latter half of the seventeenth century than today, so that quite wide belts of sea-ice developed along the coasts in 1684, and of course the mean temperature for the century was less than that for the twentieth century, accounting for the difference in absolute terms between the situation in 1683–84 and that in 1962–63.

At this point we confront the fact that the earliest instrument records extend back only just far enough to give us a tantalising glimpse of what – to judge from glaciological and other evidence – was probably the severest phase of the global climate since the end of the latest major glaciation. If we wish to reconstruct the wind and ocean circulation regimes of that time, and still more, if we wish to learn the nature of the preceding events over the centuries of decline from the early medieval warmth leading into the cold, seventeenth-century regime, appropriate methods of analysis have to be devised for taking the descriptive accounts of the weather (which are reasonably abundant for many areas of Europe and Iceland, perhaps also sufficient for the Far East) and using these to compare with the oxygen isotope results from the Greenland ice sheet and other regions, as well as the systematic analysis of tree-ring records, lake varves and so on, so that the different types of data and independent approaches can be used to check each other.

Strict analysis of a sufficient network of good

barometric pressure records has provided reliable monthly mean pressure maps, from which the prevailing wind flow can be read off, for a useful part of the middle latitudes of the northern hemisphere back to 1750 (Lamb & Johnson[17]). By making use of the very few barometer records which start earlier, and the more abundant daily observation registers for wind, weather, temperatures and rainfall from which monthly wind roses and frequency figures for other items can be constructed, it has been found possible to sketch the probable wind-flow patterns over Europe and the eastern North Atlantic as far as Iceland for each January and July back to 1680. Tests similar to those applied to the isobar analysis for later years[17] indicate that the general character of the flow is reliably portrayed. This indicates for the late seventeenth century and on to about 1712 a highly abnormal frequency of blocking situations, particularly those giving northerly winds over the Iceland–Scandinavia–British Isles region. Some of these monthly patterns in the winters of that time show a more or less continuous meridional belt of high pressure from Iceland or northern–northeastern Europe to Spain or towards the Azores, a distribution not seen on any monthly mean map since that time. (The pattern of the 1683–84 winter was a more 'orthodox' easterly situation as far as the British Isles were concerned, with a Scandinavian anticyclone.) This finding lllustrates Lamb's contention that it would be a mistake to suppose that the colder climatic regimes of the past, and particularly an ice age regime, consisted essentially of a greater frequency of patterns like those of the occasional cold winters of the present century.

As regards the regime of the medieval period and after – the warmth of the high Middle Ages and the climatic decline which followed – it is hoped to throw light on these by a research project now in progress at the Climatic Research Unit, Norwich, England. The descriptions of the weather found in annals, estate accounts, monastic chronicles and state papers and so on, from all over Europe and from Iceland, are being used to compile maps of the information available for each season of each year as far back as the information goes. These maps typically contain reports for just two or three areas in the tenth century but for 20 to 30 areas by the fifteenth and sixteenth centuries. Methods of

analysis and testing will be decided when the compilation is more nearly complete; but it seems certain that determination of the general direction of the prevailing wind flow, and cyclonic or anticyclonic character of the weather, over the central part of the well-reported region (roughly, the North Sea and surrounding lands) will be possible for the great majority of seasons of the past 1000 yr.

The past 200 years

Monthly mean sea-level (MSL) pressure charts for each January and each July for as wide an area of the world and extending back to the earliest years for which usable observational data were recorded have been constructed in the UK Meteorological Office. For Europe, the existence of a network of observing stations, some with long series of observations, provides the possibility of checking for internal consistency and January and July charts have been reconstructed back to 1750. For the North Atlantic, reasonably accurate data are available (January standard error ±2.5 mbar; July ±1 mbar) since 1790. (For discussion of the derivation of the charts and statistical reliability, see reference 17). Figs. 4.3 and 4.4 show the 40-yr MSL pressure in January and July for the earliest period over the Atlantic (1790–1829), compared with the modern, 'normal' charts based on the *Historical Daily Weather Maps* (1900–39); these figures are taken from Lamb.[9] Between the two epochs covered by these charts, mean pressure gradients increased in January, and as other data confirm,[9] the period around 1790–1829 was one of a notable minimum in the strength of the zonal circulation in January. Variations in the mean strength of the North Atlantic westerlies correlate with changes in the mean January temperature of central England, with the earlier epoch 1 °C or more cooler than the later. The colder winters of the period around 1800 and earlier are in general explained by the weakness of the mean westerly flow towards Europe. This feature of the mean maps implies that the prevailing westerly winds of middle latitudes were a less predominant pattern then than in the first half of the present century, and that other patterns, including blocking anticyclones in a great variety of positions in different years, were commoner then than in 1900–50. The tendency since about 1950 has been to-

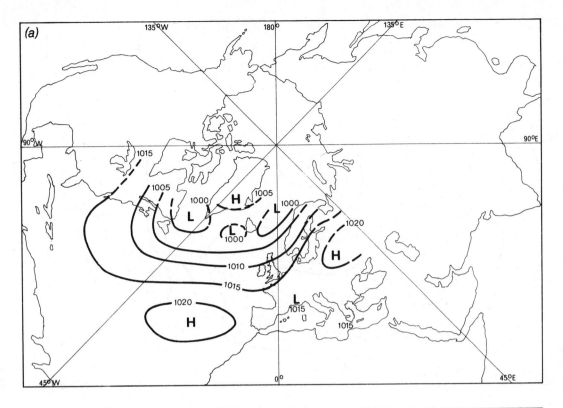

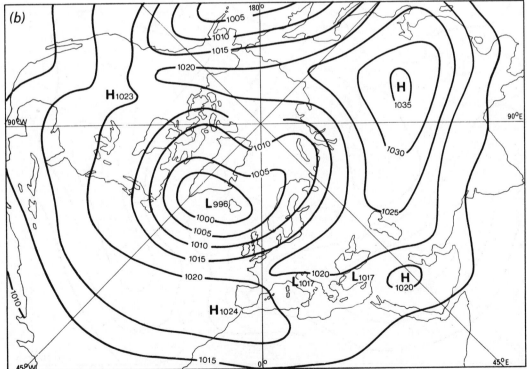

Fig. 4.3. Average monthly mean sea-level pressure distributions for the months of January in the epochs (a) 1790–1829; (b) 1900–39. Pressure gradients are weaker in the earlier epoch.

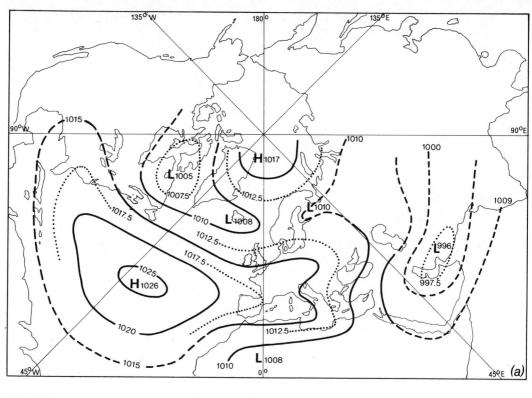

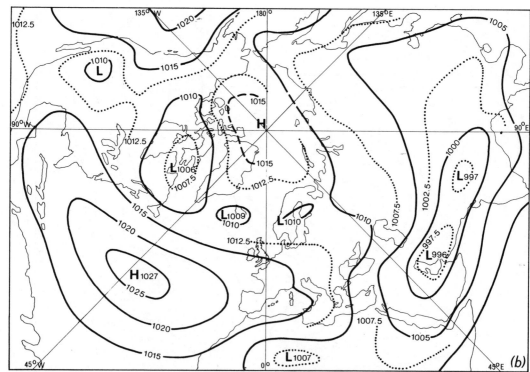

Fig. 4.4. As Fig. 4.3, but showing July average mean sea-level pressures for the same two epochs. Both Figs. 4.3 and 4.4 are based on figures in Lamb.[9]

wards a weaker circulation pattern and more frequent blocking again.

The July data show much less obvious changes between the two epochs, although it does seem that Europe had a weaker, more anticyclonic pressure field in the earlier period. Overall, the circulation data confirm the pattern in which there is a general equatorward displacement of climatic zones by 2–4° of latitude in those periods and regions where sea-ice is usually extensive. This seems particularly significant in the light of the study by Kukla and Kukla[32] (see also Section 2 of this volume) of changes in snow and ice cover of the northern hemisphere in recent years, and in view of the evidence (see, for example, reference 33) that aridity in Ethiopia has in the past been associated with cooler epochs in the northern hemisphere.

However, any comparison of the situation in the earliest part of the instrumental record with that of recent years must be treated with caution, not least because of a dramatic change in the climatic pattern in the 1830s; since 1840, at least until the early 1960s, the circulation maps do not suggest any exact analogies with the maps from before 1830, and the events of that decade seem to mark an important stage in the changing climate, at any rate in the short term. This decade is significant among those since 1750 because it produced the highest pressure generally over Europe and the Mediterranean; the most northerly position of the high pressure belt in that sector; the lowest pressure in Gibraltar associated with very high temperatures; and 'abnormal' climate bringing drought to Madeira and Malta. Further afield, during that decade great quantities of sea-ice broke up in the Arctic, so that Icelandic waters remained remarkably ice-free for a further 15 yr; Sydney and Adelaide both experienced long, dry periods; and the Antarctic sea-ice seems to have been subject to a greater than normal degree of break-up. Since 1840, the Arctic sea-ice has never regained its former extent (though it may have approached that around 1888 and 1968) which makes comparison of circulation maps from before and after that date of rather doubtful value for long-range forecasting. As far as it is possible to set a firm date on the end of the Little Ice Age, 1840 marks the most clearcut boundary; however, there is as yet insufficient evidence to decide whether the more clement

conditions in the northern hemisphere since then are likely to persist.

Variations in the longitudes of lowest and highest MSL pressure over the north Atlantic in July and January can be interpreted to provide information about upper air circulation,[9] which is of particular interest for the cool period around 1800. The wavelength L of an atmospheric standing wave is generally expressed by the Rossby formula:

$$U = \beta \, L^2/(4\pi^2)$$

where U is the prevailing zonal velocity of the upper westerlies and β is the rate of change of the Coriolis parameter with latitude at the latitude of the mainstream of the westerlies. The longitude spacing of the surface systems (ridge–trough) generally increased and decreased at times when the indices of circulation intensity were increasing and decreasing, respectively. In particular, this tendency is very clear from records of the spacing between the west Atlantic surface pressure trough and the mid-Atlantic ridge, which are closely related to corresponding features of the mainstream of the upper westerlies; the spacing of these surface features should correspond to about half a wavelength (cf. reference 34). But although the evidence suggests qualitative confirmation of changes in circulation intensity estimated from measurement of surface pressure gradients, the change in wavelength in both January and July seems too great (at 25–30 %) to be explained wholly in this way. It seems likely that some slight change of latitude of the strongest upper westerlies must also have taken place, with a poleward displacement of a few degrees of latitude between the times of weak circulation around 1800–50 and the time of strongest circulation in the earlier decades of this century.

The importance of this consistent trend is that it gives us some confidence in the reliability of estimates of probable changes in the main tropospheric westerlies at high altitude made on the basis of changes in surface patterns, particularly changes in longitude, of certain features. On this basis, there is a case for explaining the cool period around 1800 in terms of a reduction in insolation by 1–2 %, combined with associated shifts in circulation pattern. This is particularly significant to theories of climatic change based on the veiling effect of volcanic dust and on solar variations, discussed elsewhere in this volume (Section 3).

Although instrumental records are sparse from before 1750 and non-existent from before 1650, it should prove possible to extend quantitative estimates of climate to earlier centuries by comparison of various documentary records (see p. 75). The insight into climatic patterns provided by study of the records of the past 200 yr makes it possible to estimate the kind of circulation patterns which are likely to have brought the extremes of weather conditions noted in the documentary record, although there is always a need for care to avoid becoming trapped in a circular argument as this technique is used to extend our understanding of climate. For the culminating period of the Little Ice Age (1550–1700), the evidence suggests a summer flow weakened by about 30% or shifted south by 5°; the latter, or else some combination of these, is the most likely explanation. There would then have been a prevailing wave-number of five or six, with a five-wave standing pattern becoming firmly established at times; the present circulation produces a wave pattern intermediate between that of a wave-number four and a wave-number five circulation.

For the peak Little Ice Age winters, the flow seems to have been weaker by about 5–10 %, with an associated southward shift of 3–5° and the main depression track around 64° N, with considerable scatter corresponding to more frequent blocking anticyclones and many more depressions than in this century entering the Mediterranean. Unlike the present century, with a dominant three-wave pattern about the time of the most vigorous winter circulation, the Little Ice Age winters are best explained by a more common occurrence of a four-wave pattern – a feature which has tended to reappear in recent decades.

For the little optimum of 1000–1200 A.D., summer flow was most probably shifted north by 3–5° relative to the 'normal' of the early part of this century, while in winter the northward shift of the depression track may have taken it into the Barents Sea, with periods of only a two-wave circulation pattern. A cold belt across Europe near 50° N at this time can then be explained in terms of easterly winds with a long land treck, while the depressions in the Barents Sea explain the increase in the total rainfall in northern Europe.

The present situation

It is clear from the pattern of past climatic changes that naive extrapolation of any trend is no reliable guide to future prospects. In the early 1950s, there was concern lest the warming trend of the first half of this century should persist, although soon it became clear that the increase had stopped, and Thomas, for example, wrote 'the current year of 1956 has been a relatively normal year [in Ontario] and speculation is aroused as to whether or not the warming trend has been halted for the time being'.[35] Ten years later the concern about climatic change seemed based on fears that the cooling trend that had by then become clear might persist, and we find Starr & Oort in 1973 pointing out a downward trend in temperature of 0.6 °C in the northern hemisphere between May 1958 and April 1963 and speculating on the implications.[36] As those authors pointed out, such a dramatic short-term decline can only be temporary and should not be taken at face value to imply a cooling of 12 °C by the year 2000. Another short-term guide to events which may be of long-term importance comes from studies of changes in marine populations and the regions favoured by different species. Over the 25-yr period 1948–72, the total numbers of copepods and the zooplankton biomass in the North Sea have declined, the duration of the zooplankton season has become progressively shorter and the spring bloom of phytoplankton progressively later.[37] Southerly shifts in the distribution of fish have occurred, with a significant movement since 1965–67 which seems 'likely to have had its origin in climatic events' (see reference 38 and references therein). Since 1940, an average cooling of the northern hemisphere by 0.4 °C[39] has reduced the length of the growing season in England by 10 days and has undoubtedly played a part in recent agricultural troubles in other parts of the world.[40] As Dansgaard *et al.*[19] have put it 'this stresses the question of whether the future will bring back the favourable climatic conditions that prevailed in part of this century, or will remain closer to the cooler conditions that characterised most of the past six centuries or more. On the face of it, the recent warm peak was an unusual event'. Bryson[15] has summarised 'the lessons of climatic history' as:

(1) *Climate is not fixed.*

(2) *Climate tends to change rapidly rather than gradually*, with changes from a glacial to a non-glacial climate, and vice versa, perhaps taking less than a century, although full response and adjustment of the environment may take much longer; smaller but still significant changes occur over a few decades.

(3) *Cultural changes usually accompany climatic changes*, as shown so clearly by the history of the Norse colonies.

(4) *What we think of as the normal climate at present is not normal in the longer perspective of recent centuries.*

(5) *When the high latitudes cool, the monsoons tend to fail* – and the high latitudes have now been cooling for some 25 yr.

(6) *Cool periods of Earth history are periods of greater than normal climatic instability*, associated with weaker circulation, bringing extremes of drought, flood and temperature (including even occasional extreme summer heatwaves) to further disturb agriculture already handicapped by a short growing season.

We can only echo Bryson's conclusion that with these lessons in mind 'combining the nature of recent climatic change with the present narrow margin of world food-grain reserves [see reference 41], an urgent need to consider and react to the possibility of continued climatic variation is indicated.'

References

1. Chu Ko-chen (1973). *Scientia sin.*, **16**, 226.
2. Gray, B. M. (1975). *Weather*, **30**, 359.
3. King, J. W. (1974). *Nature*, **245**, 443.
4. Bowler, J. M., Hope, G. S., Jennings, J. N., Singh, G. & Walker, D. (1975). Paper presented to WMO/IAMAP climate symposium, Norwich, England, 1975; note that the paper published by these authors in the Proceedings volume (*WMO Tech. Note* No. 421, 1975) is an 'extended abstract' of a more comprehensive paper circulated by by the Australian National University team.
5. Salinger, M. J. & Gunn, J. M. (1975). *Nature*, **256**, 396.
6. Brooks, C. E. P. (1949). *Climate through the ages*, 2nd edn. Benn, London.
7. Fairbridge, R. W. (1961). *Physics Chem. Earth*, **4**, 99.
8. Zeuner, F. E. (1958). *Dating the past*, 4th edn. Methuen, London.
9. Lamb, H. H. (1963). On the nature of certain climatic epochs which differed from the modern (1900–39) normal. In *Changes of climate*, Arid Zone Research Series, 20. UNESCO, Paris. Reprinted as section 3 of reference 21 below.
10. Flohn, H. (1952). *Geol. Rdsch.*, **40**, 153.
11. Godwin, H. (1954). *Danm. geol. Unders. Raekke* II, No. 80. Copenhagen.
12. Godwin, H. & Willis, E. H. (1959). *Nature*, **184**, 490.
13. Le Roy Ladurie, E. (1971). *Times of feast, times of famine*. Doubleday, New York.
14. Griffin, J. B. (1962). In *Proceedings of the symposium on solar variations, climatic changes and related geophysical problems*. New York Academy of Science, New York.
15. Bryson, R. A. (1975). *Envir. Conserv.*, **2**, 163.
16. Harrington, H. J. & McKeller, J. C. (1958). *N.Z. Jl Geol. Geophys.*, **1**, 571.
17. Lamb, H. H. & Johnson, A. I. (1959). *Geogr. Annlr*, **41**, 94 and Lamb, H. H. & Johnson, A. I. (1961). *Geog. Annlr*, **43**, 363.
18. Cromertie, George Earl of (1912). *Phil. Trans. R. Soc.*, **27**, 296.
19. Dansgaard, W., Johnsen, S. J., Reeh, N., Gundestrup, N., Clausen, H. B. & Hammer, C. U. (1975). *Nature*, **255**, 24.
20. Bergthorsson, P. (1969). *Jökull*, **19**, 94.
21. Lamb, H. H. (1968). *The changing climate*, section 7. Methuen, London.
22. Manley, G. (1958). *Arch. Met. Geophys. Bioklim.*, *A*, **139**, 413.
23. Manley, G. (1961). *Met. Mag.*, **90**, 303.
24. Manley, G. (1974). *Q. Jl R. met. Soc.*, **100**, 389.
25. Johnsen, S. J., Dansgaard, W., Clausen, H. B. & Langway, C. C. (1972). *Nature*, **235**, 429.
26. LaMarche, V. C. (1972). *Science*, **18**, 1043.
27. Bryson, R. A. & Baerreis, D. A. (1968). *J. Iowa archeol. Soc.*, **15**, 1.
28. Wilson, A. T. & Grinsted, M. J. (1975). *Nature*, **257**, 387.
29. Friedman, I. & Smith, G. I. (1972). *Science*, **176**, 790.
30. Libby, L. M. & Pandolfi, L. J. (1974). *Proc. natn Acad. Sci. USA*, **71**(6), 2482.
31. Manley, G. (1975). *Weather*, **30**, 382.
32. Kukla, G. J. & Kukla, H. J. (1974). *Science*, **183**, 709.
33. Williams, M. A. J. & Adamson, D. A. (1974). *Nature*, **248**, 584.
34. Lamb, H. H. (1972). *Climate: present, past and future*, vol. 1. Methuen, London.

35. Thomas, M. K. (1957). Changes in the climate of Ontario. In *Changes in the fauna of Ontario.* University of Toronto Press.

36. Starr, V. P. & Oort, A. H. (1973). *Nature,* **242,** 310.

37. Glover, R. S., Robinson, G. A. & Colebrook, J. M. (1972). In *Marine pollution and sea life*, ed. M. Ruivo, p. 439. Fishing News Books, London.

38. Coombs, S. M. (1975). *Nature,* **258,** 134.

39. Budyko, M. I. (1969). *Tellus,* **21,** 611.

40. Bryson, R. A. (1974). *Science,* **184,** 753.

41. Brown, L. R. (1975). *Science,* **190,** 1053.

SECTION 2

Balancing the global heat budget

5

The heat balance of the Earth

M. I. BUDYKO

Meteorological observations provide detailed information on the contemporary state of the climate and on its changes during the period of instrumental observations, mainly for the past 100 yr. Paleogeographical investigations establish the general features of climate changes during hundreds of millions of years.

One of the main problems of physical climatology is to explain the regularities of modern climate and its past changes on the basis of the laws of physical science.

Even in antiquity it was clear that climatic conditions differ depending on latitude and season, in accordance with the magnitude of an angle of incidence of solar rays on the Earth's surface. This relationship explains the origin of the term 'climate' originating from the Greek verb αλινειν (to tilt), characterizing the angle of the solar rays.

In the nineteenth century a distinguished Russian climatologist A. I. Voeikov (1842–1916) pointed out that for understanding the regularities of the genesis of climate it is necessary to investigate the transformation of solar energy at the Earth's surface and in the atmosphere. In the monograph 'Климаты Земного шара' (*Climates of the world*, 1884), Voeikov wrote:

I believe that one of the important tasks of physical science at present is keeping an account book of solar heat received by the globe with its air and water envelope.

We want to know how much solar heat is received near the upper boundaries of the atmosphere, how much it takes for heating the atmosphere, for changing the state of water vapor added to it, then what amount of heat reaches the land and water surface, what amount it takes for heating various bodies, what for changing their state (from solid to liquid to gaseous), in chemical reactions especially connected with organic life; then we must know how much heat the earth loses owing to radiation into space . . .

The task stated by Voeikov was fulfilled in the twentieth century. Schmidt (1915), Ångström (1920) and Albrecht (1940) in their work determined the components of heat balance of the Earth's surface for individual regions of the globe. The Earth's heat balance was investigated during 1945–74 in the Main Geophysical Observatory (Leningrad) and as a result a series of world maps of the Earth's heat balance was constructed for every month and for mean annual conditions (*Atlas of the heat balance*, 1955). During further investigations these maps were refined and supplemented, and together with several maps of heat balance of the Earth–atmosphere system were published in the second edition of *Atlas of the heat balance of the Earth* (1963). Systematic observations of radiation fluxes at the upper boundary of the atmosphere by means of meteorological satellites were started in the 1960s and resulted in the construction of maps of components of the radiational balance of the Earth–atmosphere system (Raschke, Möller & Bandeen, 1968; Vonder Haar & Soumi, 1969; Raschke *et al.*, 1973). Distribution of mean values, periodical and aperiodical variability of different elements of the meteorological regime were studied in works on climatology of heat balance (Budyko, 1956, 1971), and as a result of these investigations data have been obtained on heat balance components which are related to those on temperature, precipitation and other elements of the meteorological regime observed by the world meteorological network.

Data on the heat balance are widely used in contemporary studies of the genesis of climate and, particularly, in those devoted to climatic changes. These data are the main basis of semi-empirical theories of the atmospheric thermal regime considered below.

Data on the heat balance are also used in studies on climatic theory based on modelling the general circulation of the atmosphere to validate methods of parameterizing the processes of solar energy transformation at the Earth's surface and in the atmosphere, and to check the calculation of heat fluxes (Manabe, Smagorinsky & Strickler, 1965; Manabe & Bryan, 1969).

In recent years studies of general climatic theory have resulted in world maps of the heat and water balance components (Holloway & Manabe, 1971). These maps proved to be similar to those constructed earlier from empirical data and justified the validity of the available theories on solar energy transformation at the Earth's surface and in the atmosphere.

The use of the heat balance approach enables one to obtain a good understanding of the laws of the genesis of modern climate and clarify the causes of its change.

Heat balance

Heat balance equations

Physical climatology studies the components of the heat balance of the Earth's surface, the Earth–atmosphere system and the atmosphere. They are used in the heat balance equations and present special formulations of the law of energy conservation.

The equation of heat balance for the Earth's surface is usually presented in the form

$$R = LE + P + A \qquad (5.1)$$

where R is the radiative flux of heat (radiation balance of the Earth's surface) equal to the difference of absorbed short-wave radiation and the net long-wave radiation outgoing from the Earth's surface; LE is the heat expenditure for evaporation (L is the latent heat of vaporization, E is the rate of evaporation); P is the turbulent flux of heat between the Earth's surface and the atmosphere; A is the heat flux between the Earth's surface and the lower layers of water or soil.

The heat flux A during a given period of time equals the sum of change in the heat content of the water or soil within the vertical column of a unit section going through the Earth's surface (B) and heat income resulting from heat exchange through the sides of the column with the ambient layers (F_0). F_0 can reach great magnitudes for oceans in the regions where currents exist, but is zero for land.

The equation of the heat balance of the Earth–atmosphere system has the form

$$R_s = F_s + L\,(E - r) + B_s \qquad (5.2)$$

where R_s is the radiative heat flux (radiation balance) for the system equal to the difference between the absorbed short-wave radiation and long-wave radiation going into space; F_s is the heat income entering through the sides of the vertical column of a unit section passing through the atmosphere and upper layers of the hydrosphere (or the lithosphere), where noticeable periodic or aperiodic temperature variations take place; $L\,(E-r)$ is the difference between the heat expenditure for evaporation and the heat gain from condensation in the column in question (the heat gain from condensation is assumed to be proportional to the precipitation amount per unit of time r); B_s is the change in heat content inside the column per unit time interval.

From (5.1) and (5.2) the equation of the heat balance of the atmosphere can be obtained:

$$R_a = F_a - Lr - P + B_a \qquad (5.3)$$

where R_a is the radiative balance of the atmosphere equal to the difference between the radiative balance of the system and the Earth's surface; F_a is the heat income, entering through the sides of the vertical column of a unit section passing through the atmosphere; Lr is the heat gain from condensation; B_a is the change in heat content of the atmosphere inside the column per unit time interval.

In equations (5.1–5.3) the value of the radiative balance is considered to be positive when it shows that heat is gained by the system, and all other terms are assumed to be positive when it shows a heat loss.

For annual averages B, B_s and B_a will be close to zero, R_s being also close to zero for the Earth as a whole, corresponding to the equality between the solar radiation absorbed by the Earth as a planet, and the long-wave outgoing radiation.

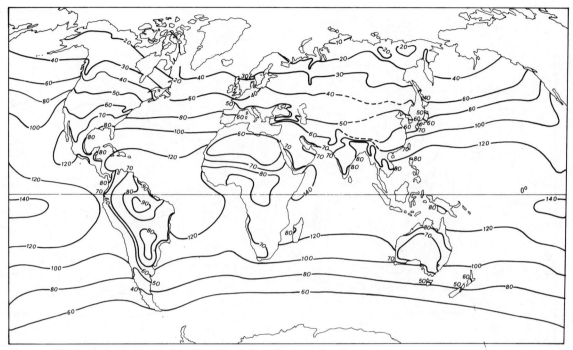

Fig. 5.1. The radiative balance of the Earth's surface (kcal cm⁻²yr⁻¹).

The geographical distribution of the heat balance components

Since the greater part of solar radiation absorbed by the Earth is absorbed at the Earth's surface (the atmosphere is comparatively transparent for short-wave radiation), study of the heat balance of the Earth's surface is of great importance for understanding the genesis of climate.

The raw materials are obtained using calculation methods relating the values of the heat balance components to meteorological elements observed at the base network of meteorological stations, that is, temperature and air humidity, cloudiness and so on. To determine the components of the heat balance of the Earth's surface the data of actinometric observations are used as well.

As is seen from the map presented in Fig. 5.1, the mean annual values of the radiation balance of the Earth's surface change from close to zero in high latitudes to 100 kcal* cm⁻²yr⁻¹ in

*1 cal = 4.1868 J.

tropical regions of continents and up to 150 kcal cm⁻²yr⁻¹ on the oceans. Zonal distribution of the values of radiation balance caused by astronomical factors is disturbed in regions where the cloud amount is considerably greater than the mean for the globe (this decreases the radiation balance). Along with this, the radiation balance is lowered in a number of desert regions of some continents due to the great albedo of deserts and their high surface temperature which increases the net long-wave radiation.

As seen from Fig. 5.2, the heat expenditure for evaporation on land changes from values close to zero in high latitudes up to 70–80 kcal cm⁻²yr⁻¹ in the wet equatorial regions. In the regions of insufficient humidification the expenditure of heat for evaporation decreases, approaching zero in most dry regions of tropical deserts. On the oceans the smallest values of heat expenditure for evaporation are observed in high latitudes near the boundary of ice cover, where they account for approximately 30 kcal

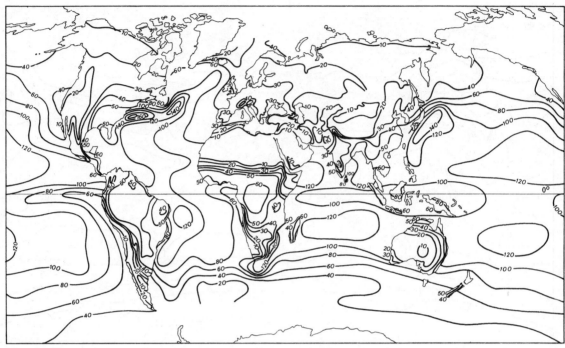

Fig. 5.2. The expenditure of heat for evaporation (kcal cm^{-2}yr^{-1}).

cm^{-2}yr^{-1}. High values of the expenditure of heat for evaporation from the oceans (over 120 kcal cm^{-2}yr^{-1}) are observed in the tropics in regions of high atmospheric pressure. Near the equator the expenditure of heat for evaporation decreases to values less than 100 kcal cm^{-2}yr^{-1} because of increasing cloud amount and air humidity.

The expenditure of heat for evaporation from oceans essentially depends on ocean currents. It increases in the region of warm currents to values exceeding 180 kcal cm^{-2}yr^{-1} (the Gulf stream) and to approximately 140 kcal cm^{-2}yr^{-1} in the region of Kuro Shio. In the region of cold currents the expenditure of heat for evaporation is lowered.

The turbulent flux of sensible heat between the Earth's surface and the atmosphere for the mean annual conditions (Fig. 5.3), as a rule, is positive and is directed from the Earth's surface to the atmosphere. Some regions of continental glaciations (the Antarctic, Greenland) and only

a comparatively small part of the ocean surface are exceptions to this.

The turbulent flux on continents, small in absolute magnitude in high latitudes, grows with decreasing latitude and increasing aridity of climate. The greatest values of turbulent flux (55–60 kcal cm^{-2}yr^{-1}) occur in the tropical deserts.

On the oceans the turbulent heat flux reaches maximum values in the regions of warm currents in the northern hemisphere (40 kcal cm^{-2}yr^{-1}). In equatorial latitudes and the southern hemisphere, where the difference between the temperature of the water surface and the air is small, the turbulent heat flux is also small. The negative values of the turbulent heat flux (the heat income from the atmosphere to the Earth's surface) occur in the regions of some cold currents.

The value of heat exchange between the oceanic surface and lower water layers for the average annual conditions equals the income or

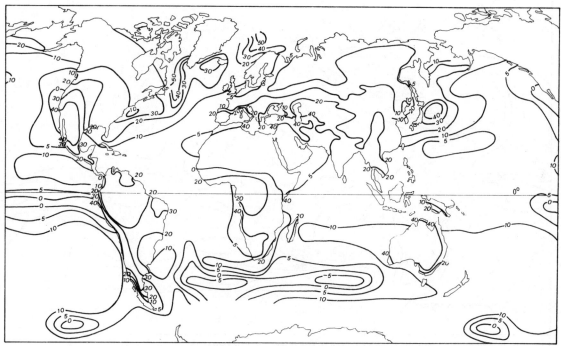

Fig. 5.3. The sensible heat flux between the Earth's surface and the atmosphere (kcal cm^{-2}yr^{-1}).

output of heat as a result of horizontal heat exchange in the vertical column, passing through the ocean. For this reason, the indicated value characterizes the redistribution of heat between different oceanic regions due to the activity of marine currents.

Fig. 5.4 shows that in low tropical latitudes the oceans accumulate a large amount of heat (nearly 20–40 kcal cm^{-2}yr^{-1}) which is transferred to the higher latitudes. The largest amount of heat is transferred from oceanic waters to the atmosphere in the regions of activity of warm currents; in some regions of the Gulf Stream it reaches 110 kcal cm^{-2}yr^{-1}. In the zones of cold currents the heat flux is usually directed from the surface of oceans to the deeper layers, its value being 50–60 kcal cm^{-2}yr^{-1}.

The above components of the heat balance of the Earth's surface in most cases change considerably during the course of the year. The radiation balance of the land and ocean surface

at all latitudes, except the equatorial zone where changes are small, reaches a maximum in summer and a minimum in winter, in accordance with variations in short-wave radiation. The heat expenditure for evaporation on the continents in the regions with a moist climate varies over the year in a similar manner to the radiation balance, whereas in arid regions its value depends considerably upon the moisture present, being lower during the period with a small amount of precipitation. On the oceans the heat expenditure for evaporation in most cases has an annual cycle opposite to that of the radiation balance. One of the causes of this regularity is an increase in winter of the difference between the temperature of the water surface and air temperature due to the great thermal inertia of oceanic water.

The annual course of the turbulent heat flux on land and oceans is often similar to that of the expenditure of heat for evaporation. In the arid

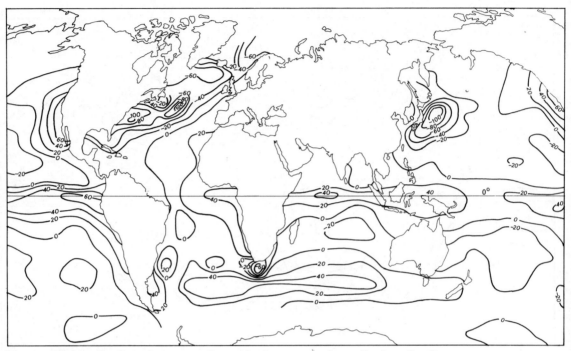

Fig. 5.4. The heat flux from the ocean surface to the deeper water layers (kcal cm^{-2}yr^{-1}).

Table 5.1. *Mean latitudinal values of the heat balance components of the earth's surface (kcal cm^{-2}yr^{-1})*

Latitude	Land			Ocean				Earth			
	R	LE	P	R	LE	P	F_0	R	LE	P	F_0
70–60° N	22	16	6	23	31	22	−30	22	20	11	−9
60–50	32	23	9	43	47	19	−23	37	33	13	−9
50–40	45	25	20	64	67	16	−19	54	45	18	−9
40–30	58	23	35	90	96	14	−20	76	65	23	−12
30–20	64	19	45	111	109	7	−5	94	75	21	−2
20–10	74	32	42	121	117	7	−3	109	95	16	−2
10–0	79	57	22	124	104	7	13	114	93	10	11
0–10° S	79	61	18	127	99	6	22	116	90	9	17
10–20	75	45	30	122	113	9	0	112	98	14	0
20–30	71	28	43	109	106	11	−8	100	88	18	−6
30–40	62	29	33	92	82	11	−1	88	76	14	−2
40–50	44	22	22	72	51	6	15	71	50	7	14
50–60	35	22	13	46	35	9	2	46	35	9	2
Earth as a whole	50	27	23	91	82	9	0	79	66	13	0

R is the radiative flux of heat (radiation balance of the Earth's surface) equal to the difference of absorbed short-wave radiation and the net long-wave radiation outgoing from the Earth's surface; LE is the heat expenditure for evaporation (L is the latent heat of vaporization, E is the rate of evaporation); P is the turbulent flux of heat between the Earth's surface and the atmosphere; F_0 is the heat income resulting from heat exchange through the sides of the vertical column of a unit section going through the Earth's surface with the ambient layers.

regions of the continents this similarity, however, is disrupted since in this case the turbulent flux in dry seasons increases.

The heat balance of the Earth

Data on the distribution of the heat balance components permit us to determine the values of these components for different latitudinal zones of the globe. Table 5.1 presents the values of the heat balance components of the Earth's surface for the latitudinal zones of land and oceans. The last row of this table contains the data on components of the heat balance of the Earth as a whole.

The above maps of the components of the heat balance of the Earth's surface used for compiling Table 5.1 and refined by data on recent investigations differ slightly from the maps in *Atlas of the heat balance of the Earth* (1963). Though several components of the heat balance (radiation balance, heat expenditure for evaporation) somewhat increased as a result of the refinement (on average by a value nearly 10 %), the basic regularities of geographical distribution of the balance of components did not change.

It is seen from Table 5.1 that in most latitudinal zones on the continents and in all zones on the oceans the heat expenditure for evaporation is greater than the turbulent flux between the Earth's surface and the atmosphere.

For the Earth as a whole the first of these components accounts for 83.5 % of the radiation balance value, the second one 16.5 %. The redistribution of heat by currents in oceans in most zones is considerably less in absolute value than the radiation balance, though in the high latitudes of the northern hemisphere oceans this component of the heat balance has come into importance.

The mean latitudinal values of the components of the heat balance of the Earth–atmosphere system for both hemispheres are given in Table 5.2 (Budyko, 1974), where Q_a is the value of radiation absorbed in the Earth–atmosphere system; I_s is the value of the outgoing long-wave radiation at the outer boundary of the atmosphere. The difference of these two values equals the radiation balance of the Earth–atmosphere system R_s. The total redistribution of energy due to horizontal motions in the atmosphere, equal to $F_0 + L(E - r)$, is designated as C_0. The values of F_0 in Tables 5.1 and 5.2, determined by different methods, are different.

Table 5.2. *Mean latitudinal values of the heat balance components of the Earth–atmosphere system for two half-years (kcal cm^{-2}month^{-1})*

Latitude	The first half-year					The second half-year				
	Q_a	C_0	F_0	B_s	I_s	Q_a	C_0	F_0	B_s	I_s
80–90°N	7.8	−4.5	0	0.8	11.5	0.1	−9.1	0	−0.8	10.0
70–80	8.2	−4.4	0	0.8	11.8	0.5	−9.3	0	−0.8	10.6
60–70	11.5	−1.8	−0.4	1.2	12.5	1.8	−6.7	−1.5	−1.2	11.2
50–60	14.6	0.0	−1.3	2.8	13.1	4.0	−4.7	−0.5	−2.8	12.0
40–50	16.9	1.0	−2.0	3.9	14.0	6.5	−2.9	0.6	−3.9	12.7
30–40	19.2	1.4	−1.7	4.3	15.2	9.5	0.9	0.7	−4.3	14.0
20–30	20.0	2.0	−0.4	2.9	15.5	13.7	0.9	0.9	−2.9	15.1
10–20	19.7	2.4	0.9	1.4	15.0	17.0	1.8	1.0	−1.4	15.6
0–10	18.4	1.4	2.2	−0.1	14.9	18.7	2.1	1.4	0.1	15.1
0–10°S	18.0	2.3	2.2	−1.5	15.0	19.7	2.4	0.8	1.5	15.0
10–20	16.2	2.6	0.9	−2.5	15.2	20.7	3.5	−0.3	2.5	15.0
20–30	13.0	1.4	0.3	−3.4	14.7	20.6	3.5	−1.2	3.4	14.9
30–40	8.9	−0.4	−0.4	−4.2	13.9	18.9	2.0	−1.4	4.2	14.1
40–50	5.9	−1.9	−1.6	−3.5	12.9	15.9	0.0	−0.6	3.5	13.0
50–60	3.3	−3.9	−2.6	−2.5	12.3	12.8	−2.6	0.6	2.5	12.3
60–70	1.0	−9.5	0	−0.8	11.3	8.1	−4.3	0	0.8	11.6
70–80	0.2	−9.8	0	0	10.0	4.4	−6.5	0	0	10.9
80–90	0.0	−8.8	0	0	8.8	3.4	−6.9	0	0	10.3

Q_a is the value of radiation absorbed in the Earth–atmosphere system; I_s is the value of the outgoing long-wave radiation at the outer boundary of the atmosphere; F_0 is the heat income resulting from heat exchange through the sides of the vertical column of a unit section going through the Earth's surface, with the ambient layers; B_s is the change in heat content inside the column per unit time interval; C_0 is the total redistribution of energy due to horizontal motions in the atmosphere.

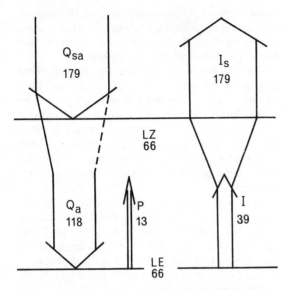

Fig. 5.5. The heat balance of the Earth (the balance components, kcal cm^{-2}yr^{-1}). Short-wave radiation absorbed by the planet Earth is represented by arrow Q_{sa}; quantity of short-wave radiation absorbed at the Earth's surface is represented by arrow Q_a; the net long-wave radiation is represented by arrow I; the total value of long-wave radiation from the planet is represented by arrow I_s; the radiative energy spent on the evaporation of water is represented by the circle LE; the sensible heat exchange between the Earth's surface and the atmosphere is represented by arrow P; and heat gain from condensation of water vapour is represented by the circle Lr.

The upper half of the left hand side of the table and the lower half of the right hand side refer to the warm half-year, and the lower half of the left hand side and the upper half of the right hand side to the cold half-year.

Table 5.2 shows that the outgoing radiation is comparable in value not only to the absorbed radiation but also to the heat transfer due to atmospheric circulation. It should be emphasized that the process of seasonal accumulation and expenditure of heat in the thickness of oceanic water is of great importance. The value of this component of the heat balance can reach 25–30 % of the value of outgoing radiation. The redistribution of heat by marine currents in several zones amounts to 15–20 % of the long-wave radiation.

It might be thought that all the above components of the heat balance exert an influence upon the mean temperature of the atmosphere, closely related to the value of outgoing radiation.

Using the data of Table 5.1 one can derive the diagram of the heat balance of the whole Earth. Assuming the solar constant to be equal to 1.95 cal cm^{-2}min^{-1} and the albedo of the Earth to be close to 0.30, we find that the short-wave radiation absorbed by the planet Earth amounts to 179 kcal cm^{-2}yr^{-1} (the arrow Q_{as} in Fig. 5.5). Of this quantity 118 kcal cm^{-2}yr^{-1} are absorbed at the Earth's surface (arrow Q_a) and 61 kcal cm^{-2}yr^{-1} in the atmosphere.

The radiation balance of the Earth's surface equals 79 kcal cm^{-2}yr^{-1} and the net long-wave radiation, corresponding to the difference between the absorbed radiation and radiation balance, equals 39 kcal cm^{-2}yr^{-1} (arrow I).

The total value of long-wave radiation from the planet Earth is equal to the amount of absorbed short-wave radiation (arrow I_s). The ratio I/I_s is considerably less than the ratio Q_a/Q_{sa}, illustrating the influence of the 'greenhouse' effect on the radiation regime of the Earth. Another characteristic of the 'greenhouse' effect is the value of the radiation balance of the Earth's surface, the radiative energy of which is spent on evaporation of water (66 kcal cm^{-2}yr^{-1} is shown in the form of the circle LE) and for the sensible heat exchange between the Earth's surface and atmosphere (13 cm^{-2}yr^{-1}, arrow P). The atmosphere gains the energy from the three sources:
(1) the heat gain from absorbed short-wave radiation, amounting to 61 kcal cm^{-2}yr^{-1};
(2) the heat gain from condensation of water vapor (shown in Fig. 5.5 by the circle Lr), amounting to 66 kcal cm^{-2}yr^{-1};
(3) the heat gain from sensible heat flux from the Earth's surface, amounting to 13 kcal cm^{-2}yr^{-1}.

The sum of these three values equals the output of heat as long-wave radiation into space which equal the difference between the values I_s and I, i.e. 140 kcal cm^{-2}yr^{-1}.

The water balance of the Earth
In the studies of the genesis of climate, along with the information on the heat balance of the Earth, it is necessary to have data on water transfer which are expressed as the components

of water balance of the continents, oceans and atmosphere.

The water balance equation for land or a reservoir can be used in the form

$$r = E + f + b \qquad (5.4)$$

where r is the amount of precipitation; E is the evaporation; f is the income or output of water as a result of its horizontal transfer; b is the change of water content in the soil or in a reservoir per unit time.

For a mean annual period the value of b is usually small compared to the other components of the balance and for the whole Earth the value of f equals zero, so for this reason for these conditions the water balance takes the simple form $r = E$.

The basic components of water balance were determined in many investigations. The latest world maps of precipitation, evaporation and run-off were published in *Atlas of the world water balance* (1974). From these maps one can obtain the mean latitudinal values of the main components of water balance presented in Table 5.3.

Table 5.3. *Mean latitudinal water balance components* $(cm\ yr^{-1})$

Latitude	Land		Ocean		Earth	
	r	E	r	E	r	E
70–60°N	57	26	86	52	65	83
60–50	71	39	111	80	88	56
50–40	64	41	109	115	85	76
40–30	58	39	92	163	77	110
30–20	61	32	71	183	67	126
20–10	90	53	137	200	124	161
10– 0	182	95	217	182	209	162
0 –10°S	194	102	166	168	172	152
10–20	112	75	129	191	125	166
20–30	59	47	116	183	103	152
30–40	66	48	123	144	116	133
40–50	138	38	131	88	131	86
50–60	188	36	135	60	135	60
Total	80	45	127	140	113	113

r is the amount of precipitation; E is the amount of water evaporated.

As is seen from the last row for the Earth as a whole the annual value of precipitation, equal to the value of evaporation, is 113 cm. Note that this value is greater than almost all similar values obtained earlier.

The data of Table 5.3 show that in different latitudinal zones the amount of water vapour

coming into the atmosphere from evaporation may be greater or less than its loss as precipitation. When this occurs the source of water vapor gain in the atmosphere is mainly the belts of high atmospheric pressure, where evaporation considerably exceeds precipitation. Removal of this water vapor surplus occurs in the zone near the equator and also in middle and high latitudes, where precipitation is greater than evaporation. The difference between precipitation and evaporation equals the difference between the income and output of water vapor in the atmosphere, resulting from the horizontal motion of air.

The values of this difference give an idea of the importance of the water vapor transfer in the atmosphere for the precipitation regime.

The semi-empirical theory of the atmospheric thermal regime

The purpose of semi-empirical theories

The problem of evolving a general theory of climate which does not contain significant simplifications presents great difficulty and to date has not been completely solved. In this connection the question arises as to whether there is the possibility of using data on heat balance for constructing schematized models of the atmospheric thermal regime which can explain a number of regularities of the genesis of climate and the causes of its change.

It was noted (Budyko, 1974) that these models had to conform to the following requirements: (*a*) the model has not to include empirical data on the distribution of individual elements of climate, particularly those which vary considerably as the climate changes; (*b*) the model has to consider realistically all the kinds of heat income noticeably influencing the temperature field, the law of conservation of energy being fulfilled; (*c*) the model has to take account of principal feedbacks between different elements of climate.

Comprehensive consideration should be given to the latter requirement.

The stability of the climate is provided by those of its feedbacks which are called negative. They aid in decreasing the anomalies of meteorological elements by moving their values towards the climatic norms. The main feedback of its kind is the dependence of outgoing long-wave radiation upon the temperature of the atmospheric air. With rising temperatures the long-wave radiation increases, which corresponds to a decrease in the temperature rise.

Another example of negative feedback is the dependence of heat transfer in the atmosphere on the air temperature gradient. Usually the heat flux in the atmosphere is directed from a zone of higher temperature to one of a lower temperature and results in levelling the temperature distribution. For climate changes, positive feedbacks are of great importance and contribute to increasing the anomalies of meteorological elements; as a result of this, they decrease the stability of the climate.

The dependence between absolute air humidity and its temperature is one of the positive feedbacks. With rising temperature, the evaporation from water or wet surfaces usually increases and this leads to a comparative stability of the relative air humidity in most climatic regions (except dry continental areas). Under these conditions the absolute air humidity increases with temperature increase, which is confirmed by numerous empirical data. Since long-wave radiation decreases with increasing absolute humidity, then the increase of absolute humidity, as temperature increases, partially compensates for the long-wave radiation increase brought about by the rising temperature.

This relationship was studied by Manabe & Wetherald (1967). They found that with constant relative humidity, changes of the solar constant influence the air temperature at the Earth's surface almost twice as much as when the absolute humidity is constant. It is evident that this feedback must be taken into account when the numerical models of climate change are used.

Another positive feedback, resulting from the influence of snow and ice cover on the albedo value of the Earth's surface, is of great importance for the regularities of changes in the thermal regime of the atmosphere for long periods of time.

The influence of snow cover on the climatic conditions was investigated by A. I. Voeikov, who established that snow cover lowers air temperature over its surface. Later, Brooks (1950) concluded that ice cover, because of its high albedo, noticeably lowered the air temperature, with the result that when ice formed or melted climate changes increased considerably.

From observational material obtained under Arctic and Antarctic conditions, as well as from satellite data, one can estimate the albedo of the Earth's surface and the Earth–atmosphere system in high latitudes and compare these estimates with those obtained for regions without snow and ice cover.

The available evidence enables us to conclude that in the summer months the albedo of the ice surface in the central Arctic amounts to approximately 0.70, whereas in the Antarctic it equals approximately 0.80–0.85. Taking into account the fact that the value of the average albedo of the Earth's surface for the snow- and ice-free regions does not exceed 0.15, it can be concluded that snow and ice cover, other conditions being equal, decreases by several times the radiation absorbed at the Earth's surface.

Snow and ice have an appreciable effect on the albedo of the Earth–atmosphere system. From satellite observations (Raschke *et al.*, 1973) the albedo of the system in the central Arctic in summer is equal to about 0.55 and in the Antarctic about 0.60 – approximately twice the value of the planet albedo, which equals 0.28.

It is evident that such great differences in the albedo values must have a pronounced effect on the thermal regime of the atmosphere. If as a result of declining air temperature snow and ice cover is formed at the Earth's surface, a sharp decrease in the absorbed radiation takes place that must contribute to further reducing the air temperature and enlarging the area of snow and ice cover. The reverse result will occur when the temperature rises, if this leads to snow and ice melting.

Allowance for this feedback in the numerical model of the thermal regime of the atmosphere showed that it exerts a marked influence on the air temperature distribution at the Earth's surface (Budyko, 1968).

To estimate the extent of this influence it is possible to give a simple example, illustrating how the mean global air temperature will change if the Earth's surface, without any clouds, is entirely covered with snow and ice. Let us imagine such a hypothetical Earth. Under such conditions, the albedo of the Earth is noticeably increased compared to the value existing now, and this will influence the air temperature. The 'effective' temperature of the

Earth, corresponding to its long-wave radiation, is proportional to $(1 - a_s)^{\frac{1}{4}}$ (where a_s is the albedo). Therefore, as the albedo changes from a value a'_s to a value a''_s, the absolute value of the 'effective' temperature changes as $[(1 - a''_s)(1 - a'_s)^{-1}]^{\frac{1}{4}}$. Taking the present-day albedo of the Earth as being equal to 0.33 and the albedo of dry snow cover as being equal to 0.80, we find that for a snow-covered Earth the mean 'effective' temperature must be reduced by approximately 75 °C.

One might think that the reduction of the mean air temperature near the Earth's surface will be greater than the value given. At present, the mean temperature of the lower layers of air is considerably increased all over the Earth's surface by the 'greenhouse' effect, which is associated with the absorption of long-wave radiation by water vapor and carbon dioxide in the atmosphere. At very low temperatures, this effect is not important, and also the formation of dense clouds, appreciably changing the radiation fluxes, becomes impossible. Under these conditions, the atmosphere becomes more or less transparent to both short-wave and long-wave radiation.

The mean temperature of the Earth's surface with a transparent atmosphere is determined by the simple formula $[S_0(1 - a_s) \times (4\sigma)^{-1}]^{\frac{1}{4}}$, where S_0 is the solar constant, and σ is the Stefan–Boltzmann constant. It follows from this formula that at $a_s = 0.80$, the mean temperature of the earth is 186K or -87 °C.

Thus, if snow and ice covered the whole surface of the Earth even for a short period of time, its mean temperature (equal now to 15 °C) would be reduced by approximately 100 °C. This estimate shows what an enormous effect snow cover can exert on the thermal regime.

The model of the thermal regime of the atmosphere
Let us present a schematic theory of the thermal regime of the atmosphere based on the equation of the heat balance of the Earth–atmosphere system and empirical data on the heat balance (Budyko, 1968). This theory takes into account the requirements listed on p. 93, specific attention being drawn mainly to the relations between the mean air temperature and other elements of the meteorological regime.

To estimate the influence of the solar radiation income and the albedo upon the mean temperature near the Earth's surface under actual conditions, it is necessary to know the relationship between the long-wave emission at the outer boundary of the atmosphere and the temperature distribution. This can be found when using the data of outgoing long-wave radiation obtained from observations or calculations.

Comparing these data with various elements of the meteorological regime, we succeeded in establishing that the monthly means of outgoing emission, in general, depend upon the temperature of air near the Earth's surface, and on cloudiness.

This relationship was expressed in the form of the empirical formula

$$I_s \doteq a + bT - (a_1 + b_1 T)n \qquad (5.5)$$

where I_s is the outgoing emission in kcal cm^{-2}month^{-1}; T is the temperature (°C); n is the cloud amount in fractions of a unit; the dimensional coefficients are: $a = 14.0$, $b = 0.14$, $a_1 = 3.0$, $b_1 = 0.10$.

It should be emphasized that this dependence of outgoing radiation on temperature in (5.5) is considerably weaker than it would be if the Earth were devoid of atmosphere. The cause of this difference is the influence on the emission of the positive feedback between the temperature and humidity of air, mentioned on p. 94.

To test the formulas relating outgoing long-wave emission to meteorological factors, it is necessary to assume a condition where the outgoing radiation for the whole globe is equal to the amount of radiation absorbed:

$$Q_{sp}(1 - a_{sp}) = I_{sp} \qquad (5.6)$$

where the values Q_{sp}, a_{sp} and I_{sp} relate to the planet as a whole. Note that in accordance with the above considerations about the effect of snow and ice cover on the thermal regime, one can show that equations (5.5) and (5.6) have at least two solutions.

Let us assume that at low air temperatures near the Earth's surface at all latitudes the globe will be completely covered with snow and ice and that the global albedo of the Earth–atmosphere system in this case will approximately correspond to that of the Antarctic, that is, 0.6–0.7. From (5.5) and (5.6) it follows that in this case the mean temperature near the Earth's

95

surface will vary in the range from $-47\,°C$ (at the lower value of the albedo) to $-70\,°C$ (at the higher value of the albedo). At these values of mean global temperature complete glaciation of the Earth is inevitable, which corroborates the above view about the ambiguous correspondence of the contemporary climatic conditions to to the external climate-forming factors. This question will be covered more comprehensively below.

From equations (5.5) and (5.6), we find the formula for the mean temperature of the Earth:

$$T_p = \frac{1}{b-b_1 n} \left[Q_{sp}(1 - a_{sp}) - a + a_1 n \right] \quad (5.7)$$

This formula is valid only within the range of the actual variation in monthly mean temperatures at the level of the Earth's surface, its accuracy depending on the accuracy of equation (5.5). It must be borne in mind that in accordance with the structure of equation (5.7), the accuracy of the temperature calculation at large values of cloudiness (with n approaching unity) decreases noticeably in comparison with calculations for conditions of small and moderate cloudiness. Although this restricts the possibility of using equation (5.7), it might still be possible to draw some conclusions from it about the effect of cloudiness on the thermal regime of the lower layers of air.

For this purpose, it is necessary to take into account the relationship between the albedo and cloudiness. This has the form

$$a_s = a_{sn} + a_{s0}(1 - n) \quad (5.8)$$

where a_{sn} and a_{s0} represent albedos of the Earth-atmosphere system with complete cloud cover and a cloudless sky, respectively.

From equations (5.7) and (5.8) this formula follows:

$$T_p = \frac{1}{(b-b_1 n)} \{ Q_{sp}[1 - a_{snp}n - a_{s0p}(1 - n)]$$
$$- a + a_1 n \} \quad (5.9)$$

Assuming, in accordance with the procedure used for the construction of maps in *Atlas of the heat balance of the Earth*, that

$$a_{s0p} = 0.66a + 0.10 \quad (5.10)$$

where a is the albedo of the Earth's surface, we find as the mean for the northern hemisphere that a_{s0p} is equal to 0.20. Then, from equation

(5.8), we obtain that at $a_{sp} = 0.33$ and $n = 0.50$, the mean $a_{snp} = 0.46$.

Note that the value of the Earth's albedo taken here is somewhat greater than the value obtained from satellite observations. The difference indicated slightly influences the results of temperature calculation if the coefficients in formula (5.5) are chosen so that in every case they give a radiation balance of the system equal to zero.

Taking into consideration the values derived here for a_{s0p} and a_{snp}, we find from (5.9) that for the mean planetary value Q_s, the effect of cloudiness on temperature is comparatively small and probably lies within the limits of accuracy of the calculations.

From (5.9) one can draw some conclusions about the effect of climate-forming factors on the mean temperature near the Earth's surface. Variations in solar radiation by 1% change the mean temperature, at cloudiness equal to 0.50 and with the existing Earth albedo, by approximately 1.5 °C. This evaluation is almost twice as great as that of the corresponding effect without the atmosphere. Thus, the radiative properties of the atmosphere intensify considerably the effect of variations in radiation on the thermal regime of the Earth's surface.

A change in albedo by 0.01 changes the mean temperature by 2.3 °C when the values of the coefficients in the formulas used are constant.

Consequently, the thermal regime depends substantially on variations in albedo, if this variation is not caused by conditions of cloudiness. As indicated above, the effect of cloudiness on the thermal regime, associated with albedo variations, is compensated for, to a considerable extent, by corresponding variations in outgoing long-wave radiation.

In individual latitudinal zones of the Earth, the thermal regime is considerably influenced by the horizontal redistribution of heat in the atmosphere and hydrosphere.

For quantitative evaluation of the effect of horizontal redistribution of heat on the thermal regime of the atmosphere, it is possible to use material on the components of the heat balance of the Earth–atmosphere system.

The equation of heat balance of the Earth–atmosphere system has the form

$$Q_s(1 - a_s) - I_s = C + B_s \quad (5.11)$$

where C equals $F_s + L(E - r)$, the sum of the heat income by horizontal movements in the atmosphere and hydrosphere.

For mean annual conditions, the term B_s, characterizing the accumulation or loss of heat for a given time period, equals zero, and the total income of heat C equals the value of radiation balance of the Earth–atmosphere system. Since the values of this balance can be determined either from observational data or by calculation, it is evident that the values of horizontal redistribution of heat can be found simultaneously.

It might be suggested that the values of C are in some way connected with the horizontal distribution of the mean tropospheric temperature. Taking into consideration the fact that the deviations of air temperature from the mean vertical distributions in the troposphere are small compared to the geographical variability of temperature, we may consider that the mean air temperature in the troposphere is closely connected with the temperature at the level of the Earth's surface. Thus, reasons exist for postulating the existence of a relation between horizontal heat transfer and the distribution of temperature near the Earth's surface.

Let us examine the relation between these values for mean annual conditions of the latitudinal zones of the northern hemisphere. For this purpose, we will calculate the radiation balance by the formula

$$R_s = Q_s(1 - a_s) - I_s \qquad (5.12)$$

The meridional heat flux in the Earth–atmosphere system can be calculated from the values so obtained of the radiation balance of the Earth–atmosphere system equal to the term C.

Taking into account that the meridional heat transfer is carried out in the form of heat transfer from warmer regions to colder ones, we may therefore consider that C depends on the value $T - T_p$, where T is the mean temperature at a given latitude, and T_p is the mean planetary temperature of the lower air layer.

To study this relationship we compare the values of C with the corresponding values of $T - T_p$, a distinct relation being established between them, which can be presented in the form of the empirical formula

$$Q_s(1 - a_s) - I_s = \beta(T - T_p) \qquad (5.13)$$

where $\beta = 0.235$ kcal cm^{-2}month^{-1} deg.C^{-1}.

The existence of such a relationship simplifies substantially the evaluation of the heat meridional redistribution in the model of the thermal regime.

From (5.5) and (5.13), we find that

$$T = [Q_s(1 - a_s) - a + a_1 n + \beta T_p]/(\beta + b - b_1 n) \qquad (5.14)$$

With the use of this formula, we can calculate mean annual temperatures in different latitudes.

To study the thermal regime of the atmosphere in various seasons, the above model should be changed to take into consideration several additional factors.

This more general model was suggested by Budyko & Vasishcheva (1971), in which the equation of heat balance of the Earth–atmosphere system is used in the form

$$Q_{sw}(1 - a_{sw}) - I_{sw} = C_w + B_s \qquad (5.15)$$
$$Q_{sc}(1 - a_{sc}) - I_{sc} = C_c - B_s, \qquad (5.16)$$

where B_s is the income or loss of heat due to cooling or warming of the Earth–atmosphere system, determined basically by the process of cooling or warming of the ocean. The values referring to the warm and cold half-years have the indices w and c, respectively.

In order to determine B_s the equations of the heat balance of the oceans were used as well as the relationships connecting the components of heat balance of the oceans to the temperature of the water and air.

The solution of these equations resulted in the formulas for the determination of the mean latitudinal temperatures of the lower air layer in different seasons. Calculations of the temperature distribution from the derived formulas gave results which agree well with observational data.

Similar theories of the thermal regime of the atmosphere were suggested by Sellers (1969, 1975), and have been discussed in many recent publications.

Testing the atmospheric thermal regime model
One of the methods of testing the theory of the thermal regime, based on the comparison of computational results of the distribution of mean-latitudinal air temperature near the Earth's surface, with observational data, is mentioned above. Another method is the comparison of the conclusions drawn from this

theory with the theories of climate evolved in other investigations.

Such comparisons have been made repeatedly and show that both the semi-empirical theory of climate, based on the heat balance data, and theories based on more or less comprehensive modelling of the general circulation of the atmosphere, give close results when studying the regularities of climate, if the theories under consideration fulfill the requirements listed on p. 93 and if they take into account the similar values of physical parameters influencing climate.

The third method of examining the model of the thermal regime consists of comparing the calculated results of climatic changes with observational data, the conclusions being discussed on pp. 105–11.

Recently, one more method of testing the theories of the atmospheric thermal regime, based on the use of data from satellite observations of outgoing long-wave emission, has been employed (Budyko, 1975). For this purpose, we use the equations of heat balance for the Earth with the solar constant S

$$(1/4)S(1 - a_{sp}) = I_{sp} \qquad (5.17)$$

as well as at its value corresponding to $S + \varDelta S$

$$(1/4)(S+\varDelta S)(1 - a_{sp} - \varDelta a_{sp}) = I_{sp} + \varDelta I_{sp} \qquad (5.18)$$

where $\varDelta a_{sp}$ and $\varDelta I_{sp}$ characterize the corresponding changes in albedo and in outgoing emission.

From (5.17) and (5.18), we find

$$\varDelta S(1 - a_{sp} - \varDelta a_{sp}) - S\varDelta a_{sp} = 4I_{sp} \qquad (5.19)$$

We shall assume that within the range of the mean air temperature variations $\varDelta T$, $\varDelta a_{sp} = A\varDelta T$ and $\varDelta T_{sp} = B\varDelta T$, and that $|\varDelta a| \leqslant 1 - a$.

In this case, we find

$$\varDelta T/\varDelta S = (1 - a) / (SA - 4B) \qquad (5.20)$$

Let the value $\varDelta T/\varDelta S$ be denoted $\varDelta T_1$ for a 1 % change of solar constant. Its value, obtained from the formulas of semi-empirical theory of the thermal regime of the atmosphere, if the albedo is constant, is equal to 1.5 °C. The value of $\varDelta T_1$ can be obtained from variations of the meteorological regime over one year for the northern and southern hemispheres, both with

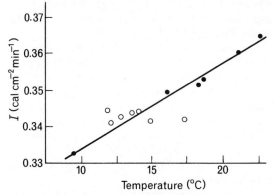

Fig. 5.6. The relation between the outgoing long-wave emission (I) and mean air temperature near the Earth's surface. Solid circles, the northern hemisphere, open circles, the southern hemisphere.

the albedo constant and with its varying as a result of changes during the year in the mean value of cloudiness for each hemisphere.

To solve this problem the values of parameters B and A in formula (5.20) must be found. We can determine B from satellite observations of the outgoing long-wave radiation.

Fig. 5.6 presents the 'Nimbus 3' (Raschke *et al.*, 1973) observational data on the outgoing radiation in the northern and southern hemispheres for corresponding seasons as a function of the mean air temperature near the Earth's surface. It is seen from Fig. 5.6 that there is a close correlation between these values, this relation being close to a linear one, which justifies the above hypothesis on the outgoing emission's dependence on temperature. Using data presented in the figure, we find that $B = 0.0024$ cal cm^{-2}min^{-1}deg.C^{-1}.

Similar data from satellite observations for 1962–66 (Vonder Haar & Suomi, 1971) give almost the same value of B.

If the albedo of the Earth–atmosphere system is equal to 0.30 and does not depend on the air temperature ($A = 0$), it follows from (5.9) that with the value of B as above, $\varDelta T_1 = 1.4$ °C. This value is in good agreement with the above estimates derived from the models of the atmospheric thermal regime for constant cloud amount.

We did not calculate the value of A from the satellite observational data since the dependence of the albedo upon the air temperature of the northern and southern hemispheres, obtained from the satellite data, can be determined to a great extent not by the influence of temperature upon cloudiness, but by the annual cycle of radiation in high latitudes (in summer, polar ice cover reflects a large amount of radiation and this results in the albedo increasing for the whole hemisphere during the warm season).

When estimating the influence of temperature upon the mean albedo for the hemispheres, one can use data on the annual cycle of cloudiness. From (5.8) we can find the variation of the albedo (Δa_{sp}) with fluctuating cloudiness:

$$\Delta a_{sp} = (a_{snp} - a_{s0p})\Delta n \qquad (5.21)$$

Using this formula, it is possible to calculate deviations of the albedo in the northern and southern hemispheres for different months from the annual means and to compare these values with the Earth's surface temperature deviations for each hemisphere from the annual means. This comparison shows that the relation between the anomalies of cloudiness and temperature is not close.

The value of A, obtained from this analysis, is equal to 0.00125 deg.C^{-1}. Using this value of the coefficient A, we can derive $T\Delta T_1$ (from (5.20)), its value being 1.1 °C.

It should be noted that the value of A, determined by this method, is not very accurate. It is possible that the main result in this case is the conclusion that the value of SA is considerably less than that of $4B$, so that the cloudiness–temperature feedback does not exert a great influence upon the relationship between the mean air temperature for a hemisphere and the solar radiation income.

The question of the effect of variations in cloud amount upon the relationship between the air temperature and heat income is comprehensively studied by Cess (1976), who also used for this purpose data from satellite observations.

From the latest collection of measurements of outgoing emission (Ellis & Vonder Haar, 1976), Cess redetermined the empirical coefficients in formula (5.5), and then he used this formula together with the heat balance equation of the Earth–atmosphere system to calculate the values of ΔT_1. The value of ΔT_1, thus obtained, equalled 1.45 °C, and the influence of the cloudiness–temperature feedback upon it proved to be small.

The calculations of Cess showed that this relationship is also not affected by individual latitudinal zones except for a comparatively small area of high-latitude regions.

From the data given here it follows that there are now several independent approaches available to elucidate to what measure the existing models of the theory of climate could be used for studying the regularities of climate genesis and the mechanism of its change. Since all these methods confirm this conclusion (for the models comply adequately with the demands indicated earlier), it seems to us that the main problem of the theory of climate is solved to a first approximation.

Unambiguity and stability of climate
Unambiguity of climate

For the study of climatic changes the question of unambiguity of climate is of great importance – can only one or several variants of the meteorological regime correspond to the given external, climate-forming conditions (solar radiation, structure of the Earth's surface, chemical composition of the atmosphere)? This question was studied by Lorenz (1968) on the basis of analyzing the equations of the theory of climate. Lorenz pointed out that two different types of solution of these equations are possible. The first of them gives one variant of stable climate resulting from averaging the instantaneous states of the fields of meteorological elements for a long period of time. Such a climate, called transitive by Lorenz, is unambiguous for the external conditions under consideration. The second type of the solution gives several variants of stable climate, every one of which refers to the same external conditions (intransitive climate).

Under the contemporary external climatic conditions one more variant of climate can exist, sharply different from the climatic conditions of present time. This is the climate of a 'white Earth', meaning complete glaciation of the Earth.

To study more thoroughly the question of unambiguity of climate one can use the equations of semi-empirical theory of the thermal regime of the atmosphere.

Let us consider, in accordance with the empirical data obtained both for marine polar ice and continental glaciations in the extratropical latitudes, that the mean latitudinal

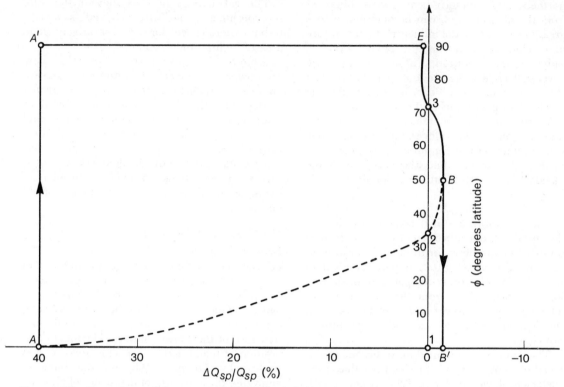

Fig. 5.7. The dependence of the mean latitude of the polar ice cover boundary in the northern hemisphere on the radiation income at the outer boundary of the atmosphere. $\Delta Q_{sp}/Q_{sp}$ is the relative change in the income of solar radiation to the outer boundary of the atmosphere, expressed as a percentage and ϕ is the mean latitude of the limit of polar glaciation of the northern hemisphere.

boundary of the permanent snow–ice cover corresponds to the mean annual temperature which equals −10 °C. When this condition is taken into consideration the dependence of average latitude of the boundary of polar ice cover of the northern hemisphere upon the income of radiation energy at the outer boundary of the atmosphere can be calculated from the model of the thermal regime for mean annual conditions.

This relation is shown in Fig. 5.7, where $\Delta Q_{sp}/Q_{sp}$ is the relative change in the income of solar radiation to the outer boundary of the atmosphere, expressed as a percentage; ϕ is the mean latitude of the limit of polar glaciation of the northern hemisphere.

The relationship between ϕ and $\Delta Q_{sp}/Q_{sp}$ is shown as a system of thick lines illustrating that this has an ambiguous character and that it

appreciably differs with an increase or decrease in the heat income to the outer boundary of the atmosphere.

Consider first of all the case of an increase in the income of radiation from its initial small value. At a small heat input, complete glaciation of the Earth takes place ($\phi = 0°$) which remains with increasing the heat income to its present value (point 1) and to the value exceeding by several tens percent (point A) the indicated one. The regime of glaciation corresponding to point A is unstable and with a small increase in the heat income transforms to the regime of complete absence of glaciation (point A'). Further increase in the heat income corresponds to retaining the ice-free regime.

With a decrease in the heat income from its initial value, considerably exceeding the solar constant, the conditions of an ice-free regime are first observed ($\phi = 90°$). On reaching point E, close to modern value of the heat income, polar glaciation arises which rapidly grows with decreasing heat income. After reaching point 3, corresponding to the present-day climatic regime, when the heat income is reduced by 2 % of the present value, the ice sheet reaches 50° N (point B).

The regime of glaciation, corresponding to this point, is unstable and with a small reduction in the heat income transforms to the regime of complete glaciation (point B'), which is retained with further lowering of the heat income.

Using the above model, we can also obtain the relation between the values ϕ and $\Delta Q_{sp}/Q_{sp}$ presented by the dashed line AB. Since this curve is characterized by an increase of ϕ with a decrease of $\Delta Q_{sp}/Q_{sp}$ (or a lowering of ϕ with an increase of $\Delta Q_{sp}/Q_{sp}$), then it might be possible that the curve corresponds to unstable regimes of glaciation passing to the regimes of complete glaciation or to its complete absence with small variations in the heat income. For this reason, the point on the curve AB corresponding to the contemporary value of the heat income (point 2), characterizes an unstable regime which cannot exist for a long period of time.

Thus, the dependence of polar glaciations upon the heat income has the form of an hysteresis loop with sections AA' and BB', corresponding to the transition from one solution of the equations used to another, designated by arrows in Fig. 5.7.

Other sections of the hysteresis loop corresponding to stable regimes of glaciation can characterize the dependence of polar glaciations on $\Delta Q_{sp}/Q_{sp}$, both with increasing and decreasing the heat income. We note that the regularities presented in Fig. 5.7 can be established on the basis of general considerations to be taken into account when building different models of the thermal regime.

As stated above, with the present available value of the heat income to the outer boundary of the atmosphere, a stable state of complete glaciation on Earth is possible with very low temperatures at all latitudes ('white Earth').

The stability of such a regime is explained by the very high albedo of a surface covered with snow and ice. In this case, climatic conditions which are characteristic for Antarctica now can exist all over the Earth's surface. It can be concluded, apparently, from any realistic theory of climate that there is a possibility of a stable existence of a 'white Earth'. Consequently, for the existing income of solar radiation a regime of complete glaciation of the Earth is possible, and is labelled '1' in Fig. 5.7. This regime is also possible with a radiation income less than the present one (this is depicted in Fig. 5.7 by a line going away along the x-axis to the left of point 1), and with a radiation income which is greater than the present value, down to the value which would correspond to the temperature of melting ice in most warm regions of the Earth's surface.

On reaching this temperature part of the Earth's surface will be free from ice cover; this will lead to an albedo decrease and an increase in the radiation absorbed. The point corresponding to the limit of the regime of complete glaciation with increasing the solar radiation is marked A in Fig. 5.7. It is natural to assume that between the conditions shown by points A and 3, there exist regimes of partial glaciation of the Earth which can be plotted as the line connecting these points. The form of this line can be established in the following way.

Data from observations, as well as general physical considerations, show that under modern climatic conditions with increasing heat income to the Earth's surface the mean air temperature near the Earth's surface rises and the polar ice recedes. With decreasing heat income there is an observed lowering of the mean air temperature and an advance of glaciation. In accordance with this regularity, the line going from A to 3 must approach the latter from the left, which is possible if this line intersects the vertical axis in at least one more point (point 2).

Thus, we can draw a conclusion about the possible existence under present conditions of a third climatic regime, presenting the second variant of partial glaciation of the Earth, with a greater ice coverage than there is in the present climatic regime. This regime, however, cannot be considered stable, because it is characterized by the possibility of a transition to a state of complete glaciation or to a state of complete absence of glaciation with negligibly small changes of climate-forming factors.

After intersecting the vertical axis at 3 the line characterizing the regimes of partial glaciation must, at some value of the radiation income, exceeding its present value, reach the horizontal line corresponding to the ice-free regime (point E). From E the line corresponding to further increase in radiation converts into a horizontal straight line going away to the right.

In several studies the calculations showed that at the present level of radiation income an ice-free regime is possible in the Arctic. If we assume this is possible, then the resultant regime would be plotted as a point at the vertical axis of the graph (at $\phi = 90°$) and in this case the line characterizing the regime of partial glaciations must reach the horizontal line corresponding to the ice-free regime not to the right of the end of the vertical axis but to the left of it. Such a position of the point of intersection of the lines is explained by the fact that if the ice-free regime is possible at the present radiation income, then it is also possible with very slightly less radiation.

Thus, in the case under consideration the line characterizing the possible regimes of glaciation passes through the above point as a horizontal line that requires the vertical axis to be cut by this line at least at one point. Consequently, since the possibility of an ice-free Arctic with the present-day level of the solar radiation income exists there must exist one more regime of partial glaciation with a smaller ice area as compared to the modern climate-forming conditions.

In addition to the relationship presented in Fig. 5.7, the relationship between the mean planetary air temperature and the heat income, calculated from the same model, is shown in Fig. 5·8. All the symbols are similar to those in Fig. 5·7. As is seen from Fig. 5.8, the relationship between the mean planetary temperature and the heat income is similar in many respects to the corresponding one between the limit of polar glaciations and heat income, and has the character of an hysteresis loop as well.

Real temperature changes during the main part of the history of the Earth are presented by sections of lines in Fig. 5.7. and Fig. 5.8, characterizing possible thermal regimes of our planet, from the values placed somewhat above point E, to the value lying somewhat above point B.

It is interesting to compare this part of the

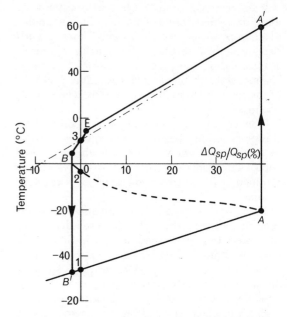

Fig. 5.8. The dependence of the mean planetary temperature upon the radiation income at the outer boundary of the atmosphere. $\Delta Q_{sp}/Q_{sp}$ is the relative change in the income of solar radiation to the outer boundary of the atmosphere, expressed as a percentage.

line in Fig. 5.8 and the chain-dotted line passing through point 3 and corresponding to variations in the mean planetary temperature in the absence of the effect of polar ice on the thermal regime. This comparison shows to what extent polar ice increases the temperature variations caused by fluctuations in the income of heat.

Thus, the contemporary climate is not unambiguous; other possibilities exist, given the existing external climate-forming factors. This conclusion was drawn both on the basis of the computations of the heat balance given in the preceding section and with the use of semi-empirical theories of climate (Budyko, 1972a; Faegre, 1972), and the theories of general circulation of the atmosphere (Wetherald & Manabe, 1975). It is notable that in all these studies the same conclusion is drawn about the possibility of 'white Earth' existing with very low temperatures at all latitudes; this suggests that such a regime of climate must represent a very stable state.

The sensitivity of climate to changes of the external factors

Climatic changes depend upon the sensivity of a meteorological regime to the fluctuations of the external climate-forming factors. The ΔT_1 parameter, suggested in the preceding section, could be used as one of the characteristics of climate sensitivity. ΔT_1 is the value of the variations in the mean temperature for the Earth (or for one of the hemispheres) of the lower air layer when the solar radiation income is increased or decreased by 1%.

The value of ΔT_1 will differ depending on the period of time to which it refers. For climatic changes over comparatively short periods of time, in the order of several years or decades, the value of ΔT_1 can be determined from theories of climate in which the albedo of the Earth's surface is considered to be constant. This is seen, for example, from data on the warming of the 1920s–30s, when the area of polar ice decreased by approximately 10 %. Simple estimates show that this change in the area of polar ice could alter the value of ΔT_1 for the northern hemisphere by approximately 0.1 °C, a comparatively small value. However, this conclusion does not mean in this case the possibility of neglecting the influence of change in polar ice area on the air temperature in high latitudes, which, as is seen from the data below, is rather substantial even with short-term climatic variations.

One of the first reliable estimates of ΔT_1 was obtained by Manabe & Wetherald (1967), who found it to be equal to 1.2 °C. It follows from the above semi-empirical theory of the thermal regime of the atmosphere that $\Delta T_1 = 1.5$ °C. The two values refer to the conditions of constant albedo, not only of the Earth's surface but of the Earth–atmosphere system as well, that correspond to the absence of changes in cloud amount.

The results of calculating ΔT_1, given in the preceding section, for the real atmosphere from satellite data of outgoing emission give a value of ΔT_1 equal to 1.1–1.4 °C (Budyko, 1975) or 1.45 °C (Cess, 1976).

In the empirical analysis of the contemporary climatic changes with the results given below, the value of ΔT_1 was found to be 1.1 °C.

All these values correlate adequately, confirming the possibility of their utilization in the studies of climatic changes. It is likely that the true value of ΔT_1 lies in the range from 1.0 to 1.5 °C. When determining the value of ΔT_1 for long-term periods, it is necessary to take into account the relationship between the albedo and the air temperature. In the first computation of this kind carried out with the use of the semi-empirical theory of the thermal regime of the atmosphere (Budyko, 1968), it was established that the thermal regime of polar ice feedback considerably increases the value of ΔT_1. This corresponds to increasing the climatic sensitivity to variations in climate-forming factors. In this calculation as well as in the calculations carried out for constructing Figs. 5.7 and 5.8 the value of the albedo of the Earth–atmosphere system for the polar zone, obtained from the first data of satellite observations, was used, which, as it then proved, was an overestimate. The calculations with more accurate values of the albedo, obtained in recent observations, gave smaller values of ΔT_1. In this case again, however, the climatic sensitivity to changes in solar constant is much greater than in the models with uniform albedo over the Earth's surface.

We note that the utilization of more accurate values of the polar region albedo somewhat changes the relationships presented in Figs. 5.7 and 5.8, the general character of which in such an event remains, however, invariable.

Several estimates of the ΔT_1 value for long-term variations of climate are given by Cess (1976), who pointed out that from the model of the general circulation of the atmosphere constructed by Wetherald & Manabe (1975) one can find $\Delta T_1 = 1.9$ °C. Taking into account the relationship between the mean albedo of the Earth and the mean air temperature presented in the work of Wetherald & Manabe and considering the relation between the air temperature and outgoing emission obtained from satellite data, Cess found that $\Delta T_1 = 2.2$ °C. Using the above semi-empirical model of the thermal regime of the atmosphere with the parameters in the model being rounded off in accordance with contemporary satellite data, Cess obtained $\Delta T_1 = 2.6$ °C. He concluded that all the values are in good agreement.

Taking into account the fact that the sensitivity of the model of Wetherald & Manabe is somewhat lowered due to a few approximate assumptions, one can suppose that the true

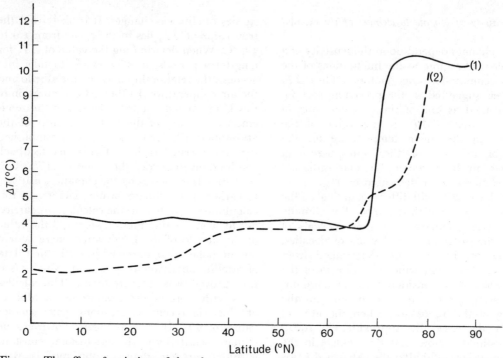

Fig. 5.9. The effect of variations of the solar constant on the mean latitudinal air temperatures. (1) the calculation using the semi-empirical model, (2) the calculation by the Wetherald–Manabe model.

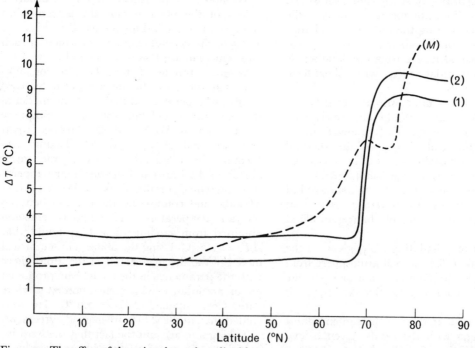

Fig. 5.10. The effect of changing the carbon dioxide concentration upon the mean latitudinal air temperatures. (1), (2) the calculations using the semi-empirical model, (M) the calculation by the Manabe–Wetherald model.

value of ΔT_1 for the given conditions lies in the range 2.0–3.0 °C.

Let us now compare the results of calculating the climatic sensitivity with the use of different models: the semi-empirical theory of the thermal regime and the theory of general circulation of the atmosphere.

Fig. 5.9 shows data on the change in the mean air temperature near the Earth's surface at various latitudes of the northern hemisphere with an increase in the solar constant by 2 %.

Curve 1 presents the computation results from the semi-empirical model, curve 2 from the model of Wetherald & Manabe (1975). In Fig. 5.10 the results of a similar calculation for the case of doubling the carbon dioxide concentration in the atmosphere are given. Curves 1 and 2 depict the results obtained from the semi-empirical model of the atmospheric thermal regime with different parameterizing of the effect of carbon dioxide on temperature for local conditions, curve M corresponds to the calculation of Manabe & Wetherald (1975) based on the theory of general circulation of the atmosphere.

Data displayed in Figs. 5.9 and 5.10 refer to the case of variable albedo of the Earth's surface, that is to the conditions of climatic changes for long-term periods of time. Adequate agreement between the results of computating the sensitivity of climate, obtained by different methods, corroborates the possibility of using the contemporary climate theories for the study of its changes.

Energy factors of climatic changes
The causes of climate changes
The question of the causes of climatic changes, until recently has been regarded as unsolvable. The hypotheses about physical mechanisms of climatic changes suggested in various investigations were frequently not confirmed by quantitative calculations and the grounds that were presented gave rise to different objections. Meanwhile, the question of the causes of climatic changes has come into practical importance.

In several investigations it has been established that man's activities have begun producing an effect on the global climatic conditions, with this effect being rapidly increased. Thus the necessity arose to develop methods of forecasting the climatic changes and ways to influence the climate in order to prevent unfavourable deterioration of atmospheric conditions.

It seems that the number of factors substantially influencing the climatic change is comparatively small. Among these are the amount of solar radiation received by the Earth as a whole and by its various latitudinal zones, the shape of the Earth's surface (the position of continents and oceans, their location relative to the poles, the height of the continents and so on) and the composition of atmospheric air (mainly the carbon dioxide concentration and atmospheric aerosol mass).

When studying long-term climatic changes, it should be mentioned that due to the close relation of the processes of heat exchange in the atmosphere, hydrosphere and cryosphere, the atmosphere, oceans and ice cover should be considered as a unified physical system. On this assumption the state of oceanic water or position of continental glaciations could not be considered as external factors determining climatic change, though their effect on the state of the atmosphere could be very great. Taking this into consideration it is probable that the influence of all other external climate-forming factors upon climate is less significant, except for those factors which are comprehensively covered below.

Solar radiation
In several investigations it is anticipated that during the evolution of the Sun, as a star, its radius gradually decreases and luminance increases. Schwarzschild (1958) suggested that during the Earth's history, solar luminosity could have increased by 60 %. This corresponds to a solar constant increase of 1 % per approximately 80 Myr. This change of the solar constant can exert a substantial influence upon climatic changes over hundreds of millions of years (Ångström, 1965; White, 1967). In several studies the assumption was used that the solar constant essentially changes over time intervals of the order of tens of thousands of years, but this hypothesis remains unproven.

Of special interest is the question of the possibility of fluctuations of the solar constant over shorter periods of time, appropriate to the modern climatic changes. Many attempts to detect such short-term variations of the solar constant, from the material of astronomical and actinometric observations, proved unsuccessful, and for this reason there was a wide-spread opinion that if these variations in the solar constant do take place, they are within the margin of precision of the available measurements (Allen, 1958; Mitchell, 1965; Ångström, 1969). Recently a scientific conference was devoted to the question on variations in the solar constant (*Proceedings of the Workshop: the solar constant and Earth's atmosphere*, 1975), where the conclusion was made that at present no data confirming the existence of such variations are available.

There is no doubt that the amount of solar radiation received by different latitudinal zones in individual seasons periodically varies due to a change in the position of the Earth's surface relative to the Sun. The opinion has been repeatedly expressed that such changes can significantly affect climate and, in particular, contribute to the development of glacial epochs. This conception, advanced, in particular, by Milankovich (1920, 1930, 1941), is based on calculation of the secular trend of the three astronomical elements: eccentricity of the Earth's orbit, inclination of the Earth's axis to the plane of the ecliptic and the time of precession of the equinoxes as a result of the Earth's axis precession (the dates in the annual cycle when the Earth is closest to the Sun). All these elements vary with time due to the influence of the Moon and other planets upon the Earth's motion, each period of variations amounting to several tens of thousands of years.

Variations of these elements do not influence the annual amount of solar radiation reaching the Earth as a whole, but they change the total radiation received by various latitudes in individual seasons. Milankovich contended that the position of continental glaciation basically depends upon the radiation regime of the warm season when glaciations melt. He calculated the changes in astronomical elements over the Quaternary period and assumed that periods when the amount of heat received in summer at the latitudes where glaciations develop (60–70° N) was low corresponded to periods of increased ice cover, while the periods which had the hottest summers in this critical region were associated with decreased ice cover.

Recently several attempts were made to construct numerical models of a glaciation developing, utilization of which could allow us to assess the changes in the area of polar ice sheets with variations of the Earth's surface position relative to the Sun. In some cases such attempts resulted in the conclusion that the influence of astronomical factors on the development of glaciations is not significant. This seems to be explained by the fact that corresponding models take no account of the feedback between ice cover and the thermal regime of the atmosphere. Taking into consideration this feedback and using the semi-empirical model of the atmospheric thermal regime, it could be shown that variations of the Earth's surface position relative to the Sun exerted a pronounced influence upon the development of glaciations (Budyko & Vasishcheva, 1971).

Some results of the calculations made are presented in Table 5.4. As is seen from the data, during the epochs of most appreciable radiation increase for the warm periods of the year in the zones of 60–70° latitude, a considerable advance of the limits of ice cover took place. In the regions where the ice sheet penetrates, a marked temperature lowering is observed, though the mean planetary temperature varies very little.

Table 5.4. *Climate change in the glacial epochs*

Period of time, thousands of years to 1800 A.D.	$\Delta\phi°$N	$\Delta\phi°$S	ΔT (°C)
22.1 (Würm 3)	8	5	−5.2
71.9 (Würm 2)	10	3	−5.9
116.1 (Würm 1)	11	2	−6.5
187.5 (Riss 2)	11	0	−6.4
232.4 (Riss 1)	12	−4	−7.1

$\Delta\phi°$N is the decrease of the mean latitude of the polar ice boundary in the northern hemisphere as compared to its present position; $\Delta\phi°$S is the decrease of the mean latitude of the polar ice boundary in the southern hemisphere as compared to its present position; ΔT (°C) is the change in mean temperature in the warm half-year at 65° N.

Comparison of the results presented in Table 5.4 with paleogeographical data shows an adequate agreement between them. This enables us to believe that astronomical factors exerted a considerable effect on the Quaternary glaciations. A similar conclusion was obtained in the studies of Berger (1973), who used a comparable method of studying the problem in question. (See also Section 3 below.)

Since polar glaciations had not existed, until the Quaternary period, for the space of approximately 200 Myr, the question arose of why variations in the Earth's surface position relative to the Sun had not led to the development of glaciations. To answer this question one should consider the effect of other factors on climatic change.

The shape of the Earth's surface

During the Mesozoic and the first half of the Tertiary period, the average level of the continents was comparatively low, therefore a considerable part of the continental platform was covered with shallow seas. In the second half of the Tertiary the intense tectonic activity led to an elevation of the level of continents and to a gradual vanishing of many interior seas. In this epoch, in particular, the vast reservoir in present-day west Siberia, which joined the tropical oceans to the Arctic basin, ceased to exist. Therefore, the north polar ocean became a more isolated reservoir, connected mainly with the Atlantic Ocean. It is probable that in the Tertiary age the process of shifting the Antarctic to the higher latitudes, which occupied a concentric position relative to the South Pole, was completed.

Such changes in the Earth's surface exerted an appreciable influence upon the meridional heat exchange in the oceans, as is discussed elsewhere in this volume.

As calculations of the heat balance show, at the present epoch the meridional heat transfer in the oceans amounts to about one half of the heat transfer in the atmosphere. It is possible that with the enhancement of the area of reservoirs joining high and low latitudes, the heat transfer in the oceans increases. The influence of increasing the meridional heat exchange on the distribution of mean latitudinal temperatures can be evaluated from the formulas of semi-empirical theory of the atmospheric thermal regime. The results of these calculations show that variations in the meridional heat exchange have no influence upon the mean planetary temperature. Relative increase in the meridional

heat exchange in the oceans reduces the mean air temperature near the equator and somewhat enhances the air temperature in middle and, particularly, high latitudes. It can be hypothesized that the decrease in the meridional heat flux in the oceans, caused by elevating the level of the continents, resulted in a temperature reduction in middle and high latitudes during the past tens of millions of years.

As is seen from the calculations, such a reduction in temperature reached several degrees. This cooling had a pronounced influence upon many natural processes; however, it seemed to be insufficient for developing great polar glaciation as long as the atmosphere contained a considerable amount of carbon dioxide.

Carbon dioxide

At the present epoch the atmosphere contains nearly 2.3×10^{12} tonnes of carbon dioxide, amounting to 0.032 % of the atmospheric air (volume percent). Carbon dioxide is used in the process of photosynthesis in order to build organic matter and is returned to the atmosphere by living organisms, in particular, by their respiration. A certain amount of carbon dioxide enters the atmosphere from the lithosphere (particularly at volcanic eruptions) and is used for forming different sediments. It has long been thought that in the past the atmosphere contained a much greater amount of carbon dioxide as compared to the present time.

Fig. 5.11 presents the results of one of the first attempts to calculate variations of the atmospheric carbon dioxide mass during the Phanerozoic (for the last 570 Myr). As is seen from this figure, during the Paleozoic and Mesozoic the carbon dioxide concentration varied in the range 0.1–0.4 %; from the end of the Mesozoic era the carbon dioxide concentration began decreasing, this process being accelerated in the Oligocene and, particularly, at the end of the Pliocene when the carbon dioxide concentration reached several hundredths of a percent.

It is possible that the reduction in the carbon dioxide concentration at the end of the Phanerozoic is explained by a decrease in volcanic activity. The dashed line in Fig. 5.11 gives some idea of the variations in volcanic activity during

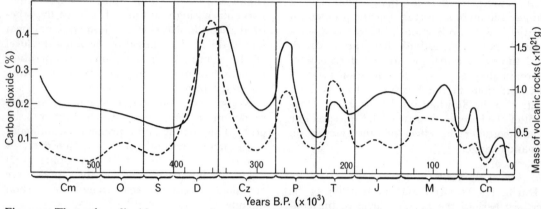

Fig. 5.11. The carbon dioxide concentration change during the Phanerozoic. '————', the concentration of carbon dioxide; '– – – –', the mass of volcanic rocks. The letters on the horizontal axis denote the geological periods of the Paleozoic and Mesozoic eras and the Cenozoic era.

the Phanerozoic and shows changes in the mass of volcanic rocks deposited over a million years. It is obvious that there is a close relationship between the mass of volcanic rocks and carbon dioxide concentration.

It has long been known that carbon dioxide exerts a specific influence on climate. This gas is more or less transparent for short-wave radiation and opaque for long-wave emission in several ranges of wavelengths, thus increasing the greenhouse effect that causes the rise in temperature of the lower air layers.

Fig. 5.12 shows the relationship between the mean for the global air temperature near the Earth's surface and the carbon dioxide concentration obtained from the semi-empirical model of the thermal regime of the atmosphere (curve 1) and from the model of the atmospheric general circulation of Manabe–Wetherald (curve 2). As is seen from this Figure, these curves are rather similar for a comparatively small increase in carbon dioxide concentration over its present level and they differ somewhat for high levels of concentration, explained by the difference of taking into account the polar ice effects on the thermal regime in the models indicated.

Calculations show that with carbon dioxide concentrations in the atmosphere above 0.1 % the air temperature near the Earth's surface does not depend on the level of concentration, being

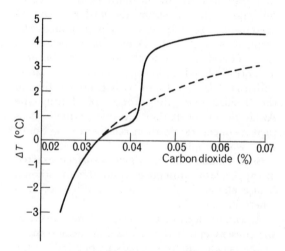

Fig. 5.12. The effect of carbon dioxide concentration on air temperature: (1) the calculation using the semi-empirical models (2) the calculation from Manabe–Wetherald model.

on average several degrees above the temperature associated with the present-day level of carbon dioxide.

It is likely that a decrease in the carbon dioxide concentration to several hundredths of a percent, as occurred in the Pliocene, considerably affected the air temperature by lowering it in high latitudes, and caused the expansion of the permanent ice cover which, with definite changes in the position of the Earth's surface relative to the Sun, extended to the middle latitudes.

The atmospheric aerosol

The study of actinometric observations shows that the amount of solar radiation reaching the Earth's surface in conditions of cloudless sky appreciably changes from year to year due to the non-stability of atmospheric transparency.

The main cause of variations in the atmospheric transparency is a change in the mass of solid and liquid particles available in the atmosphere. The so-called sub-micron large aerosol particles exert the greatest influence upon the radiation regime of the atmosphere.

Much of the aerosol mass (several tens of Megatonnes) is concentrated within the troposphere. Although the amount of stratospheric aerosol is comparatively small (it is usually below several Megatonnes), it has a noticeable influence on climate. The bulk of the aerosol mass is in the Junge layer located in the lower stratosphere, with the average altitude amounting to 18–20 km.

The stratospheric aerosol is mainly produced from sulfur dioxide that is subsequently oxidized to sulfur trioxide which almost immediately hydrolyzes to form sulfuric acid.

Due to the development of weak vertical motions in the stratosphere the sulfuric acid droplets have a prolonged residence time from several months to several years. Part of the sulfuric acid interacts with nitrogen combinations to form salt particles, but the great bulk, absorbing some water, remains as liquid particles consisting of a concentrated solution of sulfuric acid.

For global climatic changes, the comparatively uniform distribution of the stratospheric aerosol within the limits of each hemisphere that is attained due to the intensive horizontal motion of the stratospheric air is of great importance.

Volcanic eruptions have a pronounced influence upon the balance of stratospheric sulfur dioxide; with increasing volcanic activity the transparency of the stratosphere decreases. The stratospheric aerosol attenuates the solar radiation flux due to its backscattering and absorption by aerosol particles. The available estimates show that the variations of stratospheric transparency change the radiation flux coming into the troposphere by the order of several tenths of a percent. This results in anomalies of the hemispheric air temperature near the Earth's surface (relative to the temperature with the average transparency), with values reaching a few tenths of a degree.

There is some difference between the effect of the tropospheric aerosol and that of the stratospheric aerosol upon climate. The residence time of tropospheric aerosol particles is short, it amounts to 10 days on average because of the development of strong vertical motions in the troposphere and the atmospheric precipitation effect which causes the particles to settle from the air, which is one of the reasons for this difference. In this connection, variation in the total mass of tropospheric aerosol is determined by fluctuations in the rate of aerosol particle formation.

The changes in air temperature near the Earth's surface as a result of variations in tropospheric aerosol mass seem to be smaller than for the stratospheric aerosol, because the radiation absorption by the tropospheric aerosol particles results in heating the troposphere as a whole, which to some extent compensates for cooling due to radiation backscattering by aerosol particles.

Calculations of the influence of the tropospheric aerosol on air temperature near the Earth's surface, made by different authors, do not always give comparable results. Though some investigators believe that an increase in aerosol mass results in an air temperature decrease near the Earth's surface, it is suggested in some papers that in certain conditions such an increase can lead to warming. Due to the short residence time of tropospheric aerosol particle and because of the non-uniform distribution of the sources of their generation, the concentration of tropospheric aerosol particles varies considerably in space.

To understand the physical mechanism of climatic changes data on the secular trend of aerosol mass in the atmosphere, which can be obtained from actinometric observations, are of key significance.

Using observations of direct radiation under a cloudless sky at a group of actinometric stations

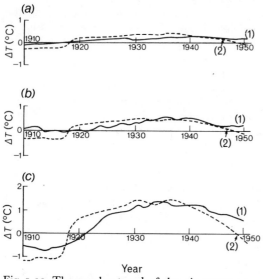

Fig. 5.13. The secular trend of the air temperature anomalies in (*a*) the northern hemisphere, (*b*) the 70–80° N belt, warm half-year, (*c*) the 70–80° N belt, cold half-year, (1) observational data, (2) calculated results.

and the formulas of atmospheric optics, variations in the solar radiation level for the two 30-yr periods 1888–1917 and 1918–47 were computed. It turned out that in the first period the total radiation was decreased by 0.3 %. Taking into account that during the first period, mean air temperature in the northern hemisphere was 0.33 °C higher than that during the second one, the value of ΔT_1 is found to be 1.1 °C (Budyko, 1969).

The air temperature variations near the Earth's surface were calculated taking into account the secular trend of direct radiation for 1910–50 and by using the formulas of semi-empirical theory of the atmospheric thermal regime. These results are presented in Fig. 5.13 as curve 2 (in each graph) which is shown to be closely related to temperature variation smoothed over the 10-yr periods, shown as curve 1.

Data presented in Fig. 5.13 corroborate the view that the warming of the first half of the twentieth century, with the peak in 1930s, can be mainly explained by a decrease in the atmospheric aerosol mass due to the volcanic activity reduction.

Anthropogenic climate changes

In recent decades and, particularly, in the last few years the global climate has been noticeably affected by Man's activities, because of a growth of carbon dioxide concentration and an increase in aerosol mass in the atmosphere.

Systematic observations of carbon dioxide concentration started in 1958 during International Geophysical Year. From the calculations Machta (1971) carried out, it follows that the present increase in carbon dioxide concentration is explained mainly by the entry into the atmosphere of constantly increasing amounts of the products of the burning of coal, petroleum and other kinds of fuel. In these conditions about half the carbon dioxide produced as a result of fuel burning remains in the atmosphere and the other half is mainly dissolved in oceanic water.

Attention must be given to the fact that Man's activities may rapidly return the atmosphere to its previous state, when the carbon dioxide mass in the atmosphere was much higher than the present-day one. Considering, in accordance with the data in Fig. 5.11, that over the last 100 Myr the carbon dioxide concentration has decreased approximately by 0.2 % of volume of the atmosphere, it might be concluded that the modern rate of increase of the carbon dioxide mass is 30 000 times higher than its rate of decrease during Cenozoic time.

Taking into account that according to calculations by Machta the carbon dioxide mass in the atmosphere has increased by 10 % due to human activities, it can be found that this would cause the mean air temperature to rise by 0.3–0.4 °C (Budyko & Vinnikov, 1976). The question arises why with the presence of a warming tendency, caused by an increase in the carbon dioxide mass, a lowering of mean air temperature took place after the 1930s. Data from actinometric observations show that this cooling could be attributed to an increase in the atmospheric aerosol mass, probably partly caused by the effect of human activities.

Recently, schematic maps characterizing the distribution of anthropogenic aerosol (Kellogg, Coakley & Gram, 1975 and Chapter 13 of this book) have been constructed. They show that the

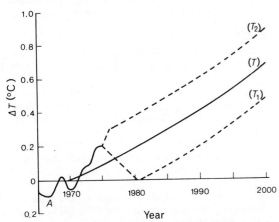

Fig. 5.14. The mean air temperature variations. (*A*) air temperature variations from observational data, (*T*) probable temperature variations due to the carbon dioxide concentration growth, (T_1) and (T_2) the limits of the probable temperature variation due to the fluctuations of aerosol mass in the atmosphere.

main mass is concentrated in the middle and, partly, tropical latitudes of the northern hemisphere. From actinometric observations it is clear that in the 1950s–60s atmospheric transparency over the continents in the northern hemisphere sharply decreased.

There is reason to believe that this reduction was to some degree explained by volcanic activities and partly caused by anthropogenic pollution of the atmosphere. The calculations show that such a lowering of the atmospheric transparency could reduce the air temperature by several tenths of a degree, more than compensating for the tendency for temperature increase due to the increase in carbon dioxide (Budyko, 1973). In the middle of the 1960s the atmospheric transparency stopped falling. Some years ago when this effect was detected the view was suggested that the cooling of the climate must soon be changed into a warming caused by the increase in carbon dioxide mass (Budyko, 1972b). In the same paper it was proposed that the warming will be accompanied by a lowering of the precipitation amount on the continents in the middle latitudes because of the decrease in the meridional temperature gradient which determines the intensity of atmospheric circula-tion which transfers water vapor from oceans to continents.

Recently it has been suggested that since the end of the 1960s the warming has started and accelerated in the first half of the 1970s (Budyko & Vinnikov, 1976). A number of disastrous droughts in Europe, Asia and North America during 1972–76 concur with this view.

Fig. 5.14 shows the variations in mean air temperature for the northern hemisphere from 1965 to 1975 (curve *A*). Also, this figure shows the probable temperature change under the effect of increasing the carbon dioxide concentration up to the end of the twentieth century (curve *T*). Curves T_1 and T_2 characterize the limits of a probable temperature change resulting from a fluctuation in aerosol mass in the atmosphere.

Data depicted in Fig. 5.14 attest that the climatic conditions of the end of the twentieth century may differ considerably from the climate of recent centuries.

References

Albrecht, F (1940). Untersuchungen über den Wärmehaushalt der Erdoberfläche in verschiedenen Klimagebieten. *Wiss. Abh. Reichsamt Wett Dienst,* **8**(2).

Allen, C. W. (1958). Solar radiation. *Q. Jl R. Met. Soc.,* **84,**(362), 307–18.

Ångström, A. (1920). Application of heat radiation measurements to the problems of evaporation from lakes and the heat convection at their surfaces. *Geogr. Annlr,* **2,** 237–52.

Ångström, A. (1965). The solar constant and the temperature of the Earth. In *Progress in oceanography,* vol. 3, pp. 1–5. Pergamon Press, Oxford.

Ångström, A. (1969). Apparent solar constant variations and their relation to the variability of atmospheric transmission. *Tellus,* **21**(2), 205–18.

Атлас мирового водного баланса (1974). Приложение к монографии 'Мировой водный баланс и водные ресурсы Земли'. М.–Л. Гидрометеоиздат.

Атлас теплового баланса (1955). Под ред. М. И. Будыко. Л. Изд. ГГО, 41 с.

Атлас теплового баланса земного шара (1963). Под ред. М. И. Будыко. М., Междуведомственный геофиз. комитет. 69 с.

Berger, A. L. (1973). *Théorie astronomique des paleoklimats,* vols. 1 & 2. Université catholique de Louvain. Louvain, 197 and 108 pp.

Brooks, C. E. P. (1950). *Climate through the ages,* 2nd edn. Benn, London. (Русск. перев.: Брукс. Климаты прошлого. М. ИЛ. (1952), 357 с.)

Budyko, M. I. (1956). Тепловой баланс земной поверхности. Л. Гидрометеоиздат. 255 с.

Budyko, M. I. (1962). Полярные льды и климат. *Изв. АН СССР. Сер. географ.* **6**, 3–10.

Budyko, M. I. (1968). О происхождении ледниковых эпох. *Метеорология и гидрология*, **11**, 3–12.

Budyko, M. I. (1969). Изменения климата· Л. Гидрометеоиздат. 35 с.

Budyko, M. I. (1971). Климат и жизнь. Гидрометеоиздат. 470 с. (Eng. trans.: *Climate and life*, Academic Press, 1974.)

Budyko, M. I. (1972*a*). Comments on 'A global climatic model of the Earth–atmosphere–ocean system'. *J. appl. Met.*, **11**(7), 1150–1.

Budyko, M. I. (1972*b*). Влияние человека на климат. Л. Гидрометеоиздат. 47 с.

Budyko, M. I. (1973). Атмосферная углекислота и климат. Л. Гидрометеоиздат. 32с.

Budyko, M. I. (1974). Изменения климата. Л. Гидрометеоиздат. 280 с. (Eng. trans.: *Climatic change*, American Geophysical Union, 1977.)

Budyko, M. I. (1975). Зависимость средней температуры воздуха от изменений солнечной радиации. *Метеорология и гидрология*, **10**, 3–10.

Budyko, M. I. & Vasishcheva, M. A. (1971). Влияние астрономических факторов на четвертичные оледенения. *Метеорология и гидрология*, **6**, 37–47.

Budyko, M. I. & Vinnikov, K.-Ya. (1976). Глобальное потепление. *Метеорология и гидрология*, **7**, 16–26.

Cess, R. D. (1976). Climate change: an appraisal of atmospheric feedback; mechanisms employing zonal climatology. *J. Atmos. Sci.*, **33**(10), 1831–43.

Ellis, J. & Vonder Haar, T. H. (1976). *Zonal average Earth radiation budget measurements from satellites for climate studies. Atmospheric Science Paper* 240. Coloradio State University.

Faegre, A. (1972). An intransitive model of the Earth-atmosphere–ocean system. *J. appl. Met.*, **11**(1), 4–6.

Holloway, J. L. & Manabe, S (1971). A global general circulation model with hydrology and mountains. *Mon. Weath. Rev.*, **99**,(5), 335–70.

Kellogg, W. W., Coakley, J. A. & Gram, G. W. (1975). Effect of anthropogenic aerosols on the global climate. *Proceedings of the WMO/IAMAP symposium on long-term climatic fluctuations, WMO Tech. Note* No. 421, pp. 323–30. WMO, Geneva.

Lorenz, E. (1968). Climatic determinism. *Met. Monogr.*, **8**(30), 1–3.

Machta, L. (1971). The role of the oceans and the biosphere in the carbon dioxide cycle. Nobel symposium 20, 16–20 August, Gothenberg, Sweden.

Manabe, S. & Bryan, B. (1969). Climate calculation with a combined ocean–atmosphere model. *J. atmos. Sci.*, **26**, 786–9.

Manabe, S., Smagorinsky, J. & Strickler, R. F. (1965). Simulated climatology of general circulation model with a hydrologic cycle. *Mon. Weath. Rev.*, **93**(12). 769–98.

Manabe, S. & Wetherald, R. T. (1967). Thermal equilibrium of the atmosphere with a given distribution of relative humidity. *J. atmos. Sci.*, **24**, 241–59.

Manabe, S. & Wetherald, R. T. (1975). The effect of doubling the CO_2 concentration on the climate of a general circulation model. *J. atmos. Sci.*, **32**(1), 345.

Milankovich, M. (1920). *Théorie mathématique des phenomenes thermique produits par la radiation solaire.* University of Paris. 339 pp.

Milankovich, M. (1930). Mathamatische Klimalehre und astronomische Theory der Klimaschwankungen. Köppen–Geiger. Handb. Klimat. I, A, Berlin. 176 S. (Русск. перев.: Миланкович. 1939. Математическая климатология и астрономическая теория колебаний климата. М.–Л. 207 с.

Milankovich, M. (1941). Kanon der Erdstrahlung und seine Anwendung am Eiszeitproblem. *R. Serbian Acad.*, **133**, 633.

Mitchell, J. M. (1965). The solar inconstant. *Proceedings of the seminar on possible responses of weather phenomena to variable extraterrestrial influences, NCAR Tech. Note*, TN–8, pp. 155–74. Boulder, Colorado.

Raschke, E. (1973). *The radiation balance of the Earth-atmosphere system from Nimbus 3 radiation measurements. NASA Tech. Note*, D–7249. National Aeronautics and Space Administration, Washington D.C. 59 pp.

Raschke, E., Möller, F & Bandeen, W. (1968). The radiation balance of the Earth–atmosphere system over both polar regions obtained from radiation measurements of the Nimbus II meterorological satellite. *Sver. Met. Hydrol. Inst. Medellanden., B,* **28**, 104.

Schmidt, W. (1915). Strahlung und Verdunstung an freien Wässerflachen ein Beitrag zum Wärmehaushalt des Weltmeeres und zum Wasserhaushalt der Erde. *Annln Hydrol.*, **43**, H.3, H.4.

Schwarzschild, M. (1958). *Structure and evolution of the stars.* Princeton University Press.

Sellers, W. (1969). A global climatic model based on the energy balance of the Earth–atmosphere system. *J. appl. Met.*, **8**(3), 392–400.

Sellers, W. (1975). The effect of solar constant variation on climate modeling. In *Proceedings of the workshop: the solar constant and Earth's atmosphere*, Big Bear Solar Observatory, 19–21 May 1975, ed. H. Zirin & J. Walter, pp. 206–31.

Vonder Haar, T. & Suomi, V. E. (1969). Satellite observations of the Earth's radiation budget. *Science*, **163,** 667–9.

Vonder Haar, T. H. & Suomi, V. E. (1971). Measurements of the Earth's radiation budget from satellites during a five year period. Part 1. *J. atmos. Sci.*, **28,** 305–14.

Wetherald, R. T. & Manabe, S. (1975). The effect of changing the solar constant on the climate of a general circulation model. *J. atmos. Sci.*, **32,** 2044–59.

White, O. R. (1967). Sun. In *The encyclopaedia of atmospheric sciences and astronomy,* ed. R. W. Fairbridge, pp. 958–63. Reinhold, New York, Amsterdam & London.

Zirin, H. & Walter, J. (eds) (1975). *Proceedings of the workshop: the solar constant and Earth's atmosphere.* Big Bear Solar Observatory, 19–21 May 1975. 332 pp.

6

Recent changes in snow and ice

G. J. KUKLA

The South Pole receives more insolation on a summer day than any other place in the world. Yet it is still one of the coldest locations on Earth. This is due to the presence of snow and ice which act as powerful cooling agents, because they: (*a*) reflect away most of the incoming short-wave radiation: (*b*) reduce the energy and moisture exchange between the underlying surface and the atmosphere; and (*c*) consume substantial amounts (80 g cal) of latent heat when melting.

As the atmosphere is principally heated at the underlying surface, the energy lost due to the presence of snow and ice represents a deficit in the Earth's heat budget. For example, bare grassland, tundra or pine forests only reflect about 15 % of incoming short-wave radiation and absorb the rest. When blanketed by fresh snow, their albedo increases to more than 80 %, rendering the incoming energy useless for heating the ground. This difference is dramatically large.

Cooling over a snow field continues at night when the cold surface releases only a small amount of water vapor into the air. In turn, the dry atmosphere allows the escape of the remaining surface heat reserves back to space. Because of the energy deficits, dry, cold air masses originate over the snow fields and progressively build up into high pressure cells which, at irregular intervals, release surges of frigid air into the lower latitudes. Related anticyclonic circulation tends to perpetuate and expand existing snow and ice fields.

The cooling impact of melting snow is particularly noticeable in spring. In the absence of wind, the air surface temperature may vary by tens of degrees over short distances between beds of melting snow and bare ground. Temperatures in the high latitudes of North America or the Eurasian continent increase from winter to summer levels within a few days once the snow melts away in May.

Snow has a particularly large impact on the climates of middle and high latitudes of the northern hemisphere. Here it greatly influences the character of the seasons, the shape of the annual temperature wave and the day-to-day weather variability in autumn, winter and spring. On the other hand, continents of the southern hemisphere, except Antarctica, are only indirectly affected by the existence of permanent snow fields over much of Antarctica and by seasonal development of pack-ice in surrounding oceans.

Snow-covered ice fields in both hemispheres have the same effect on local climate as snow cover on land. Even young ice which is still thin and dark sharply reduces the heat exchange between the ocean and the atmosphere. Furthermore, except for narrow leads, the moisture supply to the atmosphere is practically cut off. When the ice reaches a thickness of about 50 cm the heat flow from the ocean drops to negligible values.[1]

In summary, the radiative properties of land covered by snow or of the ocean covered by pack-ice differ strikingly from the bare land or open ocean. Development of seasonal snow or ice fields constitutes by far the most important variable in the energy interactions of the Earth and Sun. To understand properly the mechanism of weather and climate variability we need to understand the variability of snow and pack-ice fields.

Satellite mapping of snow and pack-ice

The practical impact of snow and ice on navigation, ground-water reserves and on the weather is such that three agencies of the US Government routinely chart snow and ice fields in both hemispheres at weekly intervals. This task was made possible by large-scale deployment of satellites. Visible light, infra-red and microwave (passive radar) sensors on board NOAA, ministration (NOAA), US Department of Commerce. They are available continuously from 1967 onwards. The maps, in a 1:50 000 000 polar stereographic projection, are based on visual photo interpretation of six to seven consecutive days of visible and infra-red satellite imagery currently provided by the scanning radiometer (SR) and very high resolution radiometer (VHRR) on board a polar orbiting

Table 6.1. *Satellite hardware used in snow and ice monitoring; after refs.* 2, 35, 36

Satellite	Sensor	Spectral band (μm)	Resolution (km)	Period of operation
Polar orbiting				
ESSA 3	AVCS	0.5–0.75	3.7	2 Oct. 1966–9 Oct. 1968
ESSA 4	APT	0.5–0.75	3.7	26 Jan. 1967–6 Dec. 1967
ESSA 7	AVCS	0.5–0.75	3.7	16 Aug. 1968–19 July 1969
ESSA 8	APT	0.5–0.75	3.7	15 Dec. 1968–present
ESSA 9	AVCS	0.5–0.75	3.7	26 Feb. 1969–15 Dec. 1973
NOAA 1	AVCS (+APT)	0.5–0.75	3.7	23 Jan. 1970–17 June 1971
	SR	0.52–0.73	3.7	
		10.5–12.5	7.4	
NOAA 2	VHRR	0.6–0.75	1–1.9	15 Oct. 1972–present
NOAA 3		10.5–12.5	1–1.9	6 Oct. 1973–present
NOAA 4	SR	0.52–0.73	3.7	15 Nov. 1974–present
		10.5–12.5	7.4	
Geostationary				
SMS-1 (GOES)	VISSR	0.55–0.70	1–7.4	17 May 1974–present
		10.5–12.5	7.4–14.8	
DMSP	VHR	0.4–1.1		
Nimbus 5	ESMR	1.55 cm	25.0	19 Dec. 1972–75
Nimbus 6	ESMR	0.81 cm	25.0	July 1975–present

Cameras and Sensors
AVCS – advanced vidicon camera system.
APT – Automatic picture transmission.
SR – scanning radiometer.
VHRR – very high resolution radiometer.
VISSR – visible and infra-red spin scan radiometer.
VHR – very high resolution sensor.
ESMR – electrically scanning microwave radiometer.

NASA and DMSP satellites provide raw data which are analyzed by automated procedures and finally visually evaluated by experienced interpreters. Table 6.1 lists the sensors used in the program and indicates their resolution.[2,3] The principal difficulty of surface mapping through satellites is the interference of clouds.

The primary source of information on snow cover in North America, Europe and Asia are weekly charts of the *Northern Hemisphere Average Snow and Ice Boundaries*, prepared by the National Environmental Satellite Service (NESS) of the National Oceanographic and Atmospheric Ad- satellite (at present, NOAA 4). Visible and infra-red spin scan radiometer (VISSR) images from a geostationary satellite (SMS-1) are also utilized.[2] The snow cover is categorized visually into three classes of relative reflectivity. Snow and sea-ice are distinguished from clouds by skilled analysts. They take into account, among other things, characteristic textures of the surface, such as dendritic patterns of forested valleys, outlines of major rivers and lakes, parallel leads in the ice and also persistence of observed features from one day to another. Running 10-day composite minimum brightness (CMB) charts are used as a complementary tool[4] and surface reports are occasionally consulted for areas with prolonged cloud cover. The charts show the snow line position as locally observed on the last cloud-free day which usually occurs in the second half of the week. While the subpoint resolution of satellite sensors currently ranges between 1–7.5 km, the scale of the charts limits the precision to about 20–50 km. Wiesnet & Matson[5] estimated the randomly distributed positioning error of snow boundaries to be 3 % in the concurrent charts and 5–7 % in the 1966–70 period. However, additional error is introduced in areas with long-lasting clouds where a local boundary may be plotted for several days earlier than the other charted features. For this reason, in autumn when the snow cover progressively increases, the charts tend to underestimate the snow extent, while the spring charts overestimate it.

The US Air Force Global Weather Center produces two parallel series of snow cover charts for the northern hemisphere. The first shows the snow depth in six categories, the highest corresponding to snow over 25 cm deep. The other shows the age of the cover in six classes ranging from fresh (24 h) to permanent (at least six months old). The mapping was started in 1976. Both types are computer printed in a scale of 1:30 000 000 at weekly intervals. The charts do not show the relative brightness or reflectivity of snow and ice fields. Satellite observations are used as the basic data input and are completed and checked by ground station reports. Information on the snow thickness in areas where direct measurements are not available is generated by computer program from given temperature and precipitation. Long-term monthly averages are used if real time reports are missing.[6]

Sea-ice in the Arctic and Antarctic waters is charted routinely in weekly intervals by the Fleet Weather Facility (FLEWEAFAC) of the US Navy. Charts are in polar stereographic projection. The Arctic basin is mapped in two segments, each in a scale of 1:18 000 000. Maps distinguish several groups of sea-ice concentration in octas. They are based on visual interpretation of satellite images obtained with NOAA, NASA and DMSP polar orbiting satellites. Visible and infra-red SR, VHRR and very high resolution sensor (VHR) data are preferred in

summer analyses, whereas the SR images in 1.55 cm wavelengths from Nimbus satellites and the infra-red images obtained by VHRR are used in winter.[7,8] These can be computer processed for enhanced contrast which greatly assists in distinguishing the characteristic pack-ice textures from clouds.

Although decisions on the sea-ice concentration are subjective and sometimes locally inconsistent, the charts, in general, are very accurate and are extensively cross-checked by ship and aircraft reports. They are available from 1971 onward and with complete winter coverage from 1973 onward.

Seasonal Change in the Extent of Snow and Ice Fields
Fig. 6.1, based on NOAA and navy charts, shows the extent of snow and ice at four characteristic intervals of 1974, namely January, April, August and October. Snow and ice conditions during that year were close to the average for the nine years of available data.

In January, the snow cover in the northern hemisphere, after repeated outbreaks of polar air into the middle latitudes, approached its maximum seasonal extent. At the same time, pack-ice in the southern hemisphere shrank close to its seasonal minimum. The global seasonal maximum of 77.7 km² (cf. Table 6.2) was reached during this month.

In April, snow cover in the northern hemisphere, although it had already begun to retreat, was still extensive due to frequent snow storms which deposited fresh snow along its southern border. Pack-ice in the northern hemisphere reached its maximum extent, while in the southern hemisphere the pack-ice started its northward advance in the zone of almost complete darkness.

Where not hindered by winds, large sections of open ocean quickly froze over and wide strips of fast ice formed along the Antarctic coast. Large oceanic segments around 65° S and 20° W and in the Ross Sea, which are affected by upwelling, developed only discontinuous or thin ice cover, a condition which persisted for most of the winter. The total global extent of snow and ice in April 1974 reached 64.9×10^6 km² (Table 6.2). At the end of April and beginning of May he pack-ice and snow in the northern hemisphere reflect the largest volume of insolation for the year.

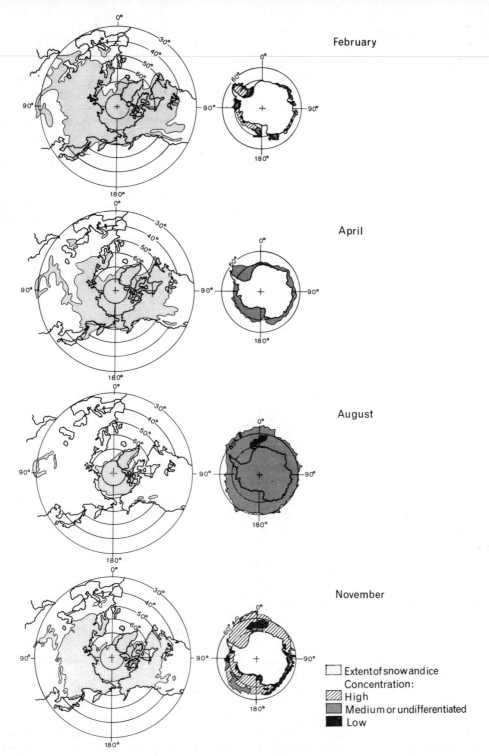

February

April

August

November

Extent of snow and ice
Concentration:
High
Medium or undifferentiated
Low

Fig. 6.1. Extent of snow and ice at four characteristic intervals of a year. From the National Environmental Satellite Service of the National Oceanographic and Atmospheric Administration weekly snow and ice boundary charts and the Fleet Weather Facility weekly ice charts. Occurrences of snow in South America and Oceania not included.

117

By the beginning of September, snow and ice in the northern hemisphere had shrunk to its minimum of 11.5×10^6 km², while the Antarctic approached its maximum. The global total of about 46×10^6 km² is very close to the lowest for the year. Global surface albedo was also at a low point. This is partly because of the striking drop of reflectivity of the pack-ice fields in the central Arctic in summer. While the albedo of the elevated Antarctic plateau remains practically as high in summer as it is in winter, averaging around 80–85 %, the

original charts prepared from satellite images. The area covered by ice was measured by a $2° \times 2°$ grid overlay equatorward of 70° S and $2° \times 4°$ of longitude poleward of 70° S. Four classes of sea-ice concentration were separately measured. The ice boundary is recognized at 4/5 octas. An index showing the net sea-ice extent minus open water areas was also calculated.

Antarctic pack-ice extent during selected seasons of the 1966–72 period was studied by Streten[13] and Sissala, Sabatini & Ackerman[14] and for 1973 and 1974 by Ackley & Keliher.[15]

Table 6.2. *Average monthly snow and ice cover in 10^6 km²*

	Jan.	Feb.	Mar.	Apr.	May	June	July	Aug.	Sept.	Oct.	Nov.	Dec.	X
A													
N. hemisphere	59.0	58.4	53.8	43.6	36.9	22.8	14.1	11.7	11.6	27.7	40.0	51.5	35.9
S. hemisphere	18.7	17.5	18.8	21.3	25.3	28.7	31.9	33.4	34.7	33.9	30.2	24.6	26.6
S. America and Oceania estimated	—	—	—	—	—	—	0.1	0.2	0.1	—	—	—	—
Globe	77.7	75.9	72.6	64.9	62.4	51.5	46.1	45.3	46.4	61.6	70.2	76.1	62.6
B													
N. hemisphere	58.5	60.1	53.7	41.5	32.0	21.5	14.3	11.0	12.4	23.8	39.6	53.5	35.2
S. hemisphere	19.6	17.3	18.6	21.6	24.6	27.6	29.6	31.1	33.9	34.0	31.9	25.6	26.3
S. America and Oceania estimated	—	—	—	—	—	—	—	0.1	0.2	0.1	—	—	—
Globe	78.1	77.4	72.3	63.1	56.6	49.1	44.0	42.3	46.4	57.8	71.5	79.1	61.5

A = 1974, B = Mean for 1967–74 for the northern hemisphere and 1973–74 for the southern hemisphere.

albedo of the permanent pack-ice fields of central Arctic drops to 60 % or less in July and August. [33] During this time, ponds of melting water and fields of moist snow cover the basin. [9,10]

Table 6.2 summarizes the changes in mean monthly extent of snow and ice fields in the two hemispheres during 1974. The data for the northern hemisphere were obtained by planimetering reflectivity classes 1, 2, and 3 in the NOAA/NESS snow and ice boundary maps.[11,12] Scattered mountain snow was taken as one-third of the measured area. The ice boundary was adjusted to fit the 4/8–5/8 ice concentration shown in the navy ice charts. Charts were measured in one to two week intervals and considered valid for the midday of the corresponding week. Monthly averages were calculated assuming a linear change of the measured parameter.

Data for the southern hemisphere were obtained from the navy ice charts and from several

Our charts do not include snow for South America, Africa, Tasmania, New Zealand and so on. Total maximum extent for these areas is about 0.2×10^6 km² in August. Their relative extent and impact on global climate is insignificant (cf. Table 6.2).

Secular variation

Considerable year-to-year differences, much larger than the possible measurement errors, were noted in the nine-year-long data set. They are represented graphically in Figs. 6.2 and 6.3.

Snow and ice cover in the northern hemisphere in 1971–73 was significantly larger than during 1967–70 and 1974–75. The difference is especially pronounced in autumn (September through November) and in late winter and early spring (February through April). The zone of largest variability is central Asia. It is interesting to note that January and July means show relatively little variation. The same con-

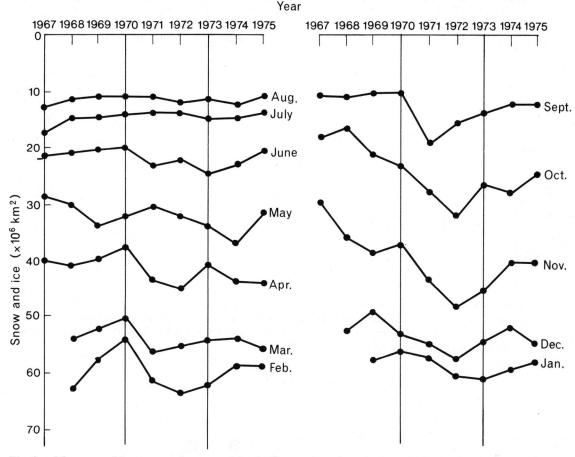

Fig. 6.2. Mean monthly extents of snow and ice in the northern hermisphere during 1967–75. Planimetered from the National Environmental Satellite Service of the National Oceanographic and Atmospheric Administration snow and ice boundary charts in two-week intervals.

clusion was reached by Wiesnet & Matson.[2] In other words, the difference between the 1967–70 and 1971–73 periods is principally due to a longer snow season and not due to more extensive winter coverage. In 1971–73 the snow cover formed considerably earlier and disappeared significantly later than in the preceding four years. A corresponding difference in autumn and spring weather patterns is also prevalent.

As a convenient climate-related index, the snow cover seasons (SC seasons) were defined by Kukla & Kukla[11] by arbitrarily selected limits of snow and ice extent. Northern SC summer is the season with less than 15×10^6 km² and SC winter with more than 55×10^6 km² of snow and

ice. SC spring and autumn are the intervening intervals.

The total pack-ice extent in the northern hemisphere did not change substantially; however, a large out-of-phase difference occurred in relative extent of ice in the North Atlantic sector and in the Bering Sea.[16] In 1972 and 1973 the ice conditions in the Bering Sea and Davis Strait were exceptionally severe, in agreement with the exceptional length of the snow season on land. But, in the Barents Sea, anomalously light sea-ice conditions were observed with particularly large areas of open water during winter and spring.[17]

Fig. 6.3 shows the 12-month running mean of snow and ice coverage in the northern

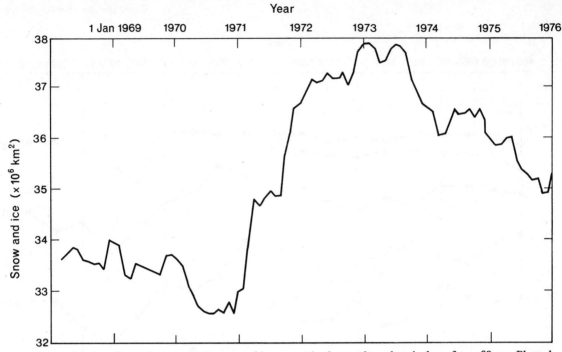

Fig. 6.3. Twelve-month running mean of snow and ice extent in the northern hemisphere for 1968–75. Plotted at the end of the month. Source: the National Environmental Satellite Service of the National Oceanographic and Atmospheric Administration charts.

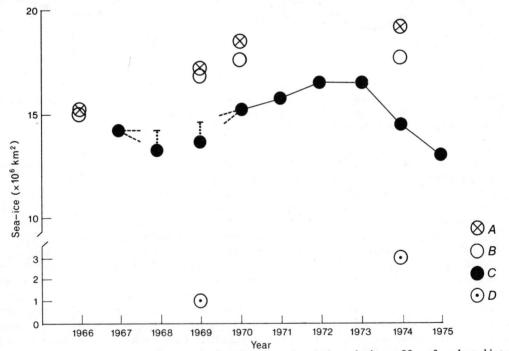

Fig. 6.4. Secular variation of pack-ice extent in the southern hemisphere during 1966–75 for selected intervals of a year. Minimum concentration: 4/8. A = August mean; B = July mean; C = end of November; D = beginning of March. Plots of A and B partly from Sissala et al.[14] and C partly from Streten.[13]

120

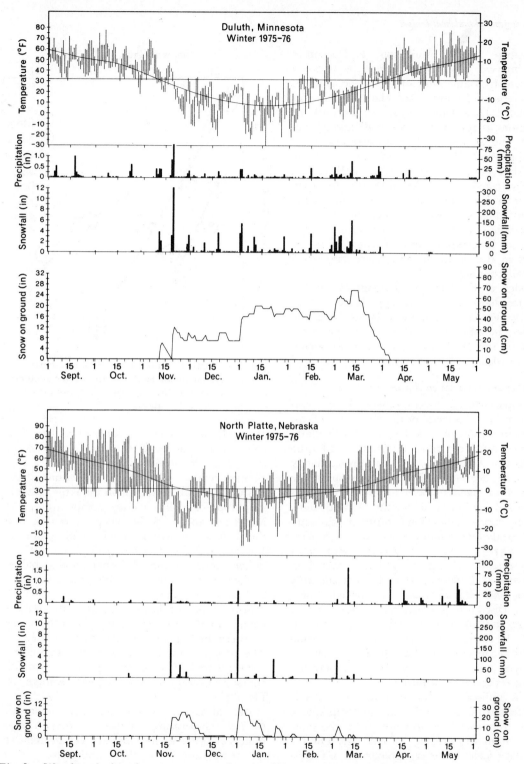

Fig. 6.5. Weather during the snow season of 1975–76 for two inland stations in North America. Duluth: 46° 50′ N 92° 11′ W; North Platte: 41° 98′ N 100° 41′ W. Daily maxima and minima of temperature, precipitation (including water equivalent of snow) reported snowfall, and snow on ground.

hemisphere for the April 1967 to December 1975 interval. The dramatic increase between 1970 and 1971 amounts to 12 % of the average.

Analysis of Antarctic ice charts is not yet completed. Data depicting the end of month ice conditions as reported by Streten[13] and Sissala *et al.*[14] and as measured by us were plotted in Fig. 6.4. While the plots do not represent the monthly averages, nor refer to the whole year, a tentative conclusion can be made that the 1972 and 1973 ice extent was larger than during the preceding and following intervals. If so, the fluctuations in the ice-covered area of the southern hemisphere and of snow and ice covers in the northern hemisphere proceeded generally in phase.

Ackley & Keliher[15], who planimetered the navy ice charts of 1973 and 1974, also noted that their values are considerably larger than the average monthly boundaries of Treshnikov[18] based on ship and aircraft observations made prior to 1960.

Impact of snow and ice on climate variability

The impact of snow cover on the weather at two locations in the interior of the North American continent is illustrated by Fig. 6.5. North Platte is located south of Duluth but is affected by the invasion of Arctic air masses to approximately the same degree as Duluth. Obviously the presence of snow in both locations correlates with relatively low air temperatures. With snow on the ground the daily temperature maxima are significantly lower than the maximum temperatures over bare ground. Maxima over snow rarely exceed 5 °C (42 °F). Duluth, located on the bank of Lake Superior, has considerably higher precipitation than land-locked North Platte. Discontinuous presence of snow cover and greater variation of daily temperature is typical for stations of the latter type. Comparing the two records it seems that with the presence of sufficient precipitation a low temperature regime tends to be stabilized.

It is difficult to demonstrate for a single station to what degree the low air temperature results from snow and to what extent the snow results from the presence of cold air. On the planetary scale, however, comparison of hemispheric and global averages minimizes this difficulty. Fig. 6.6 shows a close inverse correlation of snow and ice extent with geographically-

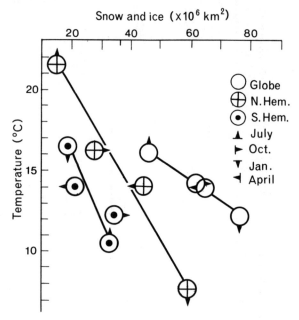

Fig. 6.6. Northern and southern hemispheric and global mean monthly air surface temperatures[34] plotted against corresponding extent of snow and pack-ice.

averaged air surface temperatures at four characteristic intervals of a year. The global total of incoming solar radiation within a year changes relatively little. The magnitude of the change is too small to explain the large seasonal amplitude of global average temperature, and is still less able to do so when we realize that the highest insolation reaches the globe in January and the lowest in July.[19] Thus it can be concluded that the hemispheric and global air surface temperatures are a function of the snow and pack-ice extent.

Several people have calculated the secular change in the air surface temperatures of the northern hemisphere for the past decade.[20–22] The 7-month running averages of Yamamoto[23] correlate well with the corresponding 7-month running averages of snow and ice coverage (cf. Fig. 6.7).

Annual averages of northern hemisphere air surface temperature calculated from Yamamoto's data[23] and the annual average relative air temperatures of the atmosphere below 500 mbar after Dronia[24], and Angell & Korshover[25] correlate with the corresponding figures of snow and ice coverage (Fig. 6.8). The

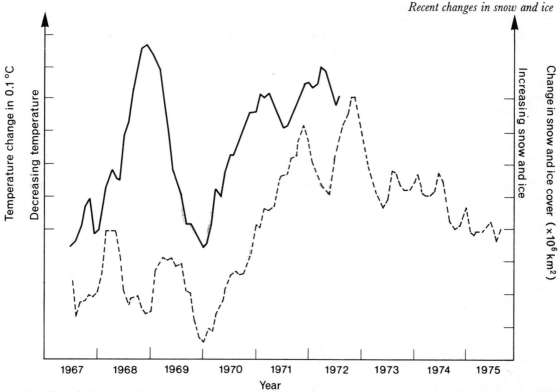

Fig. 6.7. Correlation of seven-month running mean of air surface temperatures (———) in the northern hemisphere after Yamamoto *et al.*[23] with corresponding running mean of snow and ice cover (– – – –). Plotted mid-month.

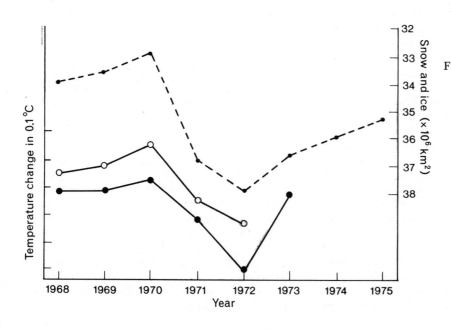

Fig. 6.8. Correlation of annual means of snow and ice cover (●– – –●) in the northern hemisphere with annual average air surface temperatures (o———o) after Yamamoto *et al.*[23] and of atmospheric column below 500-mbar level north of latitude approx. 35° N (●———●) after Dronia.[24]

correlation of temperature and snow and ice extent suggests that the latter index, obtainable from satellite imagery with little delay, could be used as an up-to-date indicator of hemispheric temperature trends.

Data sets for the southern hemisphere are still not sufficiently long and accurate for confident conclusions to be reached on secular changes of the average temperature. Stations reporting air surface temperatures used in the study of Damon & Kunen[26] represent, at most, only about one-third of the hemisphere, with most of the oceans completely unaccounted for. On the island stations, for the most part, a drop in temperature was recorded in the past pentade or decade, compared to the previous period. Damon & Kunen concentrated on the land record and interpreted the data as showing general warming of the hemisphere. Indeed, the Antarctic continent and much of South America and Australia warmed in the past decade, while most of the African stations showed a cooling. While the insufficient data base permits ambiguous and contradictory interpretations the evidence shows that large parts of the Southern Ocean were cooler in 1970–74 when the sea-ice around Antarctica was relatively extensive, and warmer before 1963 when the pack-ice, according to Treshnikov,[18] was reduced.

In this context it is important to realize that the 1971–73 cold and snowy pattern in the northern hemisphere was accompanied by a relatively warm central Arctic.[24] Warming of Antarctic stations is not incompatible with eventual cooling of the mid-latitudes of the southern hemisphere.

Snow and Pack-Ice versus Insolation

It is now widely agreed that the gross repetitive changes of climate known to have occurred during the Pleistocene were caused by changes in geographic and seasonal distribution of insolation over the Earth's surface.[27–29] The changes are due to the perturbations of the Earth's orbit around the Sun, referred to as the 'Milankovich mechanism'. Fig. 6.9 shows how the irradiation to the top of the atmosphere at selected latitudes in the past 130 000 yr was modulated by the perturbations.[30] It is known from independently dated paleoclimatic evidence that within the past 50 000 yr the warmest climate occurred over most of the globe around

6000 yr ago, and the peak cold glacial climate about 18 000 yr ago.[31] It is also known that the extent of permanent snow and ice, and temporary winter snow and pack-ice fields, was considerably larger at 18 000 yr ago than compared to today, while the opposite was true at 6000 yr ago.

If all the above observations and assumptions are valid, then some mechanism must exist linking the changing geographic and seasonal distribution of insolation during the Pleistocene with the changing extent of seasonal and permanent snow and ice covers. The chances are that this mechanism operates even today and could be detected in the response of seasonal snow and pack-ice fields to insolation. This is a reasonable expectation as the maximum difference in local irradiation in mid-latitudes due to the seasonal cycle is about 10 times larger and 10 000 times faster than the local change of insolation due to the Milankovich mechanism.

The northern hemisphere

The deposition and melting of snow closely depend on the available net radiation. An empirical formula obtained for Ellesmere Island[32] relates water melt M (in mm) to the net radiation of snow cover surface (in cal cm^{-2} h^{-1}) by:

$$M = R_n/7.74 - 0.56 \qquad (6.1)$$

The formula requires about 4.3 cal cm^{-2} of net radiation flux to initiate the melt of the snow pack. Using this formula the melt could theoretically start at the beginning of May under clear, calm conditions with a noon solar height of 17° and an air temperature close to 0 °C. Regional melt actually began in 1975 on 9 June and in 1974 on 21 June.[32] The delay is due to the influx of cold air from the north and to substantial high-altitude cloudiness even on relatively clear days.

Fig. 6.10 shows that in recent years snow in central Eurasia advanced when the surface irradiation on cloudless days dropped below 200 cal cm^{-2} day^{-1}. Snow retreated after the surface irradiation reached about 450 cal cm^{-2} day^{-1} in spring. The advancing snow line closely follows the changing insolation in autumn whereas it is considerably delayed behind the insolation shift in spring.

Lower insolation is needed in autumn to prevent deposition of snow because the dark

Mid-month insolation on top of the atmosphere (cal cm⁻² day⁻¹)

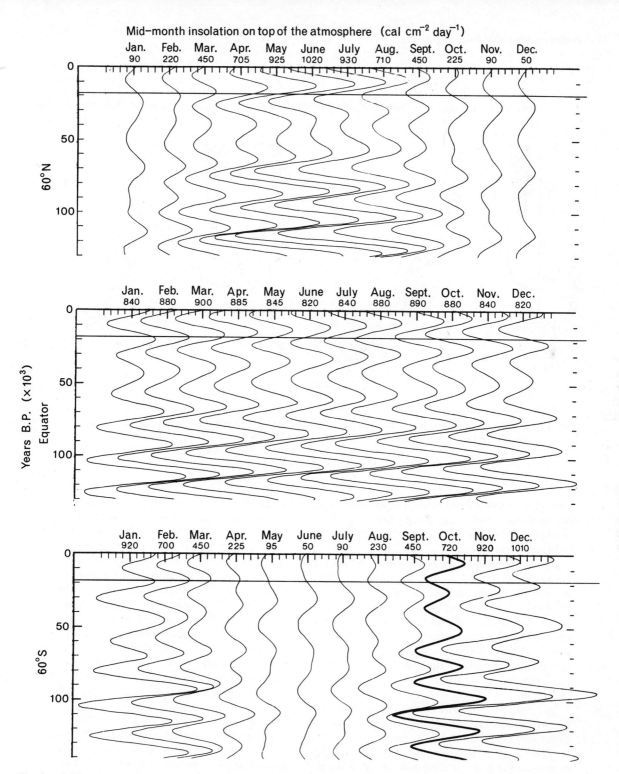

Fig. 6.9. Mid-month insolation on top of the atmosphere during the past 130 000 yr in cal cm⁻² day⁻¹. Scale in units of 10 cal cm⁻² day⁻¹. Values increase to the right and decrease to the left. Printed numeral calibrates the scale for each month. After Berger[30].

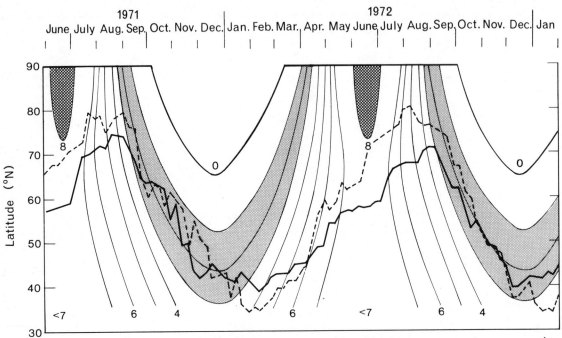

Fig. 6.10. Seasonally changing mean latitudinal position of the snow line from 1 June 1971 to 31 January 1973 in America and Eurasia, between 80° W and 100° W (solid black line) and between 50° E and 70° E (dashed line). Ground level, cloud-free insolation Q_s (10^2 ly day^{-1} indicated by vertical isolines). From September through November the snow line follows the 200 ly day^{-1} isoline.

ground with low albedo efficiently absorbs heat. An insolation level about three times as great is needed to start the melt of established, highly reflective snow cover.

Fig. 6.5 demonstrates that a low surface air temperature related to the low net radiation is one of the conditions needed for establishment of snow cover. The second condition is the availability of precipitation. The thickness of snow accumulated through winter, which closely depends on precipitation, will significantly influence the length of the melting episode. This affects the energy lost by reflection.

In summary, the duration of the snow season in the continental interior depends in a complicated manner, on insolation, and is especially sensitive to (1) irradiation and to the availability of precipitation in autumn and (2) irradiation, reflectivity, and the total thickness of accumulated snow in spring.

A past insolation regime resulting in increased

irradiation and decreased precipitation in autumn and/or spring would tend to shorten the snow season and warm the northern hemisphere. Decreased radiation and/or increased precipitation in autumn and spring would have the opposite effect. A sophisticated model able to handle in realistic detail and on a continental scale both the surface radiation balance and the precipitation is needed to solve the problem. The model should be designed to define the optimum insolation regime needed to produce minimum snow and ice coverage thus warming the northern hemisphere, and maximum cover leading to cooling. Such a model is not to hand yet.

If present knowledge on the sensitivity of the northern hemisphere temperatures to insolation changes is lamentably poor, the sensitivity of precipitation to radiation is still less understood. Land–sea interactions greatly complicate the situation. Lack of confident surface radiation measurements is another serious obstacle.

A prospective approach overcoming the latter difficulty would be to estimate the surface irradiation from data on cloudiness in two data sets on a hemispheric scale, one representing snow-rich years and the other snow-deficient years.

The southern hemisphere

Fig. 6.11 compares the average position of the ice edge around Antarctica and dissipation of the pack-ice with the seasonal march of insolation during 1973 and 1974. It also shows the opening and refreezing of coastal polynyas. The dynamics of the Antarctic pack are to a large degree associated with atmospheric turbulence. However, as the figure demonstrates, a close relation exists between the progressively increasing insolation and fast retreat of the ice edge in November and December. The advance of ice in autumn and winter proceeds at a fairly regular rate under limited insolation income and is principally controlled by the heat reserves of the ocean accumulated during the ice-free season.[1] Snow accumulates on top of the ice. After reaching its maximum extent in September and October the ice edge starts to retreat slowly and in the first half of November a spectacular break-up of the pack occurs. Large areas of broken, discontinuous sea-ice with a considerable proportion of open water form along the coast of the Antarctic continent, along the outer edge of the pack and around the polynyas in the Ross and Weddell Seas. Thus, the pack disintegrates from the inside as well as from the outside.

This is a significant difference from the pack in the central Arctic where only the top of the ice melts, without significant exposure of the ocean surface. In 1973–74 the break-up occurred under the noon solar height of about 30 ° corresponding to net radiation gain of about 12 cal cm^{-2} h^{-1} under a clear sky.

As the Antarctic ice retreats, bands of broken ice are blown out into the open ocean and the snow on top of the pack-ice melts. The average reflectivity of the pack drops by at least 20 %. While the development and dissipation of the pack is strongly influenced by atmospheric turbulence, the close dependence of its break-up on insolation is obvious. Indirectly, the early break-up also influences the onset of the ice season since the volume of radiation absorbed by

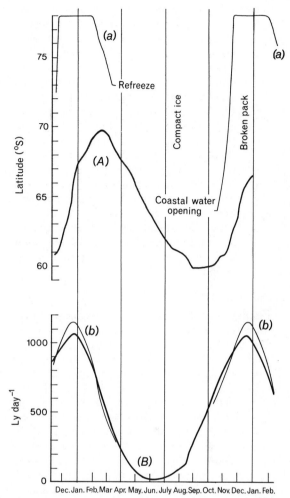

Fig. 6.11. Average latitudinal position of Antarctic pack-ice edge during 1974 (curve *A*) compared with the insolation to top of the atmosphere above the edge (*B*). Highest poleward occurrence of open water along the coast (*a*) and its corresponding insolation (*b*) is also plotted. From the Fleet Weather Facility ice limit charts. First day of a month plotted.

the ocean is larger. Thus, the length of the ice season around Antarctica seems to depend principally on insolation during November. Scarcity of meteorological stations around Antarctica prevents the testing of this assumption by measuring the climatic impact of different break-up regimes.

Summarizing the observations on the sensitivity of snow and ice fields in both hemispheres to the seasonal march of insolation we find that:

(a) In the southern hemisphere the extent of the pack-ice fields is especially sensitive to insolation in early November. Higher insolation to pack-ice fields at this time should result in shorter ice seasons and a smaller average extent of the pack.

(b) In the northern hemisphere the extent and duration of the snow fields is especially sensitive to insolation in autumn (September–November) and spring (April–May). However, it is also dependent on precipitation. Sensitivity of precipitation to insolation in the northern hemisphere is not known.

(c) Minimum global extent of snow and ice and minimum radiation losses over snow and ice areas are reached in August and September.

Paleoclimatic Evidence

Testing the above conclusions against paleoclimatic evidence of the past 30 millenia, we find maximum warmth 6000 yr ago and maximum cold 18 000 yr ago. Fig. 6.9 with plots of past mid-month insolation shows that:

(a) The average global insolation to the top of the atmosphere was highest 6000 yr ago and lowest 18 000 yr ago in September. During this month the global totals of energy lost due to the presence of snow and ice fields are at a low point and absorption of radiation by the Earth's atmospheric system is most efficient (Table 6.2).

(b) In September and October when snow fields start advancing, insolation to 60° N was high 6000 yr ago and low 18 000 yr ago.

(c) In August, insolation to 60° N and to higher northern latitudes peaked 6000 yr ago and was lowest 18 000 yr ago. During August the total heat absorbed by the ocean and surface of the Arctic pack is at its peak.

(d) During April when the snow fields in the northern hemisphere begin to retreat and pack-ice reaches its maximum extent, insolation to 60° N was low 6000 yr ago and high 18 000 yr ago.

(e) In November when the break-up of the Antarctic pack occurs, the insolation to 60° S was high 6000 yr ago and low 18 000 yr ago.

Except for (d), past insolation levels peak in phase with the paleoclimatic record at critical intervals and in critical zones. The hypothesis suggesting the mechanism through which the Milankovich mechanism modulates climate can be formulated as follows. High insolation from August to November warms global climate by:

(a) advancing the break-up of the pack-ice around Antarctica and thus reducing its average annual size;

(b) delaying the establishment of snow cover in the high and mid-latitudes of the northern hemisphere.

Low insolation from August through November would have the opposite effect.

The proposed hypothesis explains the gross climate shifts of about a 20 000-yr period but still cannot explain the 100 000-yr-long glacial/interglacial cycle. The dampening effect of ocean and glaciers may become important at this time-scale.

References

1. Pease, C. H. (1975). *AIDJEX Bull.*, **29,** 151–72.
2. Wiesnet, D. R. & Matson, M. (1975). *NOAA Tech. Mem.*, NESS, **74,** 1–21.
3. McClain, E. P. (1975). In *Climate of the Arctic*, pp. 415–29. Geophysical Institute, University of Alaska.
4. McClain, E. P. & Baker, D. R. (1969). *ESSA Tech. Mem.*, NESCTM 12, pp. 1–19. US Dept. of Commerce.
5. Wiesnet, D. R. & Matson, M. (1976). *Mon. Weath. Rev.*, **104,** 828–35.
6. Luces, S. A., Hall, S. J. & Martens, J. D. (1975). AFGWC. *Tech. Mem.*, 75–1, pp. 1–19. US Air Force Weather Service.
7. McClain, E. P. (1973). *Arctic*, **26,** 45.
8. Gloersen, P., Wilheit, T. T., Chang, T. C. Nordberg, W. & Campbell, W. J. (1974). *Bull. Am. met. Soc.*, **55,** 1442.
9. Marshunova, M. S. & Chernigovskii, N. T. (1971). In *Meteorological conditions in the Arctic during IQSY*, ed. I. M. Dolgin & L. A. Gavrilova. Jerusalem: Israel Program for Scientic Translations. (Translated from Russian.)
10. Badgley, F. I. (1961). In *Proceedings of the symposium on Arctic heat budget and atmospheric circulation,*

RM-5233-NSF, pp. 267–77. Band Corporation, Santa Monica, California.

11. Kukla, G. J. & Kukla, H. J. (1974). *Science*, **183**, 709–14.

12. Kukla, G. J. (1976). In *Proceedings on meteorological observations from space: their contribution to the first GARP) global experiment*, pp. 110–13. NCAR, Boulder, Colorado.

13. Streten, N. A. (1973). *Arch. Met. Geophys. Bioklim., A*, **22**, 119–34.

14. Sissala, J. F., Sabatini, R. R. & Ackerman, H. J. (1972). *Polar Rec.*, **16**, 367–73.

15. Ackley, S. F. & Keliher, T. E. (1976). *AIDJEX Bull.*, **33**, 53–76.

16. Sanderson, R. M. (1975). *Met. Mag.*, **104**, 313.

17. Haupt, I. & Kant, V. (1976). In *Proceedings on meteorological observations from space: their contribution to the first GARP global experiment*, pp. 179–86. NCAR, Boulder, Colorado.

18. Treshnikov, A. F. (1967). In *Proceedings of the symposium on Pacific–Antarctic sciences, eleventh Pacific science congress*. Pp. 113–23. Dept. of Polar Research, National Science Museum, Tokyo.

19. Kukla, G. J. (1975). *Nature*, **253**, 600–3.

20. Mitchell, J. M., Jr. (1963). In *Changes of climate*, Arid Zone Research Series, 20, pp. 161–81. UNESCO, Paris.

21. Willett, H. C. (1950). In *Centenary proceedings*, pp. 195–206. Royal Meteorological Society, London.

22. Budyko, M. I. (1969). *Tellus*, **21**, 611–19.

23. Yamamoto, R. T., Iwashima, T. & Hoshiai, M. (1972). *J. met. Soc. Japan*, **53**, 482–6.

24. Dronia, H. (1974). *Met. Rdsch.*, **27**, 166–74.

25. Angell, J. K. & Korshover, J. (1975). *Mon. Weath. Rev.*, **103**, 1007–12.

26. Damon, P. E. & Kunen, S. M. (1976). *Science*, **193**, 447–53.

27. Milankovich, M. (1941). *Kanon der Erdbestrahlung*. Königliche Serbische Akademie, Beograd.

28. Broecker, W. S. (1966). *Science*, **151**, 299–304.

29. Hays, J. D., Imbrie, J. & Shackleton, N. J. (1976). *Science*, **194**, 1121–32.

30. Berger, A. (1975). In: *Proc. WMO/IAMAP Symposium on Long-Term Climatic Fluctuations*, pp. 65–74. WMO, publ. no. 421, Geneva.

31. Flint, R. F. (1971). *Glacial and Quaternary geology*. Wiley, New York.

32. Woo, M.-K. (1976). *Catena*, **3**, 155–68.

33. Kondratyev, K. Ya. (1969). *Radiation in the atmosphere*. Academic Press, New York.

34. Schutz, C., & Gates, W. L. *Publs* R–915 (1971), R–1029 (1972), R–1317 (1973), R–1425 (1974). Rand Corporation, Santa Monica, California.

35. Fett, R. W. (1976). In *Proceedings on meteorological observations from space: their contribution to the first GARP global experiment*, pp. 171–7. NCAR, Boulder, Colorado.

36. Wilheit, T. T. (1975). In *The Nimbus 6 user's guide*, pp. 87–108. NASA/Goddard Space Flight Center, Greenbelt, Maryland.

SECTION 3

Astronomical influences   JOHN GRIBBIN

7

Long-term effects

A great many astronomical influences have been invoked at one time or another as possible explanations of different aspects of climatic change; they range in size from the implausible but not unproven possibility that a small black hole resides at the Sun's centre and affects its output, to the influence exerted on the Earth and Sun, according to some cosmological ideas, by the expanding Universe. The range of time-scales involved is equally great – from the age of the Universe itself down to the regular cycle of day and night – and clearly some overall scheme is needed in order to review in comprehensible fashion the less implausible of these many ideas. I have chosen to begin with the longest of these time-scales and work down towards shorter-term effects, although stopping short of any discussion of diurnal variations! The review is not intended to be comprehensive, in that the more implausible speculations have, by and large, been omitted from the discussion. It is, however, perhaps worth beginning this look at astronomical influences on the terrestrial climate by considering a theory which, although implausible in the eyes of most modern cosmologists, does emphasise clearly the very minor – indeed, insignificant – role which the Earth and our solar system have to play in the universal scheme of things. Our urgent and understandable human concern about the impact of small climatic changes on our society may thus be put in a rather different perspective, emphasising the need for constructive cooperation on the largest human scale in order to meet the challenge of even the smallest ripples of climatic change occurring on one small planet.

The idea that the 'constant' of gravity may weaken with time was put forward in a cosmological theory by Dirac, some 40 years ago,[1] and the geophysical implications of such a possibility, including the possible relevance to long-term climatic change, have been discussed in detail by Jordan,[2] although the complete mathematical formulation of the hypothesis with all its cosmological implications was presented only very recently.[3,4] Rather similar ideas have been investigated by Hoyle & Narlikar,[5] but in general the idea has not gained favour among cosmologists, not least since it implies continuous creation of matter in the Universe; in spite of Jordan's work, the implication that the Earth must have expanded over geological time is equally unwelcome in the context of modern geophysics. The climatic implications are also contentious; the higher constant of gravity in the past would require that the Sun was considerably hotter, since the luminosity of a star like the Sun is proportional to the *seventh* power of the constant of gravity, and the greater past heat implied by this dependence seems difficult to reconcile with the present understanding of the time-scale for the evolution of life on Earth. Speculation that a complete cloud cover 'in Palaeozoic or even Pre-Cambrian times' with 'a thickness of perhaps 10 km' (reference 2, p. 158) might have produced more equable conditions on the surface of the planet seems to be grasping at straws. A steady cooling of the kind required over thousands of millions of years is also difficult to explain in terms of past ice ages revealed by the geological record (see Chapter 1 of this volume) and it seems that for practical purposes we can set Dirac's hypothesis aside as an intriguing toy for the cosmologists. But although this particular proposed universal influence may not have affected the luminosity of the Sun, it is reasonable to ask just how constant the solar heat output is on this sort of time-scale.

Leaving aside fluctuations associated with the formation of the Sun and solar system, and not looking ahead beyond the remaining few thousand million years of the Sun's lifetime as a 'main sequence' star to the time when it will expand dramatically, engulfing the inner planets, the answer is that the solar furnace should, by and large, burn very steadily. Although the heat is produced by nuclear fusion reactions in the middle of the Sun, the large size of the outer layers provides a reservoir of gravitational potential energy which can be drawn on to smooth out any irregularities in the flow of energy. If the surface radiates more heat than is produced in the centre, the Sun (or any star) will contract, its internal temperature will rise and equilibrium will be restored; if more heat is being produced than is radiated, the star expands, the centre cools and again equilibrium is restored. By astronomical standards the response is quick – such adjustments take a few tens of millions of years at most, which should be compared with the many thousands of millions of years of the lifetime of a star like the Sun on the main sequence. By human standards, however, such minor adjustments could be of profound importance, and since there is some evidence that ice epochs, during which a succession of ice ages follow one another at fairly rapid intervals, last for something on the order of 10 Myr each it is hardly surprising that some astronomers have looked at the implications of minor disturbances in the solar interior for our understanding of ice epochs.

Arbitrarily varying the solar 'constant' to explain past phases of warming and cooling was in vogue before our understanding of stellar structure improved during the 1950s (see, for example, Wexler's contribution to reference 6), but this was never a satisfactory 'solution' to the problem, merely raising the further questions of why and how the Sun might vary in such a way, at more or less regular intervals if the evidence that ice epochs are separated by roughly 250 Myr[7] is to be taken at face value. Nevertheless, by making such *ad hoc* assumptions about solar fluctuations it is possible to produce outline models of the resulting climatic fluctuations on Earth, and these are of value when looking at more sophisticated theories of solar variability.

The most interesting idea to emerge from this work is that while a decrease in solar 'constant' (hereafter referred to as the solar parameter) could certainly initiate a cold epoch on Earth, a slight *increase* in the value of the solar parameter might be more effective for the spread of glaciation. The chief architect of this hypothesis was Simpson,[8] who argued that whereas a 10 % decrease in solar radiation would reduce the equator–pole temperature gradient and inhibit evaporation so that precipitation at high latitudes became insignificant, so there would be no ice epoch even though the planet cooled, a 10 % increase in solar radiation would increase evaporation to produce more cloud cover and precipitation, increase the equator–pole temperature gradient to produce vigorous circulation, and perhaps produce a decrease in temperature at high latitudes (because of the increased reflectivity of the increased cloud cover) with the extra precipitation falling as snow to build up glaciers even while lower-latitude temperatures increased. Today, the first objection to such an hypothesis is the growing weight of evidence that the most recent period of extended ice cover, at least, corresponded to cool, dry conditions (cf. references 9, 10). Perhaps, however, the hypothesis remains in contention (just!) in the modified form developed by Bell (see reference 6) as an extension of work by Willett.[11] In this version, the model involves an initial slow decrease in the value of the solar parameter to 'pre-cool' the Earth in readiness for an ice age. An enlarged polar cap in the northern hemisphere, with its strong cooling influence at high latitudes, and high albedo, could then contribute significantly to the development of an ice age as the Sun warmed up again sufficiently to produce the necessary precipitation. A further increase in solar warmth would melt the ice, producing a warm, wet interglacial, in turn succeeded by cooler, arid conditions as the heat from the Sun again declines. By invoking as many repetitions of this pattern as required, the irregular behaviour of the solar parameter can be adjusted to fit any recorded pattern of ice ages and interglacials within an ice epoch, which is seen as a period of solar flickering rather than corresponding to a single decrease (or increase) in heat output. This may be convenient for the climatologist, but poses some puzzles for the astronomer – puzzles, however, which can certainly be resolved according to at least one recent astronomical theory of ice epochs.

Öpik was the first astronomer to attempt an explanation of how the Sun might lapse into temporary phases of slightly non-equilibrium behaviour, and in its final form[12][13] his model contains several features which are particularly relevant to these recent ideas; some aspects of this work are discussed also by Bhatnagar in Shapley.[6] The basis for the model, which can usefully be carried forward into other models, is that from time to time conditions inside the Sun may become appropriate for convective mixing to occur over a large part of the interior, producing first an increase in heat and associated expansion of the outer layers as more hydrogen fuel is mixed into the nuclear burning region, then a period of cooling and contraction as conditions return towards the more normal equilibrium situation.

According to Öpik, the first phase would correspond to the onset of an ice age, since although the interior becomes hotter so much energy goes into the expansion of the outer layers that the net effect is a decrease in the luminosity of the Sun. It is when the core cools and the outer layers contract, giving up their surplus gravitational energy, that the luminosity would increase, ending the ice age rather abruptly on Öpik's model in its initial form. Bell's model, however, would suggest that the ice age sets in only when the luminosity increases after the initial decrease. The major difficulty with either version of the model is that there is no entirely satisfactory trigger mechanism to initiate the convective mixing disturbance. Öpik's own suggestion, that the residue from the nuclear burning builds up as a barrier around the core, eventually becoming convectively unstable, is at odds with more sophisticated models of stellar interiors.[14] But several astrophysicists felt compelled to take another look at the possibility of something triggering convective instability in the Sun when it became clear in the early 1970s that they were faced with very real difficulties in attempting to account for the lack of a detectable flux of solar neutrinos within the framework of standard stellar models.

This is not the place for a full discussion of the solar neutrino 'problem'; it is sufficient to say that the problem is simply that if the centre of the Sun is at the temperature required to maintain its present luminosity in a steady state then the nuclear reactions going on should produce a flux of neutrinos detectable at the Earth by suitable apparatus. In spite of the efforts of Davis and his colleagues, no flux of the appropriate intensity has been found, and the observations are consistent with the possibility that no neutrinos are being produced in the stellar interior at present (cf. reference 15). This could be explained if the temperature at the centre of the Sun is some 10 % less than the equilibrium value required by the present surface luminosity; on Öpik's model, that would imply that conditions inside the Sun are now returning to normal after a convective hiccup, and this ties in sufficiently well with the evidence that the Earth has recently undergone an ice age to have encouraged considerable speculation among the astronomical community in recent years.

Fowler[16] noted the coincidence but offered no detailed model to explain the presumed convective mixing; Dilke & Gough[17] and others[18,19] soon followed the speculation up with fuller models which showed conclusively that mixing of one kind or another could explain away the solar neutrino problem, but which did not explain satisfactorily why that mixing should recur at intervals of approximately 250 Myr, nor offered a satisfactory quantitative theory to trigger the required effect. At least one imaginative theory linked the idea with the evidence from Mariner 9 photographs of climatic change on Mars,[20] but this was certainly a premature speculation at that state of the art. Furthermore, the question remains of whether or not the associated solar fluctuations will produce ice age conditions during the warming phase or during the cooling phase, and whether the models can be fitted to the geological record. Very little work has been done on these problems recently, and the climatic aspects have not received a comparable facelift to that received by the astrophysical face of the problem since the 1950s. The need for something of the kind is particularly highlighted by the development at last of a model which provides, from the astronomical point of view, a very satisfactory trigger mechanism to initiate changes in solar luminosity at the required 250 Myr intervals, producing erratic fluctuations over a period of 10 Myr or so.[21,22]

As McCrea points out,[21] the basic idea that the position of the solar system in its orbit around our Milky Way galaxy might affect climate to produce 'cosmic seasons' was suggested long ago by Shapley,[23,24] who, unfortunately, did not then have the benefit of the more recent astronomical thought which helps to make McCrea's modern version of the idea so attractive. The first step in explaining these cosmic seasons comes from the observation that the interval between ice epochs (about 250 Myr) is, in round terms, half of the time it takes our solar system to orbit once around the centre of our galaxy. Since the structure of our galaxy is a flattened disc of stars with two prominent 'spiral arms' (rather like the galaxy shown in Fig. 7.1) this suggests that ice epochs may bear some relation to the crossing of a spiral arm by the solar system, a suspicion heightened by the fact that we are today experiencing an ice epoch at a time when we are on the edge of one of the two main arms in our galaxy, the Orion arm. In a typical spiral galaxy (Fig. 7.1) the bright arms delineated by a profusion of hot, bright stars are edged by dark lanes of cold dust and gas; the situation is the same in our galaxy, and McCrea relates the occurrence of ice epochs to passages through these dark lanes, rather than through the bright part of the spiral feature. It is thought that these dark lanes are the true, permanent features which give spiral galaxies their structure, formed as a standing wave pattern through which stars and gas move (or which moves past the stars and gas, depending on the point of view chosen); indeed, the bright stars of the spiral arms can then be interpreted as hot, young stars formed in the compression lanes of the wave,[22] as the solar system itself formed some 20 orbits of the galaxy ago.

During a passage through such a region of relatively dense interstellar material, the Sun will accrete matter and become more luminous as gravitational potential energy is released as heat. Within dust lanes, however, the distribution of interstellar clouds is itself patchy, so that passage through such a lane involves repeated encounters with dust clouds of various density. The speed of the Sun relative to each cloud encountered, as well as the density of each cloud, affects the Sun, but with reasonable assumptions about velocity and density distributions McCrea finds that each ice epoch (the passage through a dust lane) would last a few million years, while each passage through a cloud (which he equates with the most extreme period of an ice age) would last for about 50 000 yr. Because we are now on the edge of the bright Orion spiral arm, having just emerged from the associated dust lane, 'there should', in McCrea's words, 'have been an Ice Epoch during the past few million years extending up to near the present time'. The coincidences of time-scales are persuasive, but the weakest aspect of the model lies in the detailed theory of how a passage through a cloud and associated brightening of the Sun could produce ice age conditions on Earth.

As presented by McCrea[21] the model depends on an extension of Simpson's idea of a hot Sun/cool Earth situation developed by Hoyle & Lyttleton[25] in the 1930s, and it therefore is subject to the same criticism in the light of more recent developments as is that work. It should also be mentioned that Williams,[26] while also finding the coincidence of time-scales seductive, has suggested that the trigger for ice age conditions is related not to the spiral structure of the galaxy but to the flexure produced by the tidal influence of the Magellanic Clouds. But this idea has not been worked out in such detail as the dust lane hypothesis, and Williams presents no detailed explanation of how the effect might work, so that his model is little better than the *ad hoc* assumption of solar variations to match the geological record. On the other hand, there is some direct evidence from the variations in the texture of lunar soil that dust lane encounters have indeed occurred at about 250 Myr intervals,[27] and there remains the evidence from solar neutrino experiments that the Sun is not in a normal state at present. In addition, the detection of small, periodic oscillations of the solar surface[28-30] provides a new tool for investigating conditions inside the Sun (in an analogous way to the use of the techniques of seismology to provide an insight into conditions inside the Earth), and one interpretation of one of these vibration modes suggests that, in agreement with the indications of the solar neutrino detection experiments, the interior of the Sun is some 10 % cooler than it would be if it was in equilibrium with the amount of heat now being radiated from the surface.[28]

It may well be that McCrea is correct in suggesting that the effects of dust cloud encounters are responsible for ice age conditions on Earth, but that the detailed mechanism he suggests is wrong. Some guide to the effect of accreting dust on the interior of the Sun can be gained from a comparison with models of binary systems in which one star accretes material from its neigh-

Fig. 7.1. The spiral galaxy M 51 showing the double arm structure typical of many galaxies, including our own. Photograph from the Hale Observatories.

bour; it seems that there is no great influence on the interior in such a situation,[31,32] but it is very difficult to ascertain the effects of sudden *removal* of the infalling material, since the models do not converge to a unique solution when simulation of this behaviour is attempted (D. L. Moss, personal communication). This is not a situation of any great interest to astrophysicists studying binary systems, however, so that it has not been studied very intensively in the past, and there is ample scope for further investigation of the problem which may yield positive results – there is also, in the light of McCrea's work, now some interest both for astrophysicists and climatologists in such investigations. Meanwhile, it is tempting to speculate that the removal of the additional heat source near the surface provided by the infalling material ('taking off the lid') might provide the trigger for convective instability.[33] The critical factor might be not the submergence of the solar system into a dust cloud, but its subsequent emergence; either way, repeated passages through such clouds within a few million years – much less than the time needed for the Sun to return to equilibrium after a disturbance – could produce an erratic pattern of variation superficially in line with the variations of the recent ice epoch with its many individual ice ages. And both hot Sun/cool Earth and cool Sun/cool Earth models can be fitted within this framework of variations.

Although time-scales of millions of years are of little relevance to the problems of climatic change as they are likely to affect mankind in the immediate future, one clear and important implication for the short term does emerge from this haze of astronomical theory and speculation. The cosmic seasons hypothesis, the absence of a detectable flux of solar neutrinos, the 2-h 40-min vibration of the Sun and the occurrence of an ice epoch over the past few million years are all consistent with the idea that the Sun today is not in its normal state. If that is indeed the case, it could be of the greatest importance in any attempt to understand (or justify) some of the models of climatic variations on much shorter time-scales, which involve assumptions about solar activity and variability which seem abhorrent to any astrophysicist brought up to believe in the Sun as a typical, stable star at a typical stage of its evolution. One model of this kind will be discussed in detail in Chapter 8.

References

1. Dirac, P. A. M. (1937). *Nature*, **139**, 323.
2. Jordan, P. (1971). *The expanding earth*. Pergamon Press, London.
3. Dirac, P. A. M. (1973). *Proc. R. Soc.*, **A333**, 408.
4. Dirac, P. A. M. (1974). *Proc. R. Soc.*, **A338**, 439.
5. Hoyle, F. & Narlikar, J. V. (1971). *Nature*, **233**, 41.
6. Shapley, H. (editor) (1953). *Climatic change*. Harvard University Press, Massachusetts.
7. Allen, C. W. (1973). *Astrophysical quantities*, 3rd edn. Athlone Press, London.
8. Simpson, E. C. (1938). *Nature*, **141**, 591.
9. Hammond, A. L. (1976). *Science*, **191**, 455.
10. Williams, J. (1975). In *Proceedings of the WMO/IAMAP symposium on long-term climatic fluctuations*, *WMO Tech. Note* No. 421, p. 373. WMO, Geneva.
11. Willett, H. C. (1949). *J. Met.*, **6**, 34.
12. Öpik, E. J. (1950). *Mon. Not. R. astr. Soc.*, **110**, 49.
13. Öpik, E. J. (1958). *Scient. Am.*, Offprint No. 835. W. H. Freeman, San Francisco.
14. Chapman, S. & Cowling, T. G. (1970). *The mathematical theory of non-uniform gases*, 3rd edn. Cambridge University Press, London & Cambridge.
15. Davis, R. & Evans, J. M. (1973). *Proceedings of the 13th international cosmic ray conference*, vol. 3, p. 2001. University of Denver.
16. Fowler, W. A. (1972). *Nature*, **238**, 24.
17. Dilke, F. W. W. & Gough, D. O. (1972). *Nature*, **240**, 262.
18. Rood, R. T. (1972). *Nature Phys. Sci.*, **240**, 178.
19. Ezer, D. & Cameron, A. G. W. (1972). *Nature Phys. Sci.*, **240**, 180.
20. Sagan, C. & Young, A. T. (1973). *Nature*, **243**, 459.
21. McCrea, W. H. (1975). *Nature*, **255**, 607.
22. McCrea, W. H. (1975). *Observatory*, **95**, 239.
23. Shapley, H. (1921). *J. Geol.*, **29**, 502.
24. Shapley, H. (1949). *Sky Telesc.*, **9**, 36.
25. Hoyle, F. & Lyttleton, R. A. (1939). *Proc. Camb. phil. Soc.*, **35**, 405.
26. Williams, G. E. (1975). *Earth Planet. Sci. Lett.*, **26**, 361.
27. Lindsay, J. F. & Srnka, L. J. (1975). *Nature*, **257**, 776.
28. Severny, A. B., Kotov, V. A. & Tsap, T. T. (1976). *Nature*, **259**, 87.
29. Christensen-Dalsgaard, J. & Gough, D. O. (1976). *Nature*, **259**, 89.
30. Brookes, J. R., Isaak, G. R. & van der Raay, H. B. (1976). *Nature*, **259**, 92.
31. Moss, D. L. & Whelan, J. (1972). *Mon. Not. R. astr. Soc.*, **156**, 115.
32. Moss, D. L. & Whelan, J. (1973). *Mon. Not. R. astr. Soc.*, **161**, 239.
33. Gribbin, J. (1977). Appendix in *White holes*. Paladin, London & Delacorte, New York.

8

The search for cycles

It seems a fundamental facet of human nature that when we are confronted with a time series of varying data points our immediate reaction is to extrapolate the series linearly and our next reaction is to improve the extrapolation by looking for cycles in the supposed pattern of variation. Naive extrapolation of recent climatic data should have become disreputable after the realisation that the mean northern hemisphere temperature had begun to decline in the middle of this century just at the time when fears were being expressed about the consequences of a continuation of the previous warming trend (see Chapter 4 of this volume); but this has not deterred many people in recent years from expressing similar concern about the indefinite continuation of the cooling trend of the past two decades, concern based not on any physical model but simply on the existence of the trend. Of course, others have argued a case for the likelihood of continued cooling based on physical arguments, but that is another, and far more respectable, story. To some extent, in spite of the seemingly permanent presence of sensational stories of one kind or another, there is little real danger of such simple-minded extrapolations being taken too seriously, since it is obvious that any upward or downward trend of, say, temperature, cannot continue indefinitely; not even the most sensational of popularisers has, for example, jumped to the conclusion that the downward trend of northern hemisphere temperatures of 0.6 °C between May 1958 and April 1963 (reference 1) implies a cooling of the hemisphere of 12 °C by the end of this century. But cycles have a more insidious fascination. After all, if any warming trend must eventually reverse, and so must the subsequent cooling trend, what could be more natural than the discovery of a repeating pattern of ups and downs with a more or less regular periodicity? It is so easy to find quasi-periodic variations in chosen stretches of a randomly varying time series that if we are attempting to understand the time variations of some physical quantity such as temperature it is essential that every effort should be made to apply rigorous statistical tests for the presence of the suspected periodic fluctuations, and to relate these variations to physical influences wherever possible. Fig. 8.1 highlights some of the dangers involved in such a search for cycles.

This figure is taken from an article on randomness by Jacchia,[2] and was obtained by him using an electronic computer to generate random digits in the range 0 to 9 and plotting the deviation of the running mean from the expected average value, 4.5. The cumulative excess or deficiency $O - C$ is defined as $\Sigma x - 4.5n$, where x is the random digit chosen and n is the number of trials; the points in Fig. 8.1 are plotted at intervals of 200 in n, commencing at $n = 0$. Such a plot is both fascinating and instructive for astronomers and climatologists alike. Consider the section from $n = 0$ to $n = 40\,000$ in isolation – how natural it would be to see the fluctuations as evidence of some regular (even sinusoidal) variation. Or look at the plot as a whole and imagine we had plotted temperature as a function of time; the trend towards a new ice age seems clear, in spite of the short-lived fluctuations which from time to time briefly counteract the long-term trend. Yet this is simply a plot of the accumulation of random errors, containing no genuine trend useful for predicting future results of the experiment, and no genuine cyclic variations. In real life, both the astronomer and the climatologist are confronted with the more complex situation of a mixture of

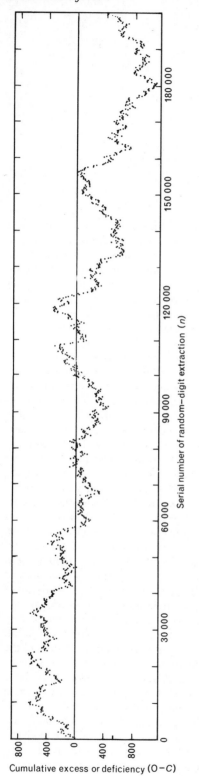

Cumulative excess or deficiency (O−C)

Serial number of random–digit extraction (n)

such random effects with genuine trends and cycles in the same data. It is hardly surprising that the search for astronomical influences on climate should provide a happy hunting ground for the cycle seeker. The discovery of a genuine cyclic variation amidst the confusion of pseudocycles can provide such a powerful tool for understanding and predicting climatic change that it is all the more important to sort the wheat from the chaff by applying tests of statistical significance and physical cause; so it is with the lesson of Fig. 8.1 very much in mind that we should begin our look at the possible cyclic influences of astronomical phenomena on climate.

The possible link between the present climatic regime and the position of the solar system in its long orbit around our galaxy can hardly be included in a discussion of cyclic effects since we have no time series extending over thousands of millions of years which can be tested to indicate the significance of the supposed effect. That model stands or falls essentially on the plausibility of the physical model presented. The longest astronomical cycles which might influence climate and which can be tested by appropriate statistical investigation of the long time series available in the terrestrial record are those involving variations in the orbit and inclination of the Earth, generally grouped under the name 'Milankovich Model' (or hypothesis) after the Yugoslav geophysicist who developed the idea some 50 yr ago, although he had been preceded

Fig. 8.1. An accumulation of 100 000 'random errors'. A variable integer x has been chosen at random in the range 0–9, so that the average expected from a large number of trials is 4.5. The cumulative figure plotted is $x - 4.5n$, where n is the number of trials, and this quantity is plotted at intervals of 200 in n, commencing with $n = 0$. The temptation to see long-term trends and even quasi-sinusoidal variations in this curve, even though we know that it is generated by a random process, provides a suitable caution to interpreters of climatic, astronomical and other time series. From Jacchia,[2] reproduced by permission of the publishers and L. G. Jacchia.

by others who speculated about such possibilities even in the nineteenth century (see discussion in reference 3). There are three astronomical cycles which formed the basis of Milankovich's model, and which remain as the basis of modern versions of the model, although details of the explanation of why these cycles should affect climate have been modified greatly over the years.

First, there is the variation in the eccentricity of the Earth's orbit, which 'stretches' from nearly circular to more elliptical and back, with a cycle time in the range 90 000 to 100 000 yr; both the degree of stretch and the exact time taken for the complete cycle circular–elliptical–circular vary slightly. Secondly, the longitude of perihelion varies relative to the equinox – in other words, the season of closest approach to the Sun varies – with a more precisely defined cycle time of approximately 21 000 yr. At present, the Earth is closest to the Sun during the southern hemisphere summer, and furthest from the Sun during the northern hemisphere summer; the situation will be the opposite in some 10 000 yr from now. And thirdly we must consider the effect of the varying obliquity of the Earth relative to the plane of its orbit. This roll effect changes the tilt of the Earth between 21.8° (more nearly upright) and 24.4° (more inclined) with a cycle time of about 40 000 yr (from 21.8 down to 24.4 and back to 21.8). A greater obliquity implies a more pronounced difference between summer and winter.

The situation today regarding these three effects is that the Earth's orbit is nearly circular and becoming more so, reducing the difference between maximum and minimum insolation a change of as much as 30 % for the most extreme orbits possible); the longitude of perihelion at present gives the coolest northern hemisphere summers (with correspondingly cold southern hemisphere winters, so that seasons differ more markedly in the south); and the obliquity is decreasing with a tendency to decrease the differences between seasons in both hemispheres. But what does all this mean in terms of the Milankovich model?

Taking first the criterion that the model must be physically realistic (or at least plausible), there have been real difficulties in explaining how the relatively small changes in insolation produced in this way could be sufficient to initiate or end glacial spread. In addition, since these astronomical variations do not change the total amount of heat received by the Earth but merely its distribution, so that when the northern hemisphere winters are cold those in the south are relatively warm, the evidence that ice ages occur essentially synchronously in both hemispheres was long taken as refuting the model. But the present distribution of the land surface of the globe seems to provide a resolution of both of these difficulties. As Calder[4] has pointed out, even with assumptions that are so simple that they 'are almost frivolous' a model can be constructed which allows the spread of ice above 50° N if the summer insolation above that latitude falls by 2 % from its present value, well within the range of Milankovich variations. Somewhat less frivolously, Kukla[5] showed that the Earth's surface temperature is most affected by seasonal changes of irradiation in the interior of North America and Eurasia, and that the highest sensitivity to insolation variations occurs in the autumn. Such models provide reasonable agreement with the patterns of major climatic changes of the recent geological past. These results are suggestive, especially when taken with the statistical evidence, and at present there seems no reason to dismiss the Milankovich model on grounds of physical implausibility. Since no definitive model of the effect of these three astronomical influences on climate has yet emerged, however, it seems inappropriate to discuss the present tentative models at length here.

The remaining nagging doubt about the mechanism – the near synchroneity of northern and southern glaciations – disappears when we consider the elegant explanation proposed by Hays[6] when presenting some of the best statistical evidence in support of the model yet available (see also reference 7). Power spectra of temperature variations revealed by analysis of the isotopic composition of marine shells in cores drilled from the Southern Ocean floor show clear peaks corresponding to the 21 000- and 40 000-year Milankovich cycles in data covering the past 150 000 yr. This time series is too short to provide a test of the physical reality of the longest climatic cycle of the Milankovich model, but other data covering much the same period have shown the 21 000-yr cycle at least.[8] Equally significantly, the Lamont-Doherty work reported by Hays also provides an indication of the varying salinity of the oceans, which depends on the amount of water locked up in ice and also influences the isotopic ratio (see

Chapter 3 of this volume). The evidence confirms that both hemispheres undergo near-synchronous glaciations, with (for the most recent example at least) an indication that the south leads the north both into and out of an ice age. It is this emphasis on *near* but not perfect synchroneity which provides the clue to explain global ice ages within the overall Milankovich model.

In qualitative terms, it is clear that in order to initiate the spread of ice in the northern hemisphere we require cool summers so that the snow cover of the land at high latitudes is not completely melted (see Chapter 6). Once the area of snow and ice cover builds up, the process may be hastened by the increased albedo. However, in the southern hemisphere there is scarcely any land at high latitudes which is not already covered by a permanent ice-cap. Snow which falls on the sea cannot remain even if summers are cool, and the only way to spread the covering of sea-ice is to have very severe winters in which large volumes of ocean water are frozen, when once again the increased albedo plays a part in hastening the development. So it is the present distribution of continents, requiring cold northern summers and cold southern winters to initiate the spread of ice, which has produced a situation in which the astronomical variations of the Milankovich cycles can produce nearly synchronous ice ages in both hemispheres. It is already clear that continental drift has played a large part in determining the present climatic regime, since it has given us a southern polar continent and a land-locked northern polar sea which can both possess ice-caps. For much of the Earth's history, one or both poles may have been ice-free and the distribution of land cannot have favoured the Milankovich mechanism except on rare occasions. This is entirely consistent with the evidence that ice epochs are rare events, and that the recent climate of the Earth has been unusual.

The overall pattern emerging from this work provides an important insight into the processes of climatic change, indicating the significance of overlapping, additive effects which operate on very different time-scales. Over hundreds of millions of years, the tectonic activity of the Earth has slowly shifted continents into a situation where polar caps are possible and full ice ages can occur given some additional stimulus.

Over hundreds of thousands and tens of thousands of years, the astronomical cycles of the Milankovich model provide several even better opportunities for ice cover to develop – but even the most ardent Milankovicher today does not claim that every dip in the Milankovich curve produces an ice age. Only, perhaps, when the astronomical influence is right and we have some additional influence (volcanic outbursts, perhaps, or one of the other triggers discussed elsewhere in this book) does the climatic regime which is already tilted towards an ice age by the other influences finally tip over into such a state.

So the Milankovich model passes both the test of physical plausibility and the statistical tests; like other good theories, it also provides a firm prediction which can be used as a further test. Although the model tells us nothing about cycles of less than 20 000 yr, it is relatively simple to determine our present position in the overall pattern obtained by combining the three effects, and to use this to predict future temperature trends for, say, northern hemisphere summers at high latitudes. Equally, we might simply extrapolate the empirical curves obtained by analysing the isotope variations in Southern Ocean sediments, or we might take at face value the suggestion that the south leads the north into an ice age, and build our prediction from the pattern of cooling in the Southern Ocean over the past few hundred years. The details differ, emphasising the still unfinished state of the Milankovich model (or family of models) – but the overall conclusion is the same, emphasising that the model is close enough to the finished state to be useful. We have already passed the peak of the present interglacial, according to all these predictions, and we can expect the beginning of a return to full ice age conditions within about 1000 yr, persisting perhaps for a further 100 000 yr.

This prediction takes no account, of course, of the implications of Man's interference with the environment, which could have a pronounced short-term effect (see Section 5). In addition, the probability of the trigger for the next ice age occurring in the next 100 yr must be very small, and there is no reason to fear the imminent onset of a full ice age in the immediate future (by human time-scales). Perhaps the most valuable interpretation of this evidence is obtained by looking at the other side of the coin: we are certainly not likely to experience a pronounced warming or 'climatic optimum' over the next few hundred years, and should rather expect the conditions of the past few hundred years (see Chapter 4 of this volume) to be a good guide to the best we can expect, unless the natural pattern is altered either deliberately or inadvertently.

In view of this prediction, it is particularly unfortunate that the search for astronomical cycles which might influence the terrestrial climate reveals few possibilities in the range from 20 000 yr down to a few score years. Such variations within that range of time-scales could only be produced by changes in the Sun's output, and this has not been monitored reliably for the necessary length of time. Attempts have been made to argue on the basis of the climatic record that solar fluctuations with a regular cyclic behaviour do take place on a time-scale of thousands of years (see, for example, reference 9), but these are fraught with the dangers of circular argument and depend on the assumption that certain climatic changes in the past few thousands of years were produced by changes in solar activity. Although such investigations may be important in the debate about whether or not the Sun is at present in a 'normal' state, they are of very little direct value in the debate about climatic change on Earth and will not be discussed further here.

This brings us to consideration of the claimed solar cycle influences on climate with periods in the range from a few years to 200 yr. Sunspot/weather correlations are a fashionable area of research in the 1970s, just as they were, indeed, in the 1870s, soon after the roughly 11-yr solar cycle was first recognised, when among the many comments on the phenomenon found in the pages of *Nature* and other journals was the report that work by 'a German physicist Dr Küppen' related temperature cycles on the Earth to the sunspot cycle. 'Dr Küppen proposes' our counterparts of 100 years ago were told, 'to examine the influence of periodic weather changes . . . on some phenomena of organic nature';[10] but alas neither Dr Küppen nor other nineteenth-century researchers were able to provide a definitive model of the link between weather and solar activity that would have been practicable for forecasting purposes, and the whole subject was, by and large, neglected after the initial burst of interest, until recent years. The reasons for the present upswing of interest in solar–terrestrial links is twofold.

First, we now have more data and more sophisticated statistical techniques with which to analyse those data; secondly, the arrival of the space age has made it possible to investigate conditions in interplanetary space, providing the essential physical insight into the nature of the links between the Sun and Earth.

Even so, the subject was not totally ignored during the first part of the present century, and several of the contributors to the book *Climatic Change* (see reference 3), published a quarter of a century ago, drew on the available body of data indicating even at that time the reality of the solar cycle influence on weather not just in the recent past but at different geological periods. Zeuner[11] found evidence of roughly 11-yr and 23-yr cycles in varve data both ancient and modern, as well as a suggestion of the 21 000-yr Milankovich cycle. Korn[12] found essentially the same cycles in layers of slate dated as 275 Myr old, and there are many other examples. Even where the ages of the deposits have been revised in the light of more recent interpretations of the geological record, this does not affect the evidence of cyclic variations within the varves or other strata. In addition, cycles seemingly produced by sunspot cycle influences on weather are clear in many tree-ring samples, including the classic work by Douglass.[13] One of the chief problems in interpreting this wealth of data – even without the additional data gathered and presented since the 1950s – is the large number of cycles which have been offered as solar influenced. The simple sunspot cycle revealed by counting the number of spots visible on the solar disc averages just over 11 yr in length, but can vary by several years either side of the mean; the 22- or 23-yr 'double sunspot' cycle associated with reversals of the solar magnetic field is regarded by solar physicists as a more fundamental cycle in terms of the changing conditions in the Sun (and the magnetic link may make this cycle particularly important for the Earth); and cycles of both 80–90 yr and about 180 yr have also been claimed for sunspot variations, although it should be pointed out that since we only have good sunspot data for the 370 yr or so since the time of Galileo the statistical basis for the last claim, at least, is far from good. Other sunspot cycles have also been invoked from time to time, but these four principal contenders are more than enough for us to consider in the present state of the art.

Selecting a few examples from the recent literature, we find sunspot number correlated with the length of the agricultural growing season in Scotland, rainfall in Beirut and good cricketing summers in England;[14] with rainfall in Brazil, South Africa and Adelaide, and temperature in central England;[15] with surface air temperatures in the North American continent;[16] and with changes in the pattern of distribution of southwest monsoon rainfall over India.[17] The roughly 22-yr cycle of drought in the American Great Plains has long been suggested as a 'double sunspot' cycle effect (cf. reference 18), and it would merely be tedious to list further the various climatic indicators which have been linked to solar influences associated with the 11-yr and 22-yr cycles in particular. There is no doubt that some data from some parts of the globe show variations which match the pattern of sunspot variations for at least some of the time. But sometimes the matching is in phase and sometimes in anti-phase; sometimes (and in some places) there is no matching; and often the statistical significance of the match is not enough to suggest that any solar signal *drives* the climatic change, merely that it *modulates* the noise already present in the natural climatic fluctuations. It is small wonder that meteorologists concerned with forecasting weather variations should be tempted to wash their hands of the whole business. But the concensus of opinion now is that there is a real, if small, sunspot cycle/weather correspondence in much of the available data;[19] at present this is of doubtful value in short-term forecasting, but of very great value in providing an insight into the workings of the atmosphere of our planet and the physical links between Sun and Earth, while perhaps indicating the basis of one kind of long-term trend.

A good indication of both the potential understanding of the atmosphere and the problems of interpretation provided by the sunspot/weather correlations is given by a study of South African rainfall variations reported by Dyer and his colleagues.[20-22] This work shows that rainfall variations in South Africa during this century have been oscillatory, and the initial emphasis of the work[20] was that it thus dispelled the apparently widespread notion that the southern part of Africa has been experiencing progressive desiccation with decreased rainfall since the time of Livingstone. The periods found in these oscillations, however, correspond closely to those of the sunspot cycle – although this only becomes apparent when rainfall figures for separate areas are considered separately.

The subcontinent was divided into seven homogeneous rainfall regions (see also reference 23), with each group of monitoring station records analysed in terms of both temporal and spatial regional effects and local effects. Only the regional temporal effects obtained are of interest in the context of sunspot cycle correlations, which emerge in two clearly distinct ways. For most of the northeast region of South Africa dominated by summer rainfall, the fluctuations are in phase with those of the double sunspot cycle; but further south, in a band roughly paralleling the coast of the Cape but slightly inland, the rainfall fluctuations are in almost perfect anti-phase with the single sunspot cycle. Between these two regions, where summer rainfall is present but not as dominant as in the northeast, are regions where the pattern of rainfall fluctuations is more confused, and again no strong solar cycle signal is detectable in the coastal regions of the extreme south.

Dyer's statistical evidence for this pattern is compelling,[21] but as yet we lack a satisfactory physical explanation. The best that can be said is that evidence that it is the *distribution* of rainfall, rather than the global total, which varies with solar cycle variations seems more satisfactory in terms of models in which the solar effects are seen as modulating natural atmospheric variations rather than driving the circulation into a new pattern. The latitudinal effects found in South Africa are clearly related to the very different kinds of rainfall pattern found in different parts of that country, which experience respectively strong winter maxima, strong summer maxima, equinoctial maxima (six-monthly peaks) and all-seasons rainfall. It seems that the best use of sunspot cycle correlations with meteorological data in the immediate future will be to determine the global distribution of stations where the different correlations – with single or double cycle, in phase or anti-phase – are found. If these echo on a global scale the latitudinal or regional distributions found on the smaller scale of South Africa we may then be in a position to understand the changes in overall circulation which produce these effects and are in turn, it now seems, produced by variations in solar activity.

Even now, however, established patterns like those found for rainfall in some regions of South Africa can be of use in empirical forecasting, provided there is some satisfactory means of predicting the next peak (or trough) in the cycle of solar activity. A variety of more or less bizarre methods of predicting solar maxima (see discussion in reference 24) has not yet provided us with a successful prediction, although many explanations can be fitted to past patterns of solar activity. Now, however, there is some evidence that the size of a solar activity maximum is determined at the beginning of the cycle, the time of the previous solar minimum. According to Brown,[25-27] the occurrence of days of abnormal phase in the solar diurnal variation of the horizontal component of the Earth's magnetic field (abnormal quiet days, or AQDs) shows an inverse correlation with solar cycle so that the number of AQDs at a solar minimum is a guide to the strength of the coming maximum. This raises interesting questions regarding the physics of the solar interior; but if Brown is correct we will shortly have a firm prediction of the strength of the coming solar maximum, when AQD data for the present solar minimum are analysed. In terms of physical understanding of the processes involved, however, Brown's work is important in that it presents evidence of yet another solar–terrestrial link, and that this link involves magnetic variations.

It is not yet clear just how close the links between terrestrial magnetism and climatic effects are. At one extreme we have the view that the shape of the Earth's magnetic field may play a dominant part in determining the overall circulation pattern of the globe, including such features as the dumb-bell shaped double low of the northern polar regions;[28] at the other extreme is the view that 'it is unnecessary, and probably misleading, to postulate any causal relationship between the geomagnetic field and the 500-mbar contours, because the main features of the maps of 500-mbar contours can be explained by direct calculation from physical and dynamical principles without consideration of magnetic effects' (reference 29; but see also references 30 and 31). The truth, as usual, probably lies somewhere between the extremes of present theoretical thought, and there do seem to be real links between magnetism and climate which, although not strictly cyclic effects, should be mentioned briefly here. Wollin *et al.*[32] have mentioned earlier

claims that there is a link between variations in the Earth's climate and magnetic changes, and give evidence 'that a close relationship links changes of the Earth's magnetic field and climate', using data for the period 1925-70. Harrison & Prospero[33] suggest links between geomagnetic reversals and major climatic changes; this is a theme to which we shall return in the next chapter, but within the context of solar cycle variations the most relevant aspect of magnetism/climate studies is that there is an inverse correlation between mean annual temperature and magnetic field intensity over both the short term[32] and the long term.[34,35] If we were to speculate that the cyclic influences of solar activity on the terrestrial atmosphere respresent the sum of discrete events – solar flares and storms – which affect the atmosphere by changing the nature and strength of the solar wind of particles reaching the magnetosphere and being funneled towards the polar regions by the Earth's magnetic field, it is at least an intriguing working hypothesis to suggest further that the link between magnetism and climate operates because variations in the magnetic field also affect the arrival of particles in the atmosphere. A weaker magnetic field may, in some circumstances, be equivalent to a higher level of solar activity. Whether or not this is the case, an understanding of the physical links between Sun and Earth seems to depend not on further discussion of cyclic effects, but on the growing body of observational data and theory relating to the short-term influences of specific solar events on the circulation of the Earth's atmosphere. Such studies go hand in hand with observations of the solar wind, which links the Sun and Earth, made by spacecraft; it is hardly surprising that a satisfactory understanding of the whole phenomenon of solar–terrestrial links is only now beginning to appear, when we note that it was only in 1974 that Intriligator[36] was able to report 'the first evidence for long term variations in the solar wind associated with changes in the solar cycle', with another report later in the same year[37] uncovering a variation in the spiral angle a of interplanetary magnetic field lines observed near the Earth which depends on both the polarity of the field and the phase of the sunspot cycle.

Before we move on to look at these short-term influences in detail, however, two further influences relating to cyclic variations should be mentioned. Rosenberg & Coleman,[38] in describing a 27-day cycle in the rainfall of Los Angeles and relating this[39] to the 27.3-day lunar cycle in declination (during which the Moon moves north for 13.65 days and south for 13.65 days) have drawn attention to the shortest astronomical cycle of any practical relevance to our present discussion. Bryson has noted that this lunar cycle also pulls the Pacific high cell along with it,[40] and Lamb[41] has confirmed that result. Such an effect is small in the context of the global situation, but significant as an example of a small astronomical influence which does affect terrestrial weather. But our chief concern here is with climatic change, and Schneider & Mass[42] have presented a model in which many features of climatic change over the past 300 yr can be understood in terms of the averaged effect of the cyclically varying level of solar activity.

The Schneider–Mass model is used to calculate a global surface temperature history from 1600 to 1975, using a combination of Lamb's volcanic dust veil data[43] and the controversial but intriguing suggestion of a link between sunspot number and the amount of insolation arriving at the top of the atmosphere, implying a variation of the solar 'constant' – which, following Schneider & Mass, it seems more sensible to refer to as the solar parameter. Two independent studies provide the basis for an empirical relation between the value of the solar parameter (S) and annual sunspot number.[44,45] Both suggest that the value of (S) increases as the annual sunspot number (N) increases from zero, but reaches a maximum for moderate sunspot number (about 80) and then decreases, reaching a value similar to that for zero sunspot number at high activity (N equals approximately 200); the range covered is about 2 % in S.

To an astronomer, the suggestion of such a large variation in S over a few years seems almost heretical. Observations of other stars thought to be similar to the Sun do not show this kind of variation, and the implication seems to be that the Sun is not a normal star. That may be the case – or, at least, the Sun may not be in a normal state today. But it is equally possible that it is not the quantity but the quality of solar radiation which varies sufficiently with sunspot number to produce this effect. Lockwood[46] has reported observations of the brightness changes of Uranus,

Neptune and Titan which suggest a cyclic variation linked with variations in sunspot number. Although such variations could be a result of changes in the Sun's brightness, with the outer planets simply reflecting the changing solar output, the explanation most favoured by astronomers at present is that variations in the solar wind of particles and/or variations in the solar flux at ultra-violet and shorter wavelengths may produce changes in the planetary albedos, perhaps by photochemical effects. It is interesting that the magnitude of variations in S required by Lockwood's data if there are *no* albedo changes is also about 2 %, but it is equally interesting that changes in the Sun's ultra-violet flux, in particular, seem likely to be important for the ozone layer of our atmosphere, and we certainly should not rule out the possibility that it is the transmissivity of the atmosphere which varies with sunspot number, not the total energy output of the Sun. Either way, however, the implications for surface temperatures are the same, as interpreted by Schneider & Mass.

Using the Kondratyev & Nikolsky data[45] for variability of the solar parameter and standard astronomical records of sunspot number, along with Lamb's dust veil index calibrated using data from studies at Mauna Loa of changes in atmospheric transmissivity following recent volcanic eruptions,[47] Schneider & Mass derive the 'predicted' global temperature history shown in Fig. 8.2. The agreement of many features of the derived curve with the actual temperature record of the past three centuries is striking, and among other features the recent cooling trend is seen as the effect of the high sunspot numbers of recent cycles. It is unfortunate, therefore, that there is no proven reliable guide to the levels of solar activity which we might expect during the next few cycles, although such evidence as there is (see reference 24) suggests that these recent high levels of peak activity are unlikely to be maintained. Brown's work[25–27] is perhaps the most interesting recent development in this field of sunspot prediction.

The most obvious disagreement between the Schneider–Mass model and the real temperature history of the globe is that the latter does not include pronounced variations following the '11-yr' sunspot cycle. This may perhaps be attributable to the noisiness of the data, gathered from many different stations, or it may be that the

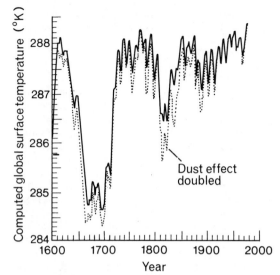

Fig. 8.2. Global surface temperature variations computed by Schneider & Mass[42] using a model incorporating the effects of volcanic dust and an empirical solar cycle variation, as described in the text. The dashed curve, indicating a computation in which the dust effect was arbitrarily doubled, indicates the small sensitivity of the model to this parameter. Fig. 8.2 was provided by S. H. Schneider and used here with his permission.

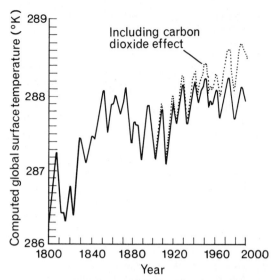

Fig. 8.3. Global surface temperature computed by the method of Schneider & Mass[42] with an extrapolation to 1989. The dotted curve was obtained by including Broecker's estimate[47] of the influence of anthropogenic carbon dioxide. Fig. 8.3 was provided by S. H. Schneider and used here with his permission.

thermal relaxation time of the Earth–atmosphere–ocean system is too long for such short-term variations to develop, so that there is a natural smoothing effect. But the absence of the 11-yr signal in the data does not, of itself, imply that the model is invalid.

As Schneider & Mass point out, 'the causes of climatic change are far from explained by this simple exercise', but their work does stress the urgent need for monitoring of the solar parameter from space over a long period, with an accuracy better than 2 %. If the climatic system is being forced by some solar influence as the model suggests, then a thorough investigation of the forcing mechanism and its strength could be of great value in improving the present understanding of the response of the system to other forcing factors, such as the increasing burden of carbon dioxide in the air. Fig. 8.3 shows the effect of adding one recent prediction of the effects of increasing atmospheric carbon dioxide[48] to the Schneider–Mass model, with a tentative extrapolation; the relationship between Man and climate is discussed more fully in Section 5 of this book.

References

1. Starr, V. P. & Oort, A. H. (1973). *Nature,* **242,** 310.
2. Jacchia, L. G. (1975). *Sky Telesc.,* **35,** 371.
3. van Woerkom, A. J. J. (1953). In *Climatic change,* ed. H. Shapley, p. 147. Harvard University Press, Massachusetts.
4. Calder, N. (1974). *The weather machine.* BBC Publications, London.
5. Kukla, G. J. (1975). *Nature,* **253,** 680.
6. Hays, J. (1975). Unpublished talk at WMO/IAMAP Symposium, Norwich, England, August 1975.
7. Hays, J. D., Imbrie, J. & Shackleton, N. J. (1976). *Science,* **194,** 1121.
8. Bloom, A. L., Broecker, W. S., Chappell, J. M. A., Matthews, R. K. & Mesolella, K. J. (1974). *Quaternary Res.,* **4,** 185.
9. Bray, J. R. (1968). *Nature,* **220,** 672.
10. Anon. (1874). *Nature,* **9,** 184.
11. Zeuner, F. E. (1946). *Dating the past.* Methuen, London.
12. Korn, H. (1938). *Neues Jb. Miner. Geol. Paläont.,* **74A,** 50.

13. Douglass, A. E. (*Carnegie Inst. Wash. Publ.*, **289,** I(1919); II(1928); III(1936).
14. King, J. W. (1973). *Nature*, **245,** 443.
15. King, J. W. (1975). *ESA Bull.*, No. **3,** 24.
16. Currie, R. G. (1974). *J. geophys. Res.*, **79,** 5657.
17. Jagannathan, P. & Bhalme, H. N. (1973). *Mon. Weath. Rev.*, **101,** 691.
18. Roberts, W. O. & Olson, R. H. (1975). *Nature*, **254,** 380.
19. Discussion Meeting on *Solar–terrestrial relationships*, held by Royal Meteorological Society (London, 1976); see also report in *Nature*, **259,** 367 (1976).
20. Tyson, P. D., Dyer, T. G. J. & Mametse, M. (1975). *Q. Jl R. met. Soc.*, **101,** 817.
21. Dyer, T. G. J. (1975). *South African J. Sci.*, **71,** 370.
22. Dyer, T. G. J. (1976). *Q. Jl R. met. Soc.*, **102,** 157.
23. Dyer, T. G. J. (1975). *Q. Jl R. met. Soc.*, **101,** 817.
24. Gribbin, J. R. & Plagemann, S. H. (1974). *The Jupiter effect*. Macmillan, London.
25. Brown, G. M. (1974). *Nature*, **251,** 592.
26. Brown, G. M. (1975). *J. atmos. terr. Phys.*, **37,** 107.
27. Brown, G. M. (1976). *Mon. Not. R. astr. Soc.*, **174,** 185.
28. King, J. W. (1974). *Nature*, **247,** 131.
29. Sawyer, J. S. (1974). *Nature*, **252,** 368.
30. King, J. W. (1974). *Nature*, **252,** 370. (This is a reply to reference 29.)
31. Green, J. S. A. (1974). *Nature*, **252,** 343. (A review of references 29 and 30.)
32. Wollin, G., Kukla, G. J., Ericson, D. B., Ryan, W. B. F. & Wollin, J. (1973). *Nature*, **242,** 34.
33. Harrison, C. G. A. & Prospero, J. M. (1974). *Nature*, **250,** 563.
34. Wollin, G., Ericson, D. B., Ryan, W. B. F. & Foxter, J. H. (1971). *Earth Planet. Sci. Lett.*, **12,** 175.
35. Wollin, G., Ericson, D. B. & Ryan, W. B. F. (1971). *Nature*, **232,** 549.
36. Intriligator, D. S. (1974). *Astrophys. J.*, **188,** L23.
37. Svalgaard, L. & Wilcox, J. M. (1974). *Science*, **186,** 51.
38. Rosenberg, R. L. & Coleman, P. J. (1974). *Nature*, **250,** 481.
39. Rosenberg, R. L. (1975). *Nature*, **258,** 457.
40. Bryson, R. A. (1948). *Trans. Am. geophys. Un.* (EOS), **29,** 473.
41. Lamb, H. H. (1972). *Climate: present, past and future*, vol. 1. Methuen, London.
42. Schneider, S. H. & Mass, C. (1975). *Science*, **190,** 741.
43. Lamb, H. H. (1970). *Phil. Trans. R. Soc.*, **266,** 425.
44. Abbott, C. G. (1966). *Smithson, misc. Collns*, **448,** 7.
45. Kondratyev, K. Ya. & Nikolsky, G. A. (1970). *Q. Jl. R. met. Soc.*, **96,** 509.
46. Lockwood, G. W. (1975). *Science*, **190,** 560.
47. Ellis, H. T. & Pueschel, R. F. (1971). *Science*, **172,** 845.
48. Broecker, W. (1975). *Science*, **189,** 460.

Addendum

After this chapter was written, significant new evidence in support of the Milankovich model was presented by Dr B. J. Mason, Director-General of the UK Meteorological Office, in the 1976 Symons Memorial Lecture of the Royal Meteorological Society. As well as reporting a new power spectrum analysis of marine core isotope data, which shows significant peaks corresponding to each of the three Milankovich periods (N. J. Shackleton, unpublished work), Mason provided a persuasive argument for the effectiveness of these astronomical variations of northern hemisphere insolation in changing the amount of ice cover.

The first importance of this work, as Mason has stressed, lies in the link between statistically significant periodic variations in the isotope data and a known physical mechanism, contrasting with the situation in so much of the shorter-term data where seemingly significant peaks often have no physical explanation, while peaks expected on physical grounds (such as the 11-yr solar cycle signature) often are not significant. With this incentive, Mason has reworked some aspects of the Milankovich model in terms of the variations in the amount of heat received at different northern hemisphere latitudes in different seasons, using both the original astronomical data of Milankovich and modern improvements on those data. (Mason, B. J. (1976). *Q. J. R. met. Soc.*, **102,** 473.)

North of 45° N the variations due to these effects cover about 4×10^{18} cal day^{-1} (1.5×10^{21} cal yr^{-1}), which is about 1 % of the total heat received by the polar cap, and in his Symons Lecture Mason commented that he was 'staggered' to find the very exact agreement over the past 150 000 yr between times of minimum insolation and times of maximum advance of northern hemisphere ice.

This impressive agreement encouraged him to look at the integrated effect of insolation variations during periods in which the ice cover advanced and retreated. Again for the region north of 45° N, when ice cover developed during the period from 83 000 B.P. to 18 000 B.P. the integrated deficiency in insolation produced by Milankovich variations amounted to some 1000 cal for each g of ice formed using Milankovich's figures, or some 556 cal g^{-1} with the modern figures. Since the latent heat of the vapour/snow transition is 677 cal g^{-1} the figures are impressive even without allowing for the possibly crucial significance of the fact that the variations in insolation are much greater in summer than in winter.

From 18 000 to 6000 B.P., when the ice was melting into water, the integrated 'excess' heat received was 4.2×10^{24} cal on Milankovich's figures and 10×10^{24} cal with the modern astronomical figures, while the latent heat required (at 80 cal g^{-1}) for the known decrease in volume of the ice was 3.2×10^{24} cal; a coincidence which stretches across 24 orders of magnitude to agree so closely is certainly not to be dismissed lightly, and work is now in progress at the Meteorological Office to produce a more detailed model taking account of the capacity of the ocean to act as a reservoir of heat and the feedback mechanisms between ice and ocean. Commenting on likely future climatic fluctuations, Mason noted that we are now shifting from a situation in which the Milankovich effect has produced excess insolation in the north (above the mean for the past 150 000 yr or so) to one in which there is a deficit. The simple astronomical calculations show, once again, the dip reaching a northern hemisphere insolation minimum in about 10 000 yr, but bottoming out before reaching conditions quite so extreme as those which prevailed at the height of the recent ice age. Even though it remains to be seen how this model can be related in detail to the global picture of synchronous ice ages in both hemispheres (including the Lamont-Doherty work referred to above), the whole situation is best summed up by Mason's own words: 'this effect cannot be laughed away . . . certainly all the energy quantities are in the right ball park'.

9

Short-term effects

It is now more than 10 yr since Mitchell commented[1] 'I think it is a rather curious thing – I am tempted to use the word shocking, but I don't think I need to say that – that in the year of our Lord 1965 and the year of our satellite 7, going on 8, we have yet to put an instrument into space above the atmosphere to find out once and for all whether the solar constant is changing along with all the other features on the sun that we know are changing. Obviously this is a very crucial question.' Obvious, perhaps; but even now we lack the dedicated satellites to investigate this kind of influence on climate, even though many satellite experiments primarily serving other purposes do provide data relevant to studies of such climatic influences. A wealth of information of this kind is provided in Bandeen & Maran,[2] where almost 10 yr after Mitchell's remark quoted above we find Rasool commenting in his summary of the symposium being reported 'I was horrified [to discover here] that, after ten years in space, we don't know what is the variation [of the Sun] in the near ultraviolet or the solar constant . . . it is almost criminal not to know this.' (reference 2, p. 423). Rasool went on to point out that in looking for mechanisms to explain observed influences of solar variations on weather it seems logical to start as close as possible to the troposphere, in order to shorten the hypothetical chain of cause and effect; the ozone region just above the troposphere can be affected by ultra-violet variations, cosmic rays or solar protons, and there seems little need to look further out but rather to concentrate on developing models which can show how the variations likely to be produced in the ozone region can influence circulation in the troposphere below. 'The stratosphere is the region where either the source is put in or the

energy is transferred . . . unless you have a solar constant variation in the visible' (reference 2, p. 425). The lack of detailed models of the energy transport from the stratosphere downward during changes caused by discrete solar events seems to be the major stumbling block to developing a satisfactory understanding of short-term astronomical influences on weather and climate at present. But the reality of these influences is stressed by a wealth – almost an embarrassment – of observational evidence and there is scope, as we shall see, for some entertaining (if incomplete) speculation about the large-scale changes resulting from an accumulation of such short-term effects.

The penetration of solar protons into the magnetosphere at high latitudes has now been investigated directly with instruments on board the satellite HEOS 2[3] and the behaviour of the ionosphere and plasmasphere during geomagnetic storms caused by solar disturbances has recently been reviewed by Jones,[4] but the complexity of these processes is only likely to begin to be unravelled after the series of interrelated studies of the International Magnetosphere Study in the second half of the 1970s. Meanwhile, the best picture of short-term solar influences on the circulation of the atmosphere comes from the work of Roberts and his colleagues, which first linked the development of low pressure systems at high latitude with geomagnetic storms and ionospheric disturbances, and now includes observations of the influence of the solar magnetic sector structure on these phenomena.

Wilcox has reviewed much of the background to this work in his contribution in Bandeen & Maran[2] (p. 24) where he summarises the three principal pieces of observational evidence as:

(1) Meteorological responses tend to occur two or three days after geomagnetic activity.

(2) Meteorological responses to solar activity tend to be most pronounced during the winter season.

(3) Some meteorological responses over continents tend to be the opposite from the responses over oceans.

Throughout the past 30 yr, since Duell & Duell[5] reported surface pressure variations in northwest Europe after geomagnetically disturbed and geomagnetically quiet days, such observations have been dismissed by some meteorologists in the absence of an accepted physical mechanism to explain them, although statistical tests of such data have long showed their reliability (cf. reference 6). The lack of a detailed model is certainly not to be dismissed lightly, but it is the duty of theorists to construct models which explain the observations, not the duty of observers to produce results in line with currently accepted theory. And by the early 1970s the observational evidence had accumulated to show the overwhelming (but still not yet satisfied) need for better theories of Sun–Earth interactions.

In 1973, Roberts & Olson[7] reported a study of the development of low pressure troughs at the 300-mbar level in the North Pacific–North America area in wintertime. They found that troughs which enter or are formed in the Gulf of Alaska between two and four days after a sudden increase in geomagnetic activity deepen more strongly than corresponding troughs formed at other times. This extended and confirmed the earlier findings of Macdonald & Roberts[8] and Twitchell[9] using 500-mbar data, and it was not long before further evidence was produced relating these changes to changes in the solar magnetic sector structure.[10,11] The atmospheric parameter used in the investigation was the vorticity area index (VAI) of Roberts & Olson,[7] defined as the sum of the area (km^2) over which the absolute vorticity (curl of the velocity vector per unit area) exceeded $20 \times 10^{-15} s^{-1}$ plus the area over which the vorticity exceeded $24 \times 10^{-5} s^{-1}$. (The concept of VAI is also discussed in detail in Svalgaard.[12]) The solar parameter to which changes in VAI were related, the solar magnetic sector structure, was first observed in the mid-1960s (cf. reference 13); the solar magnetic field which is extended outwards from the Sun by the flow of the charged particles of the solar wind can be

divided into sectors in each of which the polarity of the field is either towards or away from the Sun. There are usually four such sectors, separated by very narrow boundaries and each occupying a width of about 90° in solar longitude. As the Sun rotates, the whole pattern is carried around, sweeping past the Earth with a period of about 27 days. The Earth remains within each sector for several days, but sector boundaries pass the Earth within a few hours. At times near solar maximum a pattern with two sectors superimposed confuses this simple picture, and as yet we only have data from one solar maximum to study, but even the simple picture is sufficient to provide a new insight into the solar–terrestrial link investigated – since long before the sector structure was known – by Roberts and his colleagues.

First, the VAI concept was extended to cover the entire northern hemisphere north of 20° N with calculations made twice for each day between 1964 and 1970; the vorticity data were then compared with the dates of 'key days', when a sector boundary passed the Earth, more than 100 of which were recorded in the same interval. The passage of the sector boundary provides a clear timing signal, but it is not suggested that the meteorological response found is produced by the passage of the boundary alone, but by changes in the large-scale sector structure over a period of a few days before and after the key date on which a boundary passes.

The integrated VAI reaches a minimum roughly a day after the passage of a sector boundary, and then increases by about 10 % over the next two or three days, for such passages which occur in the winter months (November–March). Confirmation of the reality of the Sun–weather correlation came from the initially sceptical Hines & Halevy,[14] who not only convinced themselves that the effect is real but also showed that the solar 'signal' is stronger in the weather data during periods of relatively strong meteorological 'noise', which is greater in winter. This suggests a qualitative model in which the solar influence acts as a modulator of the meteorological noise, but there is still no satisfactory detailed physical explanation of the phenomenon, even though the evidence continues to accumulate.[15,16] Since geomagnetic disturbances tend to occur near the time of sector boundary crossings, Olson, Roberts & Zerefos[16] have also carried out a similar study using geomagnetic key days and the occurrence of major solar flares as their initial solar signal. The latter effect is particularly striking in the present context, with an average sequence of events roughly as follows.[16]

With the day of a large solar flare as 0,

(1) by day 1 or 2, the Northern Hemisphere VAI increases sharply, by 5–10 % above its background level;

(2) by day 2 or 3, a geomagnetic storm has started, which may persist for several days;

(3) by day 3 or 4, the VAI has decreased to a value 5–10 % below its value before the flare;

(4) by day 5 or 6 the VAI has recovered back to its original level.

Interesting though these studies are, they are primarily concerned with short-term influences of solar activity and related phenomena on weather, not climate. But it may be that the insight which is now being provided by this work has some bearing for climatic change. We saw in Chapter 8 how the model of Schneider & Mass can 'explain' many features of recent climate in terms of solar variations, and this kind of an effect could, perhaps, be associated with the average influence of many solar flares and many sector boundary crossings. But these are small effects, important though they may for mankind, and astronomy has a tradition of grand speculation which leaves tinkering with mere fractions of 1 % (of, say, the solar parameter) as something almost beyond the pale (which may, perhaps, be why we still do not know if the Sun varies by 1 or 2 % over 11 yr or so); so to close this review of astronomical influences on climate I shall, in keeping with that tradition, mention the possibility of a far more dramatic event, the complete removal of the Earth's magnetic field, which has occurred several times in the past 2.5 Myr and which may on more than one occasion have allowed the short-term influence of solar activity to influence the terrestrial climate dramatically, as well as making conditions distinctly uncomfortable for land-based life.

Since it was realised that the Earth's magnetic field undergoes complete and, by geological standards, frequent reversals, several attempts have been made to link this phenomenon with the breaks in the fossil record which often occur at around the same time (see reference 17 and references therein). Deep-sea cores from several

locations suggest the reality of this link, since during the past 2.5 Myr eight species of radiolaria found in these cores became extinct, and six of these extinctions occurred in near coincidence with polarity reversals.[18] One school of thought held that the link might operate through the mutagenic influence of the increased flux of cosmic rays penetrating to the Earth's surface while the protecting magnetic field was absent during a reversal, but this now looks to be too small an effect to account for the changes; another speculation is that the changing magnetic field has a direct biological effect; but the most attractive theory is that climatic changes (in the broadest sense) associated with the magnetic reversal play a part in producing the faunal extinctions.

Harrison & Prospero[17] have developed this idea simply on the basis of the link between magnetic field intensity and climate (mentioned in the previous chapter) and the links between geomagnetic disturbances and climate referred to above. But the link between solar flares and the pattern of terrestrial events listed by Olson *et al.*[16] can also be included in the overall picture.

Many solar flares produce large fluxes of particles, notably protons, which are detectable at the Earth; these events occur sporadically, although the general frequency of such solar proton events depends on the overall level of activity of the roughly 11-yr solar cycle. The range of energies of individual particles produced in this way is from 10^4 eV to more than 10^9 eV, and the intensities and distribution of energy among the particles vary considerably from one event to another. Crutzen, Isaken & Reid[19] showed that ionisation produced by these particles in the stratosphere causes the formation of large quantities of nitric oxide, so that only a few intense events in a year will suffice to produce as much nitric oxide as is produced by direct oxidation of nitrous oxide. Through the interaction of nitric oxide with ozone this is 'directly affecting the efficiency of the ozone shield that protects the surface of the Earth from potentially harmful solar ultraviolet radiation. Although each solar proton event lasts only a few days, the lifetime of nitric oxide in the stratosphere is long, and its effect on the ozone layer is likely to persist for several years' (reference 20).

The importance of ozone in the stratosphere for

climate and the general circulation of the atmosphere needs no elaboration here, and even without developing this work further we already have a powerful new insight relevant to the occurrence of solar cycle effects in climatic variations. But it requires only a small further step to gain insight also into one aspect of the relationship between geomagnetism and climate.

Today, the presence of the geomagnetic field acts to steer the solar protons to high latitudes, so that the direct effects on the atmosphere are significant only above about 60° (either north or south of the equator). The nitric oxide is spread by atmospheric transport, but according to Reid *et al.*[20] it is diluted by a factor of seven in proportion to the fraction of the Earth's surface lying at latitudes above 60°. During polarity reversals, however, the geomagnetic field disappears, and is at best very weak for a few thousand years, during which time both solar protons and cosmic rays from outside the solar system can penetrate the atmosphere at low latitudes. In terms of the effect on ultra-violet transmission, the implications are clear. A large solar flare of the kind normally encountered during the solar cycle of activity would remove enough ozone to produce ultra-violet radiation levels 15 % greater than those normal on the Earth's surface today, while a flare 100 times larger than usual (a likely prospect sometime during a period of several thousand years) could lead to an increase of ultra-violet radiation at the surface of 160 %. Such effects are likely to be significant hazards for some forms of life on Earth; even a flare 10 times larger than usual would increase the ultra-violet they receive by 55 %.

Removing ozone and changing the transmissivity of the atmosphere are in themselves events of climatic importance, and it is possible that an increased flux of ions in the upper atmosphere could lead to increased cloud cover (see reference 17). The theories are, as yet, speculative; there is even room to suggest that the climatic effects produced as a result of the penetration of solar protons at all latitudes may have contributed to the faunal extinctions which occurred at the times of geomagnetic reversals, which certainly seem to have been times when many species would have been exposed to additional stress factors. The idea applies equally to events more remote than those of the past 2.5 Myr, and Reid *et al.* furnish their grand speculation when considering events some 65 Myr ago: 'the mechanism we have described may have had the dominant role in some of the massive faunal extinctions that have occurred in the distant past. For example, roughly one-third of all living species became extinct at the close of the Cretaceous, which was a period marked by a resumption of polarity reversals, following a very lengthy period of normal polarity.' The dinosaurs, it seems, may have been ill-adapted to cope with sudden bursts of ultra-violet radiation (or associated climatic events) to which they had been unable to adapt because the long absence of geomagnetic reversals failed to provide them with a taste of things to come. The message echoes that of McCrea mentioned in the first chapter of this section; we may owe our very existence to the effects of astronomical influences on the atmosphere. But the greatest significance of the Reid *et al.* model today remains its application to the influence of solar proton events occurring in the normal way at different times in the solar cycle. Here, perhaps, is a physical basis for some of the effects discussed by, for example, Schneider & Mass; as our understanding of the influence of discrete, short-term solar events on the weather grows, the idea that longer-term solar influences, cyclic and otherwise, may simply be the aggregate effect of many short-term influences becomes increasingly attractive.

References

1. Mitchell, J. M. (1965). *Proceedings of the seminar on possible responses of weather phenomena to variable extraterrestrial influence, NCAR Tech. Note* TN–8. P. 157. Boulder, Colorado.
2. Bandeen, W. R. & Maran, S. P. (eds) (1974). *Possible relationships between solar activity and meteorological phenomena*, Preprint X–901–74–156. NASA, Washington.
3. Domingo, V., Page, D. E. & Wenzel, K.-P. (1974). In *Correlated interplanetary and magnetospheric observations*, ed. D. E. Page, p. 507. D. Reidel, Dordrecht.
4. Jones, T. B. (1976). *Nature,* **259,** 9.
5. Duell, B. & Duell, G. (1948). *Smithson. misc. Collns,* **110**(8), 34.
6. Craig, R. A. (1952). *J. Met.,* **9,** 126.
7. Roberts, W. O. & Olson, R. H. (1973). *J. atmos. Sci.,* **30,** 135.
8. Macdonald, N. J. & Roberts, W. O. (1961). *J. Met.,* **18,** 116.
9. Twitchell, P. F. (1963). *Geophys. Bull., Dubl.,* **13,** 69.
10. Wilcox, J. M., Scherrer, P. H., Svalgaard, L.,

Roberts, W. O. & Olson, R. H. (1973). *Science,* **180,** 185.

11. Wilcox, J. M., Scherrer, P. H., Svalgaard, L., Roberts, W. O., Olson, R. H. & Jenne, R. L. (1974). *J. atmos. Sci.,* **31,** 581.

12. Svalgaard, L. (1973). Paper presented to Seventh ESLAB Symposium (Saulgav, 1973); reprinted as *SUIPR Report* No. 526. Stanford University Institute for Plasma Research.

13. Wilcox, J. M. (1968). *Space Sci. Rev.,* **8,** 258.

14. Hines, C. O. & Halevy, I. (1975). *Nature,* **258,** 313.

15. Wilcox, J. M., Svalgaard, L. & Scherrer, P. H. (1975). *Nature,* **255,** 539.

16. Olson, R. H., Roberts, W. O. & Zerefos, C. S. (1975). *Nature,* **257,** 113.

17. Harrison, C. G. A. & Prospero, J. M. (1974). *Nature,* **250,** 563.

18. Hays, J. D. (1972). *Bull. geol. Soc. Am.,* **83,** 2215.

19. Crutzen, P. L., Isaksen, I. S. A. & Reid, G. C. (1975). *Science,* **189,** 457.

20. Reid, G. C., Isaksen, I. S. A., Holzer, T. E. & Crutzen, P. J. (1976). *Nature,* **259,** 177.

SECTION 4

Modelling the changing climate

I O

The role of the oceans in the global climate system

TIM P. BARNETT

Most serious workers in climate research agree that the oceans play vital roles in the global climate system. Many reasons have been given for this conclusion, but two obvious facts bring the point home. (1) The atmosphere is in contact with the oceans over 72 % of the Earth. If the atmosphere is at all sensitive to 'bottom boundary conditions', then it must be aware of, and perhaps reflect, the large oceanic changes known to occur. That the lower boundary conditions are important follows from the facts that atmosphere is essentially a huge heat engine and at any one time the amount of heat (fuel) stored in a vertical column of air extending from the Earth's surface to the fringes of space is approximately the same as the heat contained in an equivalent column of ocean extending from the surface to a depth of *only 3 m.* Thus it is the ocean that initially accumulates and stores incident solar radiation for later release to the atmosphere. It is little wonder, then, that many feel the oceans can initiate changes in the atmospheric climate. (2) Perhaps even more important, the oceans, because of their great heat capacity, can act as a stabilizing influence on an otherwise frenetic atmosphere. These two aspects of 'the ocean problem' could fill a textbook.

In view of the limited space available here, it will be necessary to limit severely discussion of the exchanges between oceans and atmosphere. It is only possible to note the farsighted contributions of early workers in the field.[1-5] The major contributions of isolated climate researchers in the past three decades, a period when they were out of a meteorological mainstream otherwise preoccupied with the construction of large numerical models of the atmosphere, can also only be acknowledged.[6-11] In spite of the range of climatic

time-scales discussed in this volume it has been necessary to focus on the temporal interval between one month and one decade. It is at these time-scales that ocean/atmosphere interactions are best documented and perhaps partially understood. More important, these time-scales are of most immediate significance to mankind (see Section 5). Finally, although this is a section on modelling, no extensive discourse on the strengths and weakness of present ocean and ocean/atmosphere coupled models will be undertaken since it is available elsewhere.[12-16] The present ocean modelling effort is perhaps a decade removed from present atmospheric modelling capability. Unfortunately, the present ocean observations and observational effort needed to verify the models are not dissimilar to their meteorological counterparts 25 to 50 yr ago!

Links between oceans and atmosphere
General physical relationships

The oceans drive the atmosphere above them principally through the transfer of latent and sensible heat energy. It is this energy that partially sustains the major features of the general circulation. In addition, the large heat capacity and much greater density and mass of the oceans can provide the atmospheric features with the qualities of self-enhancement, quasi-stability and, perhaps, long-term predictability. The atmosphere responds by imparting momentum to the ocean. This momentum is the vital sustenance upon which the major oceanic current regimes depend.

Of major importance to ocean/atmosphere coupling is the fact that the sites of maximum exchanges of heat and momentum are generally

Fig. 10.1. Global distributions of wind stress and latent heat flux. The wind stress data for winter (December, January, February) and summer (July, August, September) come from Evenson & Veronis[18] and Hellerman.[17] The estimates of heat flux come from the work of Budyko.[19] The respective units are 10^{-1} dyne cm^{-2} and kcal cm^{-2} mo^{-1}.

not congruent in time and/or space.[17-19] There is also a dissimilarity between hemispheres in the general distributions of momentum and heat flux (Fig. 10.1). The momentum flux (taken here as proportional to wind stress) which drives the currents is put in along zonal bands within the westerlies and the trades. Significantly more momentum is put into the southern oceans than their northern neighbors. The major northern hemisphere *latent* heat exchange,* on the other hand, is rather closely confined to the continental margins being associated with the (warm) western boundary currents. Secondary maxima lie under the trade winds while minima occur in the oceanic upwelling regions along the equator. In the southern hemisphere, the main energy sources lie under the trade wind belts since there are no good equivalents to the continental boundary conditions. Thus, the heat added differentially to the atmosphere must be redistributed internally if the major circulation systems are to be sustained.

Just as the direction of the wind stress is reversed between mid- and low-latitudes, so the direction of *total* heat exchange is reversed. The apparent deficit/surplus in oceanic heat content is balanced out, in large part, by an internal redistribution of heat by the (wind driven) ocean currents. Thus the spatial character of the forcing functions which, at least partially, drive each field are different and thus the principal interactions between the fields occurs at different locations.

The relative phases of the exchanges of heat and momentum also differ between the two media. The momentum flux to the ocean is quite high in mid-latitude during the hemispheric winter. During the hemispheric summer the flux is almost negligible in the northern hemisphere and substantially reduced in the southern oceans. At lower latitudes a seasonal signal exists, but in general it does not have the amplitude of its mid-latitude counterpart (the Indian Ocean monsoons are clearly an exception). The *total* heat exchange is also quite different over the course of a year. At mid-latitudes, the latent and sensible exchanges dominate so that the ocean gives up heat during the late autumn, winter and early spring. During the remainder of the year

*The exchange of latent heat is shown since it is generally much larger than the sensible heat exchange and since these two fluxes together represent the *direct* ocean input to the atmosphere.

radiation dominates the heat budget and the oceans accumulate heat. By contrast, the heat flux in the equatorial oceans is relatively constant over nearly the entire year (the Indian Ocean again excepted). Thus, there is a distinct, but dissimilar, latitude-dependent phase relation between the seasonal forcing functions which drive the respective members of the ocean/atmosphere system.

In summary, the oceans and atmosphere act as a (non-linearly) coupled system, each with a set of strong internal and integrally personal interactions affecting the degree of coupling. Furthermore, the interactions between the systems occur in large measure in different regions of a common physical space. It is little wonder that even the most rudimentary models of the ocean/atmosphere system are difficult to construct.

Mechanisms of coupling

Many of the physical processes that determine the essence of both the ocean and the atmosphere are virtually identical. The differences, however, may be a key to 'the climate problem'. For instance, the oceans are heated from above making them somewhat stable while the atmosphere is heated from below thereby being somewhat more unstable. Also, many of the major space–time characteristics of each physical process, indeed, each medium, are related as the ratio of the media density 1000:1; others to their relative mass 1 300 000:1. These facts help to explain why the ocean is relatively sluggish in its reaction to outside perturbations. This ability of the ocean to 'remember' past influences, coupled with its importance as an atmospheric energy source, suggests to some that oceanic 'events' may be precursors of long-term, large-scale atmospheric 'events'. From a different point of view, the oceans act as a low pass filter on the high frequency perturbations of the atmospheric field. This feature, plus the energy source argument, suggests to others that the ocean is a stabilizing influence in the global climate picture rather than an initiator of change.[20] Without this source of negative feedback the global climate might, indeed, be highly unstable as some recent (but oversimplified) models suggest.[21 22]

The actual processes most important in the coupling are well known, or so we think. The fluxes of sensible and latent heat from the oceans provide energy to the atmosphere. But only a set of complicated measurements can determine the exchange rates at present. Since such methods are impractical for climatic work, it is the common approach to use 'bulk formulas' (which depend on simple, observed variables) to estimate the exchanges.[23] Unfortunately, these formulas and the crude data that are used with them are generally fraught with so many problems, as to be of questionable value for estimating the year-to-year *differences* in heat exchange implied by changes in oceanic heat content.[24]

Estimating the input of momentum to the ocean presents equal difficulties. Again, the complicated measurements required to estimate the flux directly are generally replaced with a simple formula for wind stress that depends on the square of the wind speed.[23] But at what altitude should the wind be observed? Given a quasi-logarithmic wind profile and the fact that the wind speed is squared in the stress calculation, this is certainly not a trivial question. Furthermore, the 'constant' of proportionality between (wind²) and stress is not accurately known and may depend on atmospheric stability,[25] roughness and so on. The fundamental problem is that it is not known how the momentum passes into the ocean! Some data suggest the momentum goes first into the ocean surface wave field and is then transmitted (via wave breaking?) to the oceanic flow field.[26] Such a picture, if true, is not consistent with the assumption from the Ekman theory of tangential stress on a flat sea surface. Yet this theory is routinely used by modellers and others in estimating the response of the near-surface ocean layers to wind stress.

In summary, the actual physical processes that affect the ocean/atmosphere coupling are not well understood and/or difficult to estimate accurately numerically. In the former situation, for instance, oceanographers do not now have a good model that can describe the variations of heat content in the ocean's mixed layer.[16] In the latter situation, it is not at present known if, in fact, the ocean/atmosphere systems are sensitive to fluctuations in the processes comparable with the magnitude of the errors in their evaluation. Fortunately, there are now several scientific programs aimed at ameliorating these problems.[27] But for the immediate future we must reconcile ourselves to the crudest estimates of ocean/atmospheric coupling on climate scales.

Ocean climate changes

The point of this section is to illustrate that the oceans experience larger inter-annual changes in some of their important properties, that is, 'the oceans have their own climate'. Particular emphasis will be placed on the space scale and timescale of these inter-annual (anomalous) fluctuations to show that they are similar to those observed in the atmospheric climate. The causes of the oceanic fluctuations are not well established and so will not be discussed at length.

Of the many properties that could have been presented, it seemed appropriate to single out those which are best defined, somehow reflect or affect the oceans' *near-surface* heat content and which are known to be related to *short*-term climatic changes in the atmosphere (see below). The reader should note that on *longer* time-scales such as 1000 yr, the characteristics of *bottom waters* formed near the poles may be of climatic importance (see Appendix). At any rate, the discourse, while not complete, does illustrate the variability that may be associated with the oceans' ability to affect the atmosphere.

Equatorial temperature changes

What is perhaps *the* largest climatic fluctuation in the ocean's near-surface temperature field occurs over a huge area of the equatorial Pacific Ocean. These inter-annual changes, often referred to as El Niño, were recognized decades ago. By contrast, the inter-annual changes in the equatorial Atlantic and Indian Oceans, which are more poorly documented, seem to be smaller in both magnitude and areal extent.

Recent results have indicated that the occurrence of El Niño events, usually acknowledged to be an abnormal increase in surface water temperature off the South American coast, is associated with major changes in the ocean/atmosphere system across the entire tropical Pacific.[28-34] Furthermore, the largest events seem to be associated with global perturbations of climate and perhaps changes of climatic regime.[35-37] More will be said about the atmospheric implications of equatorial warming later. Here I simply concentrate on the characteristics of the ocean temperature changes.

The lateral extent and magnitude of the inter-annual changes in equatorial sea surface temperature (SST) are vividly illustrated in Fig. 10.2.[38] During the spring and summer of 1972 one

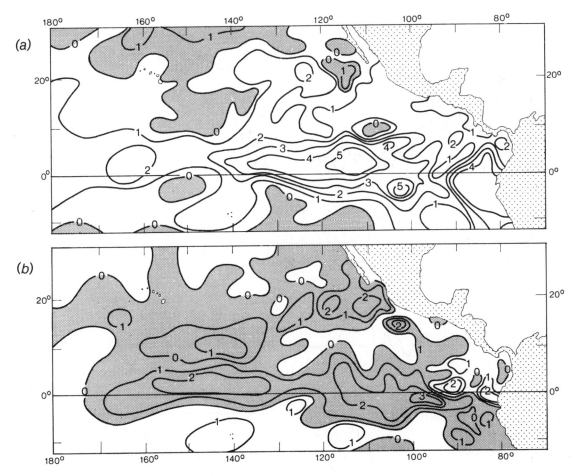

Fig. 10.2. Sea surface anomalies (°C) during warm (El Niño) and cold epochs in the equatorial Pacific. (*a*) July 1972, (*b*) July 1973. Data from the National Marine Fisheries Service, Fishing Information.[38]

of the larger warming events of this century was at its height. Water temperatures were up to 6 °C above their long-term mean values. The warm water extended from South America to the limit of the data; more than 25 % of the Earth's circumference. Recent studies[34] show that on average the warmings do not extend west beyond the date line. Warm water was somewhat asymmetrically distributed about the equator with the center of the distribution being south of the equator near South America and north of the equator in the central ocean.

One year later the tremendous lens of warm water had been replaced by water that was 2–3 °C colder than the long-term mean. The general

distribution of the anomalous cold water is much the same as that of the anomalous warm water during the previous year. The inter-annual change of 8–9 °C in the central ocean is at least three times the seasonal signal.

A detailed analysis[30,34] of time histories of monthly SST anomalies from the equatorial zone shows that all stations along the South American coast between 5° N and at least 10° S are significantly correlated with each other. Highly significant correlations exist in the zonal direction from the South American coast (80° W) to Canton Island (172° W). Beyond Canton lies a 'quiet zone' wherein SST anomalies are not significantly related with fluctuations to the east (or anywhere

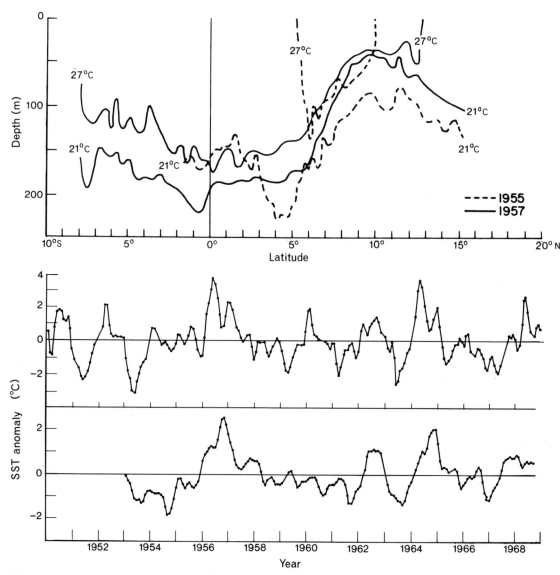

Fig. 10.3. Upper panel: meridional sections of temperature across the central equatorial Pacific during cold (1955) (– – –) and warm (1957) (——) events. Note in particular the distribution of the 27 °C isotherm. The data were taken from Austin.[39] Lower panel: time series of monthly sea surface temperature (SST) anomalies (°C) from Talara, Peru (upper graph) and Christmas Island (lower graph). Note that the low frequency components of these curves account for most of the observed variance.

else for that matter).[34] The correlation analysis thus suggests that SST anomalies in the zone 80–180° W and within at least ±10° of the equator fluctuate coherently. Thus the horizontal dimensions of the largest equatorial changes are some 2000 km by 10 000 km.

While the SST fluctuations are much discussed in the literature, the variable of most importance for present purposes is the change in oceanic heat content. Limited information is available on these changes during warm/cold epochs at the equator. But this evidence suggests drastic dissimilarities in oceanic thermal structure between the two situations.

The large contrast between epochs is illustrated in the upper panel of Fig. 10.3.[39] The data come from nearly identical bathythermograph sections taken in the central Pacific before and during the large, warm event of 1957–58. Similar sections at different longitudes during the 1972 event show the same general features (D. McLain, personal communication). In the earlier section the meridional gradients of the isotherm depths (considered as Margules surfaces) can be used to pick out the 'normal' equatorial surface currents as defined in most text books, that is, the eastward flowing North Equatorial Countercurrent between 4–10° N surrounded by the westward flowing South and North Equatorial Currents. The 'thermal equator', is associated with the North Equatorial Countercurrent and centered near latitude 8° N, its long-term mean position.

During the warm period of 1957 the thermal structure was greatly altered. The meridional gradients of isotherm depth suggest that *eastward* flow existed from 10° N to 10° S, that is, the South Equatorial Current, if it can still be called by that name, reversed its direction of flow. The thermal equator expanded in meridional extent and its centroid shifted some 700–800 km to the south, coinciding approximately with the geographic equator. The amount of heat contained in the upper 200 m increased tremendously. The size of the change and its depth of penetration both suggest that a simple argument about an altered rate of air/sea heat exchange and vertical mixing in the ocean cannot account for the observations. They must be due, in large part, to major internal adjustments in the oceanic density field.[40,41] The ocean is determining, to some extent, its own destiny.

There is reasonable evidence to suggest that similar changes occur over much of the coherent equatorial zone during warming events, particularly near the eastern boundary.[31] The resulting dislocation and strengthening of the thermal equator over some 10 000 km of the Earth's surface does have some very significant effects on the atmosphere, as we shall see later.

The time-scales which characterize the anomalies are described in two different ways. It is informative simply to look at a 20-yr time history of monthly anomalies from two representative stations in the equatorial coherent zone (Fig.

10.3, lower panel). Typical intervals between smaller events are in the order of 2–3 yr, with the larger events occurring perhaps once per decade. The anomaly signal seems noisier near South America than in the central ocean. This may be due to the fact that the mixed layer is shallower near the eastern margin than in the central ocean. So the 'thermal inertia' of the ocean is smaller near the continent.

Another measure of the characteristic time of the temperature changes may be obtained from the study of their autocorrelation functions.[30,34] In the equatorial coherent zone these functions resemble exponentially damped sinusoids. The first zero crossing of the correlations is about 8 month at South American and 18 month at Christmas Island, values which give time-scales in concert with the figures cited above.

Cross-correlation analysis between stations in the equatorial zone show[30,34] that the central ocean stations *lagged* those at the continent by about 6 month. Thus the major warming/cooling begins first at South America and 'progresses' westward. Its inferred speed of propagation, if in fact it is a propagating feature, is oceanic in nature and is 0.5–1.0 m s⁻¹, again suggesting internal interactions are the prime mechanisms for effecting the change in the oceanic thermal structure.[40,43]

Mid-latitude temperature changes

Large-scale SST anomalies, sometimes as large as 40 % of the normal seasonal signal, occur in the mid-latitudes of the northern hemisphere ocean. These fluctuations have been extensively documented in both the North Pacific and North Atlantic. Limited data from the Indian Ocean suggest that significant inter-annual variations occur there also. Documentation of similar occurrences in the southern hemisphere ocean is generally lacking.

The mid-latitude SST anomalies are thought to represent an abnormal heat source/sink for the general atmospheric circulation, thereby affecting its character. Although the anomalies are not as large in magnitude as their equatorial counterparts they cover huge areas of the ocean and have long lifetimes. Thus, their integral effect could be of importance to the short-term climate.

I shall concentrate on describing fluctuations in the North Pacific Ocean's temperature field. This restriction makes it possible to include some recent measurements revealing the sub-surface character of the SST anomalies and some results on air/sea coupling in this area. For those readers with a strong preference for the North Atlantic story, references 10, and 44–49, will provide an entreé to the many significant results that have been obtained for that region. These results are similar in many respects to those that will be shown for the North Pacific.

The horizontal extent of North Pacific SST anomalies are most easily described statistically since there is a wide range of variation in the patterns. One of the first such descriptions[50] simply took the SST anomaly at a base-point in the ocean and *contemporaneously* correlated it with surrounding data, the averaging being done over many years of data. The isolines of correlation coefficient thus describe the spatial coherence of the anomalies. Working from base-points located at positions of maximum variance in the anomaly field, the authors demonstrated regions of highly significant (1 %) correlation that measured roughly 3000 km in the zonal direction and 2000 km in meridionional extent. The results also suggest that one or two anomalies of opposite sign exist concurrently in the Pacific.

A more general description of the horizontal scales of the SST anomalies[51,52] and their frequency of occurrence is available through the use of empirical orthogonal functions (EOFs).[52,53] This analysis technique extracts from a data field the major coherent patterns beginning with the one that accounts for the most field variance and progressing to less and less significant patterns. The first three EOFs of the SST anomaly field, computed for all months of a 28-yr record, are shown in Fig. 10.4. These three functions together account on average for 43 % of the field variance. The principal feature appears in mid-ocean. After the first function, however, the patterns show an increasing preoccupation with the western boundary region.

The principal scale is confirmed to be of order 2000–3000 km. The amount of variance at the western boundary and the number of functions needed to account for it can be explained since this is the area where the Kuroshio and Oyashio Currents collide. This interaction and the different water types associated with these currents are known to produce highly complex, small-scale spatial patterns in the oceanic properties of that area.

In summary, the horizontal dimensions of SST anomalies in the North Pacific are comparable with the scale of major atmospheric features. The geographic distributions of variance and scale analysis suggest that the observed anomalies in the central ocean are of a different character to their counterparts to the west. Unfortunately there is not a clear spatial separation between the two types of anomaly. Thus, while low frequencies dominate the spatial field, the characteristics and dynamics of SST anomaly behavior may be strongly dependent on position.

Recent results have provided the first description of the structure *beneath* SST anomalies. This description is crucial to understanding their generation and destruction. Are they confined to the upper (mixed) layer of the ocean thereby suggesting generation by local atmospheric processes? . . . Or do they extend to depths greater than seasonal processes can penetrate thereby suggesting an oceanic origin? The following evidence shows that *both* possibilities seem likely.

Long-range US Naval aircraft have been used to monitor the thermal structure through the heart of the principal mid-ocean SST anomaly regions[24] as shown in Fig. 10.4. Preliminary estimates of the vertical structure of SST anomalies are illustrated in Fig. 10.5. These typical examples show (upper panel) a strong, cold anomaly located almost entirely above the mixed layer depth (about 90 m). One month later (middle panel) the anomaly has apparently weakened but now extends to depths greater than those to which the seasonal signal penetrates. The hydrostatic stability at the latitudes of the maximum vertical extent of the anomaly is known to be very low, suggesting that penetrative convection has caused the anomaly to deepen. A completely different (summer) situation is shown in the lower panel. Here a large, positive anomaly exists in the upper 100 m but is superficially disguised over much of its meridional extent by a thin surface layer of unusually cool water. This masking effect clearly demonstrates the problems associated with characterizing the upper ocean heat content during summer from SST alone.[54]

Another view of the vertical structure of anomalies has been developed[55] from long time histories of temperature data from weather ships NOVEMBER (30° N, 140° W) and PAPA (50° N, 145° W), positions away from the areas of

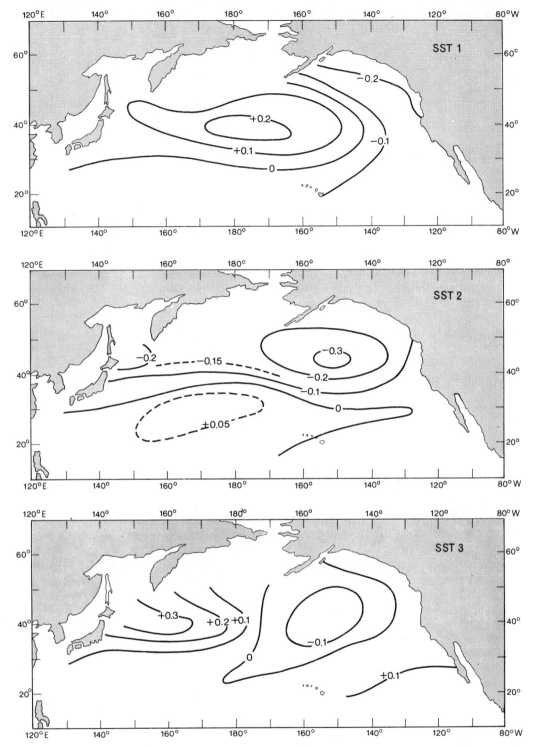

Fig. 10.4. First three empirical orthogonal functions of the sea surface temperature (SST) anomaly field in the North Pacific Ocean irrespective of season. Note the large (atmospheric) scale associated with the first pattern and the tendency for substantial variance along the western boundary of the ocean. After Barnett & Davis.[51]

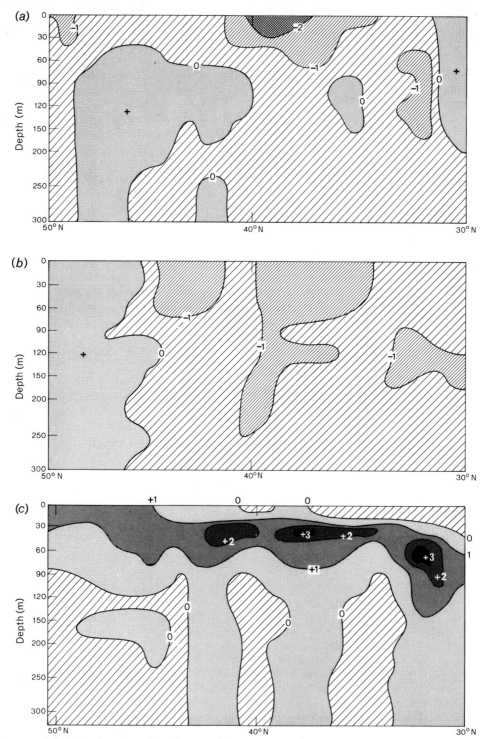

Fig. 10.5. Meridional sections through sea surface temperature anomalies. The data illustrate that the anomalies can be confined to the mixed layer or penetrate to a considerable depth in the ocean. Thus there appear to be different classes of anomalies (after Barnett).[24] (a) November 1974 along 158° W; (b) December 1974 along 158° W; (c) September 1975 along 158° W.

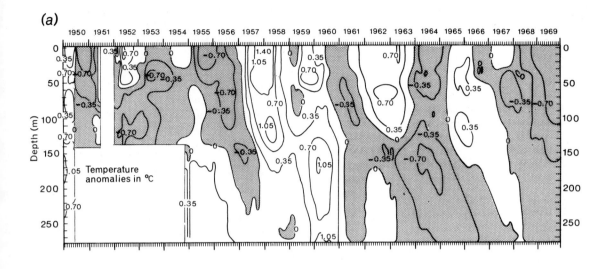

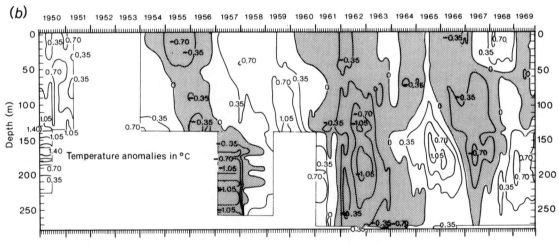

Fig. 10.6. Time series of water temperature anomalies (°C) at depth from Ocean Stations (*a*) PAPA and (*b*) NOVEMBER. The original data has been low pass filtered with a 12-month running mean. The figure further demonstrates that anomalies extend below the limit of seasonal penetration and have relatively long time-scales (after White & Walker[55]).

high SST variance. The resulting pictures (Fig. 10.6), which have been low pass filtered, show major features penetrating to depths well in excess of those associated with the seasonal signal. The anomalies thus seems well 'rooted' in the main oceanic structure. Note the tendency of the anomalies to 'originate' at the surface and appear

later at depth. A good explanation of this feature has not been offered.

The characteristic time-scales of SST anomalies have been computed in several ways. Estimates from the autocorrelation function[54] give scales in the range 3–24 months, depending on the type of function one assumes to fit the computations. The

167

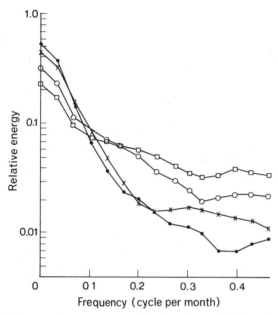

Fig. 10.7. Power spectra of selected amplitudes of empirical orthogonal functions of sea surface temperatures anomalies (see Fig. 10.4): n is the function mode number (after Barnett & Davis[51]). ●—● is $n = 1$, x — x is $n = 3$, o—o is $n = 7$ and □—□ is $n = 16$.

matter can be put in better perspective by investigating the power spectra (Fig. 10.7) of the amplitude functions that modulate the EOFs (Fig. 10.4) of the SST anomalies. These spectra are of the 'red' variety, and most of the energy is associated with frequencies below one cycle per year. These frequencies are not well resolved by the 28-yr record. There is a tendency for relative energy levels in the higher frequencies to rise as the order (n) of the EOF increases; that is, as smaller and smaller spatial scales are considered, the temporal variations approach those expected for 'white noise'.

The time variability of anomalies at depth is not so well known. But the weather ships' results, shown in Fig. 10.6, give a clue to at least one principal scale. The undulations of the deeper isotherms (and thus the time between major anomalies) are typified by intervals of around 3–6 yr. It is doubtful if these long periods are characteristic of the near-surface anomaly field but they may represent a significant contribution to the low frequency energy seen at the surface (Fig. 10.7).

Unfortunately, the problem of assigning a time-scale to mid-latitude anomalies is complicated in the upper layers of the ocean by the fact that the physical processes that dominate the heat budget are seasonally dependent. Thus the persistence, even the existence, of certain classes of anomalies, is a function of season. This point is brought home by inspection of a time-dependent correlation function[51,54] of the SST anomaly field. This calculation shows the extremely high persistence in the anomaly field at one to three-month time lags. It also clearly shows that an anomaly which exists in winter generally vanishes by the summer (the masking effect?), with a slight (not statistically significant) suggestion that it may arise the following winter. By contrast, anomalies present in the summer seem capable of persisting for 1–1.5 yr, although they may briefly 'wink out' during the subsequent summer.

In summary, mid-latitude SST anomalies and their sub-surface manifestation have time-scales between several months and several years. Thus, while the spectra of monthly values of, say, sea-level pressure are rather independent of frequency (=white), similar spectra of oceanic fluctuations are associated strongly with low frequencies (= red). There is clearly a mismatch in time-scales between the ocean and atmosphere at the lower frequencies. The oceanic features also show a seasonal dependence that complicates estimates of their general character.

Ocean/atmosphere coupling

My intent in this section is to demonstrate the close, perhaps symbiotic, relationship between climatic changes in the ocean and atmosphere. The previous examples of oceanic variability will form the basis of this section. However, I will concentrate here on the atmospheric changes that preceded and/or followed the oceanic changes. The key question of causality will not generally be addressed – indeed the answer is not known. The main point, for the present, is to show that ocean/atmosphere events are intimately coupled at climatic scales, emphasizing that major advances in understanding the mechanisms of climatic change will not occur without consideration of the oceans. This point is emphasized in other sections that describe numerical simulations of the atmosphere's response to oceanic variability.

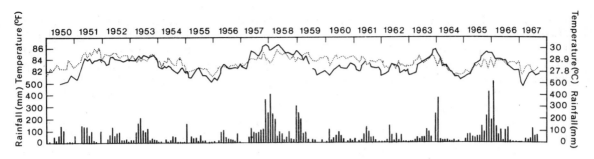

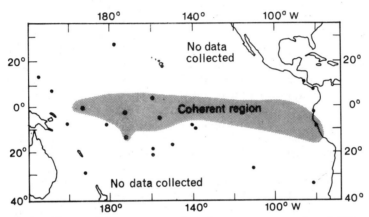

Fig. 10.8. Upper panel: time history of air temperature (.......), sea temperature (——) (°F) and rainfall (mm) at Canton Island (after Bjerknes[36]). Lower panel: regional distribution of coherent relationships between rainfall anomalies (after Dobritz[56]).

Equatorial interactions

It was shown earlier that large, anomalous changes in the SST and heat content of the equatorial Pacific Ocean occur over scales of thousands of kilometers and several years. These fluctuations are associated with massive changes in the equatorial atmosphere. Again, many examples are possible. The few that have been picked are felt to be of major importance in determining the features of the general circulation in the tropics, and perhaps elsewhere.

Equatorial rainfall/Intertropical Convergence Zone. The heat energy that fuels the tropical circulation comes largely from condensation of water vapor which is equated here to rainfall. This latter field has its maximum over much of the Pacific in the vicinity of the Intertropical Convergence Zone (ITCZ). So the amount and areal distribution of rainfall in the region is a gross measure of one of the principal forcing functions for the atmospheric system. How does the distribution and magnitude

of rainfall vary during warm/cold epochs in the equatorial ocean?

The anomalies of equatorial rainfall have been studied with cross-spectral analysis.[56] One of the prime results, which was only partially factual, was the inferred existence of a narrow strip extending along the equator from South America to 170° E wherein all the low frequency, interannual rainfall fluctuations are coherent (Fig. 10.8). More recent data from satellites indicates that rainfall at both ends of the strip may be coherent but that the center of the region may have a mind of its own (reference 32 and C. Ramage, personal communication). In the historical sense, this zone is referred to as the equatorial dry zone since it is usually an area of low rainfall. This zone is nearly identical with the coherent zone of equatorial SST anomaly (shown in Fig. 10.2). It has been suggested that this is no coincidence.[36,57]

Under normal conditions, the water in the

zone is colder than the air, a result of upwelling (Ekman divergence) at the equator and advection of cool water from the east. The moisture flux to the atmosphere is then inferred to be low. During warming sequences, the water temperature can exceed the air temperature. The moisture flux then greatly increases and so does rainfall.[36] This argument appears to be substantiated by the data from Canton Island (172° W) (Fig. 10.8). Since Canton lies in the coherent zone this could imply heavier rainfall throughout the whole zone. But there are several alternative hypotheses that can also be substantiated with the same data! Satellites show that the increased rainfall at Canton results from an eastward shift of some 2000–3000 km in the doldrum rainfall maximum (C. Ramage, personal communication). Further to the east the 'increase' of rainfall is apparently due to a southward shift of the ITCZ (see below). Thus there is no real increase in rainfall, rather just a geographical dislocation of where the rain falls (C. Ramage, personal communication). For present purposes either possibility is acceptable.

A major increase in rainfall along the equator suggests, as mentioned above, that the ITCZ will also be closer to the equator than normal ... particularly if, as it seems, the position of the oceanic thermal equator and the ITCZ are indeed congruent.[58] Recent analysis of observed winds in the tropical Pacific shows the latitudinal position of the ITCZ between longitudes 180 and 100° W is very well correlated with equatorial water temperatures in the central ocean ($r = 0.83$).[58] Thus during warm events the ITCZ over the central ocean is some 400–600 km south of its position during cold ocean events, thereby supporting the 'dislocation' hypothesis.

It is clear that major spatial changes in the equatorial rainfall pattern and ITCZ position accompany large, warm water events. Whether one is causing the other, or whether each is being caused by yet another mechanism, is a question to be debated elsewhere. For now we merely conclude that a major displacement of a principal atmospheric forcing function occurs over a significant fraction of the global circumference; in conjunction with major changes in ocean climate. The magnitude of the forcing may also increase if, indeed, more rainfall occurs during warm events.

The trade winds. There are massive differences in the trade wind field between warm and cold water epochs in the eastern/central ocean. This is illustrated (Fig. 10.9) by contrasting ship-reported wind observations[58,59] for January/February 1956 (cold) with those from January/February 1958 (warm). These times correspond to those of Fig. 10.3 which showed upper ocean heat content. The distribution of SST for these times was similar to that previously shown for the 1972–73 period (Fig. 10.2).

During the cold period the zonal winds (*easterlies*) are particularly strong near the date-line at and south of the equator. No clear equatorial minimum in the zonal component is evident in mid-Pacific. The tropical high-pressure ridge ($u = 0$) in the northern hemisphere is from 25–30° N. During the warm period, however, *westerlies* instead of easterlies prevail from the western Pacific out to the date-line along the equator. A clear minimum in zonal flow exists at the equator in mid-ocean. The northern hemisphere high-pressure ridge now runs from 20–25° N. Note that the zonal winds in the eastern ocean have changed little between epochs although the changes in water temperatures are maximum in that region.

The meridional component of the wind also changes dramatically. Northerly winds existed in the southwestern Pacific during the cold period although they reverse and become southerly below 10° S during the warm period. North of 10° S, air from the northern hemisphere is now crossing the equator where it did not before. The convergence at the ITCZ ($v = 0$) which was diffuse and far north during the cold period has shifted south by some 4–6° latitude and intensified significantly. The strong meridional component in the eastern Pacific weakens during the warm period near 135° W, suggesting the 'coherent' rainfall area may not be as depicted in Fig. 10.8. But the wind *increases by more than 50 %* off the coast of South America as does the strength of the convergence at the eastern end of the ITCZ. Also increased is the meridional component of the northeast trades across virtually the entire ocean.

Such large changes in trade wind field constitute major perturbations in the oceanic forcing function. So it is little wonder that large oceanic changes accompany the wind changes. By the

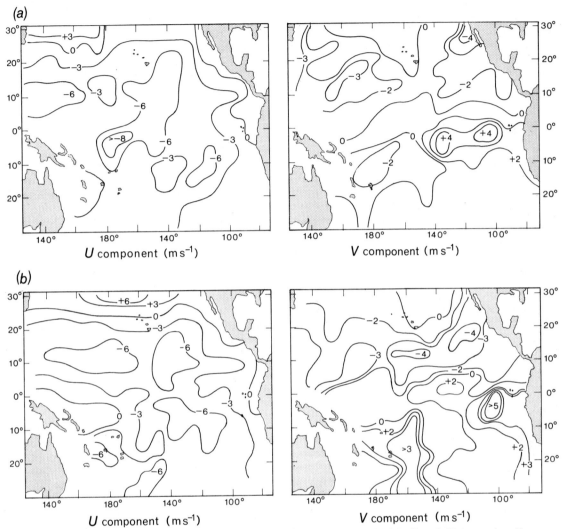

Fig. 10.9. Reconstruction of the trade wind fields during conditions of (*a*) cold (1956) and (*b*) warm (1958) water events in the equatorial Pacific. The illustrations show the *u* and *v* components of the wind (m s^{-1}) obtained from actual ships' observations during the months of January and February of the noted years. The major differences btween the warm/cold sequences exceed three standard deviations.

same token, the distribution of air/sea heat exchange must be drastically altered, if for no other reason than the root mean square (rms) wind speed has changed over much of the tropical ocean and the fluxes of sensible and latent heat flux are roughly proportional to wind speed (as well as SST). It seems clear that the fates of the trade wind systems and the equatorial oceans are inextricably linked.

The Hadley and Walker Cells. Bjerknes[36,57] has described how large-scale atmospheric circulation cells link various regions of the tropics and spread tropical effects to higher latitudes. The variability of these cells is hypothesized to be a major factor in short-term climatic change. The first system, which he called the 'Walker' cell, consists of a zonal circulation confined to equatorial regions and driven principally by the oceanic temperature gradient. In the Pacific, then, air flows westward from the colder, eastern area to the warm, western ocean where it acquires warmth and moisture, and subsequently rises. A return flow aloft and subsidence over the eastern ocean complete the cell. The second system, the 'Hadley' cell, carries momentum, sensible heat and potential energy from the tropics to the mid-latitudes. The northernward transport aloft is complemented by subsidence in the subtropical high-pressure ridge and a surface return flow. The surface manifestation of both of these hypothetical cells is, in large measure, the trade wind systems.

Bjerknes concluded that significant variations in the distribution of equatorial water temperatures would cause major changes in the large-scale circulation cells. The large changes in the trade wind system, discussed above, substantiate this early conclusion. Additional confirmation suggests that the extra-tropical effects envisioned by Bjerknes may also occur.[58,60-62] For instance, the atmospheric pressure field over the North Pacific Ocean changed markedly between the winters of 1956 and 1958. Statistical evidence suggests that the observed changes are strongly correlated with increased fluxes of heat and potential energy from the tropics. These latter quantities are in turn strongly correlated with the variations of ocean temperature in the eastern tropical Pacific.

Undoubtedly there is a complex interplay between extra-tropical and tropical systems. We know that both regions are highly variable at inter-annual time-scales, yet it is not clear at present whether both systems mutually interact or if the flow of 'information' is uni-directional. In any event, it is clear that the tropical ocean is involved in the communication link. Whether the ocean's role is that of the initiator, collaborator or recipient is to be discovered.

Mid-latitude interactions

Large inter-annual changes in SST and heat content of the mid-latitude oceans were described above. Although the anomalies were not as large in magnitude as their tropical counterparts, their space scale and time-scale were large enough to suggest that they could significantly influence atmospheric affairs. Sensible and latent heat flux are the mechanisms by which the perturbation is presumably affected. This subsection summarizes past work which demonstrates that a close coupling between mid-latitude atmosphere and ocean does exist. It also shows that the oceanic field can be used to 'predict' some aspects of atmospheric field, as has been suggested by Namias and others. This discussion is also relevant to the examples of oceanic variability in the Pacific given previously, and several new, exciting results are available for this area. Otherwise, it would have been just as easy to make a case using examples from the North Atlantic Sector.[44, 63-65]

If the ocean-to-atmosphere flow of heat is strong then one might expect a high correlation between anomalies of SST and the heat content of the atmosphere's lower layers. On the other hand, if the flow of heat is low or nil – if the air is warmer than the water – then there should be little communication between the media. So the heat fluxes should depend partly upon the static stability of the air, and the heat content of the planetary boundary layer on the ocean temperature.

The degree to which these ideas are operative has been investigated by Namias.[66] Utilizing a 25-yr time history of monthly SST and 1000–700 mbar-thickness anomalies over the North Pacific, he obtained *contemporaneous* point value correlations between the two variables on a 5° grid. The spatial distribution of the correlation show clearly that significant relations exist during all seasons (Fig. 10.10).

The high correlations existing south of the Aleutians imply that heat is really provided to the lower atmosphere by virtue of the fact that normally cold continental air masses are continually fed into this area, particularly during the cold season. The region beneath the northeast trades is also a continual heat source. This results from the flow of cool (northern) air over warm surface water. The zone of minimum correlation lies in the northern flank of the Pacific high-pressure ridge. This is an area of subsidence whose stabilizing effects are enhanced by the surface flow of warm air (from the south) over the colder water.

These results seem to demonstrate that inter-annual variations in SST are related to the heat content in the planetary boundary layer and indicate the regions where the relationship is strongest. In Namias' words, 'These findings suggest the strong influence of air mass modification by air advection over water masses of different temperature and point to varying atmospheric

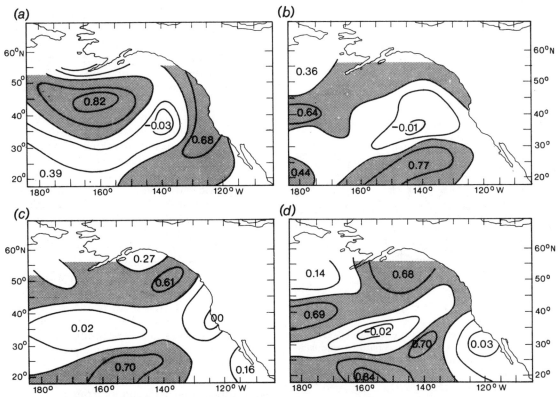

Fig. 10.10. Isopleths of correlation between the anomalies of sea surface temperature and 1000–700-mbar thickness for the four seasons (*a*) winter, (*b*) spring, (*c*) summer and (*d*) autumn over the period 1947–71. The isopleths are drawn for each 0.2 with centers indicated by the numbers. The correlations in the shaded areas exceed the 5 % level of significance (after Namias[66]).

stability as a principle component of large-scale heat exchange.'

Some recent model results suggest the mid-latitude ocean is just a slave to the atmosphere and that SST anomalies are merely manifestations of atmospheric anomalies, the ocean-to-atmosphere feedback being nil.[15,67] At the other extreme are model results that suggest the oceans dictate in large measure the behavior of the atmosphere.[68–70] The truth may include both possibilities. In fact, both points of view can be illustrated using actual data, as opposed to the synthetic model results cited above. The reader is cautioned that the results to be discussed are quite recent and, therefore, not sufficiently reviewed by the scientific community. Also, the results were obtained by advanced statistical methods that need adequate interpretation themselves before the full import of the answers can be established in the reader's mind.

The inter-annual variations in the SST and sea-level pressure (SLP) field over the North Pacific has been examined to see if one of the variables can be predicted from the other.[52] Twenty-eight years of monthly data on these fields was represented with a set of empirical orthogonal functions. These functions were used as linear estimators to determine, among other things, the amount of variance in the SLP field that could be accounted for with the SST field. This relative

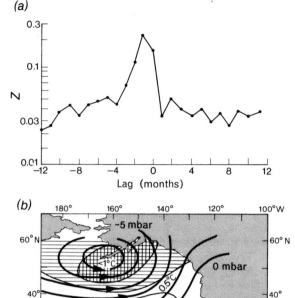

(a)

(b)

Fig. 10.11. (*a*) skill of estimating sea-level pressure in month *t* + lag using as data 10 sea surface temperature modes from month *t*. Skill index Z measures the fraction of the entire spatial field correctly estimated. The 'artificial' skill expected if there is no true skill is = 0.04 (after Davis[52]). (*b*) prognostic construction of a typical autumn SLP anomaly field from the SST anomaly field of the *anticedent* summer. The prognostic technique accounts for 26 % of the variance in the autumn SLP field over the 28-yr record (R. Davis, unpublished).

amount of variance was taken as a measure of the skill in the forecast (actually a hindcast) with 100 % being a perfect reproduction. The effects or 'artificial predictability' were accounted for in the analysis.[52,53]

The significant result is displayed in the upper panel of Fig. 10.11. The inability to predict future SLP variability from the current SST field is clear. Most striking is the positive ability to predict *past* SLP variations from *present* SST data. This suggests that, in the region studied and on the time-scales of a month to a year, the observed

connection between SST and SLP variations is the result of the atmosphere driving the ocean. This conforms to standard oceanographic dogma and is no surprise in itself. But the reverse interaction (ocean-to-atmosphere), which could not be detected must be very weak and perhaps not relevant to the SLP field.[67] Note also that the amount of variance accounted for is relatively low. Other unknown factors must be responsible for much of the observed variation.

Before these conclusions are readily accepted it must be noted that the entire ocean area was included in the SST (and SLP) descriptors. The large variance in the western part of the ocean, due presumably to the interaction of major current systems, was included in the SST field, although it presumably has little to do with local atmospheric variations. Also, no seasonal discrimination is made in the model, although it is well established that the fluxes and other processes affecting the SST and SLP field are quite seasonably dependent. This implies that the processes causing the anomalies are also seasonally dependent. The matter of seasonality was studied by developing separate prediction equations for the cold and warm seasons of the year.[52] The EOFs for the entire Pacific were still used to describe the SST field. The results were largely the same as noted above: the atmosphere appears to be forcing the ocean.

Other investigations[71] of seasonal predictability have been made, limiting the study area to the northeast Pacific, thereby excising variance associated with western boundary currents.* This region has appeared in earlier illustrations (Fig. 10.10) to be one of maximum air/sea interaction. One of the most interesting results to come from the studies was the demonstration of a significant hindcast skill (26 %) in predicting *autumn* SLP anomalies from *summer* SST anomaly patterns.

A typical field reconstruction of this finding is shown in the lower panel of Fig. 10.11. Warm water (1 °C) existing in the Gulf of Alaska during the summer leads to a −5-mbar pressure anomaly in the autumn. This success has been attributed[71] to the fact that the first autumn storms will bring cold air from the continent out over the warmer than usual water. The increased cyclogenesis resulting from unusual transfers of sensible

*R. Davis has kindly made his unpublished results available for this article, including Fig. 10.11.

and latent heat will result in lower than usual pressures. It is interesting that the increased storm activity will destroy the SST anomaly which created it in the first place. The same processes work in reverse during the existence of a cold, summertime SST anomaly. It thus seems that there is sound physical reasoning and statistical evidence to support the idea that the ocean does effect the atmosphere.

The results of this section could be viewed as contradictory. However, they actually only infer a causal connection since they are statistical in nature. Furthermore, each of the studies concentrated on a different set of space/time scales. So no real dichotomy need exist. In any event, the results show a significant time-lagged relationship between SST and SLP on climatic scales. The physics underlying the relationships have yet to be fully demonstrated.

The present state of the art and future prospects

The oceans and atmosphere are two closely coupled systems that partially depend on each other for sustenance. The forcing functions that drive one or the other system are non-homogeneous, strongly time-dependent and characterized by a large mismatch in time-scales between media. Furthermore, strong internal interactions shape to some extent the degree of external coupling between the systems. It is also true that the mechanics of external coupling are either poorly parameterized and/or difficult to evaluate accurately. It is little wonder that adequate models of the ocean/atmosphere system are not at hand. Fortunately, much attention is at present being directed toward the difficulties mentioned above. The future in this area of research holds much hope for substantial advancement.

The past decade has seen a superexponential increase in our descriptive knowledge of climatic changes occurring in the oceans. Much of the description rests on relatively crude and/or limited data. Nonetheless, the resulting picture of inter-annual change shows the oceans to be highly variable in many of their principal properties. In this article, I have concentrated on the changes in SST and near-surface heat content since these variables relate to the air/sea heat exchange (external interactions). But other oceanic variables (salinity, current systems, sea-level and so on) related to internal interactions show equally impressive changes. Both the tropical and mid-

latitude oceans are not quiescent at climatic timescales; however, good explanations for these variations are at present unavailable. The oceanographers will, hopefully, be filling this void, as well as bettering the description of oceanic climate change, over the next decade. Given the present level of support available and the fractious nature of the current oceanographic community, these expected advances may not occur with great rapidity.

The changes in ocean climate were shown to be closely linked with large-scale atmospheric changes. Both media seem capable of inducing changes in the other. Therefore, it does not seem profitable to champion one medium and argue causality from our present limited base of data and theory. Rather, our resources might be better spent understanding the mechanics of the *coupled* system. Perhaps this is an unreasonable request in view of the facts that (1) the educational systems do not now train scientists equally adept at meteorology and oceanography, (2) the levels of funding and organization between the two study areas are as disparate as the media density and (3) the levels of modelling, observation base and observational programs between the two disciplines are at very different levels of sophistication. Still, as has been shown here, the climate problem requires the joint efforts of both oceanographers and meteorologists. This cooperation may be one of the more substantial benefits derived from studying the 'climate problem'.

I am indebted to Will Kellogg for suggesting the possibility of this article and for commenting on a first draft. J. Namias and J. Barnett also provided helpful comments. G. Johnston faithfully decoded the numerous manuscript revisions while Fred Crowe did an excellent job of developing the art work. This work was carried out under the auspices of the Climate Research Group, Scripps Institution of Oceanography.

References

The references cited in the text either provide an overview of work in a particular area or represent some of the most recent contributions in the area. They thus serve as starting points from which the interested reader can delve into the rich history of climatic studies. To cite all of the interesting and important work in an article of this size would be impossible.

1. Teisserenc de Bort, L. (1883). *French Central Meteorological Bureau, Annales,* **1881, 4,** 1762.
2. Helland-Hansen, B. & Nanson, F. (1920). *Miscellaneous collections, Smithson. Ins. Publ.* **2537, 80,** 408.
3. Brooks, C. E. P. & Glasspoole, J. (1922). *Q. Jl R. met. Soc.,* **48,** 139–66.
4. Berlage, H. P., Jr (1933). *Met. Z.,* **50,** 41–7.
5. Walker, G. (1923). *Mem. India met. Dept.,* **24,** 75–131.
6. Namias, J. (1972). In *The changing chemistry of the oceans,* Nobel Symposium 20, ed. D. Dryssen & D. Jagner, pp. 27–48. Wylie Interscience Division, also Almqvist & Wiksell, Stockholm.
7. Namias, J. (1959). *J. geophys. Res.,* **24**(6), 631–46.
8. Namias, J. (1969). *Deep Sea Res.,* **16**(Suppl.), 153–64.
9. Mitchell, J. M. (1963). In *Changes of climate,* Arid Zone Research Series, 20, pp. 161–81. UNESCO, Paris.
10. Lamb, H. H. (1972). *Climate: present, past and future.* Methuen, London. 602 pp.
11. Flohn, H. (1971). *Bonn. Meteorol. Abh.,* **15,** 1–49.
12. Anon (1975). *Numerical models of ocean circulation.* National Academy of Sciences, Washington D.C.
13. Somerville, R. (1976). Proceedings of the NATO Advance Study Institute on modeling and prediction of the upper layers of the oceans. Also National Center for Atmospheric Research, MS. No. 0501–76/2.
14. Manabe, S., Bryan, K. & Spelman, M. (1975). *J. phys. Oceanogr.,* **5,** 3–29.
15. Chervin, R., Washington, W. & Schneider, S. (1976). *J. atmos. Sci.,* **33**(3), 413–23.
16. Niiler, P. P. (1976). In *The sea,* vol. 6, ed. E. Golberg *et al.,* pp. 97–115. Interscience, New York.
17. Hellerman, S. (1968). *Mon. Weath. Rev.,* **96,** 67–74.
18. Evenson, A. J. & Veronis, G. (1975). *J. mar Res.,* **33**(Suppl.), 131–44.
19. Budyko, M. (1955). *Heat balance atlas.* Clearing House for Federal Scientific and Technical Information, US Dept. Commerce, No. TT63–13243.
20. Hasselman, K. & Frankignoul, C. (1975). *International Union of Geodesy and Geophysics* XVL *General Assembly Abstracts,* I. S. 19, p. 161. IUGG, Grenoble.
21. Budyko, M. (1969). *Tellus,* **21,** 611–19.
22. Schneider, S. & Dickinson, R. E. (1974). *Rev. Geophys. Space Phys.,* **12,** 447–93.
23. Friehe, C. & Schmitt, K. F. (1976). *J. phys. Oceanogr.,* **6**(6), 801–9.
24. Barnett, T. P. (1976). *Naval Res. Rev.,* **29**(3), 36–51.
25. DeLeonibus, P. (1971). *J. geophys. Res.,* **75**(27), 6506–27.
26. Hasselman, K. H. *et al.* (1973). *Ergänzungsheft zur Deutschen Hydrographischen* Zeitschrift, Reihe *A* (8°), **12.**
27. NORPAX, POLYMODE, JASON and other subprograms of the Global Atmospheric Research Program. NORPAX, Scripps Institute of Oceanography, La Jolla California, USA; POLYMODE, Woods Hole Oceanographical Institute, Woods Hole, Massachusetts, USA; JASON, Institute of Oceanographic Sciences, Natural Environment Research Council, Wormley, Godalming, Surrey, England.
28. Bjerknes, J. (1966). *Bull. inter-Am. trop. Tuna Commn,* **12,** 3–62.
29. Miller, F. & Laurs, M. (1975). *Bull. inter-Am. trop. Tuna Commn,* **16**(5), 403–17.
30. Hickey, B. (1975). *J. phys. Oceanogr.,* **5,** 460–75.
31. Wyrtki, K. (1975). *J. phys. Oceanogr.,* **5**(4), 572–84.
32. Ramage, C. (1975). *Bull. Am. met. Soc.,* **56**(2), 234–42.
33. Namias, J. (1976). *J. phys. Oceanogr.,* **6**(2), 130–8.
34. Barnett, T. P. (1977). *J. phys. Oceanogr.* (In Press.)
35. Namias, J. (1963). *Mon. Weath. Rev.,* **91**(10, 12), 482–6.
36. Bjerknes, J. (1966). *Tellus,* **4,** 820–9.
37. Barnett, T. P. & Isaacs, J. D. (1975). Proceedings of the American Association for the Advancement of Science, Pacific Coast Environment Meeting, San Francisco, 1975.
38. *Fishing Information,* National Marine Fisheries Service, Southwest Fisheries Center, La Jolla, California, USA.
39. Austin, T. (1960). In *Symposium on the changing Pacific Ocean, 1957 and 1958, California Cooperative Fisheries Investigation,* **7,** 52–5.
40. Hurlburt, H., Kindle, J. & O'Brien, J. (1976). *J. phys. Oceanogr.,* **6**(5), 621–31.
41. McCreary, J. (1976). *J. phys. Oceanogr.,* **6**(5), 634–45.
42. Dickson, R. R. & Lamb, H. H. (1971). *International Commission for Northwest Atlantic Fisheries Environmental Symposium Contribution* No. 1.
43. Rodewald, M. (1971). *International Commission for Northwest Atlantic Fisheries Environmental Symposium Contribution* No. 1.
44. Neumann, G., Fisher, E., Pandolfo, J. & Pierson, W. J., Jr (1958). *Final report AFCRCTR–58–236, (ASTIA–AD–15255),* AF 19(604)–1284. New York University, Dept. Met. and Oceanogr. 78 pp.
45. Timonov, V., Smirnova, A. & Nepop, K. (1970). *Oceanology,* **10**(5), 586–9.

46. Valerianova, M. & Seryakov, Y. E. (1970). *Oceanology*, **10**(5), 590–5.

47. Schell, I. (1970). *Mon. Weath. Rev.*, **98**(11), 833–50.

48. Smed, J. (1965). *International Commission for Northwest Atlantic Fisheries, Spec. Publ.* No. **6**, 821–5.

49. Ratcliff, R. (1971). *Met. Mag.*, **100**, 225–32.

50. Namias, J. (1972). *Fishery Bull.*, **70**(3), 611–17.

51. Barnett, T. P. & Davis, R. E. (1975). *Proceedings of the symposium on long-term climatic fluctuations, WMO Tech. Note* No. 421, p. 439, WMO, Geneva.

52. Davis, R. (1976). *J. phys. Oceanogr.*, **6**(3), 246–66.

53. Lorenz, E. (1958). Report No. 1, Statistical Forecasting Project, Dept. Met., Massachusetts Institute of Technology.

54. Namias, J. & Born, R. (1970). *J. geophys. Res.*, **75**(30), 5952–5.

55. White, W. B. & Walker, A. E. (1974). *J. geophys. Res.*, **79**(30), 4517–22.

56. Dobritz, R. (1969). *Bonn. Meteorol. Abh.*, **11**, 1–42.

57. Bjerknes, J. (1969). *Mon Weath. Rev.*, **97**(3), 163–72.

58. Barnett, T. P. (1977). *J. atmos. Sci.*, **34**(2), 221–36.

59. Wyrtki, K. & Meyers, G. (1976). *J. appl. Met.*, **15**(7), 698–704. Also Hawaiian Institute of Geophysics Report HIG–75–2.

60. Rowntree, P. (1972). *Q. Jl R. met. Soc.*, **98**(416), 290–321.

61. Kreuger, A. & Gray, T., Jr (1969). *Mon. Weath. Rev.*, **97**(10), 700–10.

62. Barnett, T. P. (1974). *Report of the proceedings of the IUGG/IAMAP first special assembly*, Melbourne, Proces-Verbaux No. 13, p. 182.

63. Rodewald, M. (1963). In *Changes of climate*, Arid Zone Research Series, 20, pp. 97–107. UNESCO. Paris.

64. Lee, A. (1967). *Red book*, part 4. Issued from the Headquarters of the International Commission for Northwest Atlantic Fisheries, Dartmouth, Nova Scotia, Canada.

65. Murray, R. & Ratcliff, R. (1969). *Met. Mag.*, **98**, 201–19.

66. Namias, J. (1973). *J. phys. Oceanogr.*, **3**(4), 373–8.

67. Salmon, R. & Hendershott, M. (1977). *Tellus.* (In Press.)

68. Houghton, D., Kutzbach, J., McClintock, M. & Suchman, D. (1974). *J. atmos. Sci.*, **31**, 857–68.

69. Simpson, R. & Downey, W. (1975). *Q. Jl R. met. Soc.*, **101**, 847–67.

70. Huang, J. (1975). *Proceedings of the WMO/IAMAP symposium on long-term climatic fluctuations, WMO Tech. Note* No. 421, pp. 399–401. WMO, Geneva.

71. Namias, J. (1976). *Pres. second conference on ocean/atmosphere interactions.* American Meteorological Society, Seattle, Washington D.C.

11

The use of numerical models in studying climatic change

JILL WILLIAMS

As described by the Global Atmospheric Research Programme (GARP),[1] the processes of the climate system may be expressed in terms of a set of dynamical and thermodynamical equations for the atmosphere, oceans and ice, together with appropriate equations of state and conservation laws for selected constituents (such as water, carbon dioxide and ozone in the air). The equations describe the processes which determine changes in temperature, velocity, density and pressure. Also, processes such as condensation, precipitation and radiation are treated. These equations can be used to model climate, but various physical and numerical approximations must be made because of the lack of detailed knowledge of both the observed climate and the methods of computing the processes.

A hierarchy of climate models therefore exists, each model involving different approximations and thus simulating processes on a particular time-scale or spatial scale. The models have several applications. The atmosphere, hydrosphere and cryosphere are considered as the internal climate system and the Earth's crust and the space surrounding the Earth as the external system. Thus, models can be used to investigate the sensitivity of the climate system to external changes, for example, a change in the amount of solar radiation.[3-7] Likewise, the models can be used to study the sensitivity of the climate system to internal changes, for example, to an atmospheric carbon dioxide level twice that of the present[8] and to sea surface temperature (SST) anomalies.[9,10] The models can also be used to investigate specific climatic phenomena, such as monsoon circulations,[11,12] and in particular to see how these phenomena depend on boundary conditions. Lastly, one can use models to begin to assess the degree to which climate may be predictable on various time-scales and space scales.[1]

In this chapter experiments with two kinds of climate model will be considered. The first type of model is called a general circulation model (GCM) which is from the more detailed end of the climate model hierarchy. A GCM treats the equations governing the dynamics and thermo-hydrodynamics of the atmosphere for a finite number of grid points, and by integrating the equations forward in time from a given initial condition, a model climate is determined after a period of 40 or more days. The other type of model which will be considered is an energy balance equilibrium model, from the less detailed end of the hierarchy, in which atmospheric motion and state is determined as a steady-state solution of equations governing the energy balance of the atmosphere.

Schneider & Dickinson[2] have given a detailed description of climate modelling, stressing the physical basis of each kind of model and its contribution to the understanding of the climate system. In a report on an international study conference[1] which discussed the physical basis of climate and climate modelling, the design and use of climate models are discussed in detail. Another comprehensive review of the physical basis of climate and the development and use of climate models is given by the US Committee for GARP.[13] Others[14,15] have reviewed observational and theoretical studies of the general circulation of the atmosphere and discuss the use of numerical models of climate. The use of GCMs in paleoclimatic reconstruction has been described by Barry.[16]

Present climate models have several recognized shortcomings. For example, almost all models assume that the ocean surface temperature is fixed at the appropriate climatological value, rather than involving a joint atmosphere–ocean model. The latter[17–19] are at an early stage of development. Arakawa[14] suggests that the parameterization* of the time-dependent clouds and their interactions with other processes, radiation processes in particular, is perhaps the weakest aspect of existing GCMs. This aspect has also been discussed elsewhere.[20] Other weaknesses of such models are in the treatment of subgrid-scale motions, parameterization of vertical transports of momentum by orographic and other subgrid-scale gravity waves and the poor response to steep mountains. Further shortcomings are found in the treatment of the hydrological cycle.

In spite of the acknowledged shortcomings, however, it is still considered that climate models represent the best, and possibly the only, approach to understanding climate sensitivity. In the rest of this chapter emphasis will be placed on the use of climate models from the more detailed end of the hierarchy, to investigate the response to glacial period conditions, as an example of how we can get as much information as possible out of the models, to aid in our understanding of climate.

Use of models to investigate the response to glacial period boundary conditions

Over the past few million years the Earth's climate has alternated aperiodically between glacial periods, in which ice sheets expanded over North America, Scandinavia and elsewhere in the globe, and interglacial periods, such as the present day. The most recent glacial period reached its maximum 18 000–22 000 yr ago, at which time the ice extended over presently unglaciated land as illustrated in Fig. 11.1. Conditions at the maximum of that glacial period can be deduced from palaeobiological evidence (see Section 1 of this

*Many models at the more detailed end of the hierarchy solve the equations for a grid of points. The National Center for Atmospheric Research (NCAR) model, for instance, uses a grid of 5° in latitude and longitude for the horizontal plane and has six layers of 3 km in the vertical plane. Atmospheric phenomena with scales near to or less than the gridlength are usually referred to as subgrid scale. These small-scale phenomena can have a significant influence on the atmospheric general circulation. An attempt must therefore be made to include the effects of these subgrid-scale processes and this is done by approximating or 'parameterizing' the statistical effects of these subgrid processes in terms of the large-scale parameters.[21]

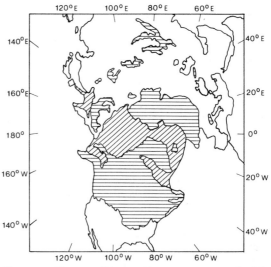

Fig. 11.1. Extent of ice at the most recent glacial maximum (shaded area) (about 18 000 yr ago). Adapted from reference 13.

volume and Flint[22]). The most recent compilation of evidence for conditions 18 000 yr ago is described by the CLIMAP project members[23] and will be discussed in more detail later.

Before the development of numerical models of climate, reconstructions of the climate during glacial periods were based either on analogue techniques, using extreme events at the present day as representative of past conditions, or on diagnostic relationships between conditions at the Earth's surface and the atmospheric circulation.[16]

The availability of physical, numerical models of climate introduced a new tool for palaeoclimatic reconstruction. The models used so far for investigation of glacial period conditions are those from the more detailed end of the hierarchy of climate models and they are not concerned with the problem of long-term climatic change. Instead, they are used to investigate the response of the atmospheric circulation, with no feedback to the ocean, to certain boundary conditions, in this case glacial period conditions. We are therefore solving the equations which govern atmospheric processes to find a climate in equilibrium with a certain set of boundary conditions. The results of such model experiments can tell us nothing about the origin or development of the glacial period.

The experiments with the climate models follow a general pattern. The first work involves collection of boundary conditions appropriate to the latest glacial maximum. The boundary conditions required are usually the extent and height of ice sheets and pack-ice, continental geography, the albedo of land surfaces (vegetation and snow cover) and the distribution of ocean surface temperatures. The model is run first of all to simulate the present-day climate using present-day boundary conditions, the results usually being expressed as 30-day means of meteorological fields for the month (usually January or July) which is being modelled. The latter results can then be compared with those from an equivalent run using glacial period boundary conditions. The problem of determining whether the difference between the climatological statistics of a present-day (or control) case and a glacial period (or perturbation) case is a significant one is treated in Chapter 12 of this book.

Models used to simulate present-day and glacial period climate

Five sets of experiments using different models have been performed. Four of these will be compared here and the fifth, which was made by Manabe, is only reported in brief.[24]

The first experiments using a numerical model to study the atmospheric circulation during the maximum of the latest glacial period were made by Alyea.[25] Alyea developed a quasi-geostrophic, two-level, spectral model of the northern hemisphere circulation. The author does not compare his model in any detail with present-day observed climate, although the model's ability to simulate the northern hemisphere summer monsoon has been evaluated.[26] Alyea performed two experiments, one using present-day and the other glacial period boundary conditions. Each simulation was integrated forward in time for 60 days using mean July initial states. The 30-day period between days 15 and 45 was used for representing the 'climate' of each simulation.

Williams, Barry & Washington[27] and Williams[28] used the NCAR GCM, which has been described in detail elsewhere.[29,30] The model has a horizontal resolution of 5° in latitude and longitude and the vertical coordinate is height, with

the atmosphere divided into six layers of 3 km each.

Saltzman & Vernekar[31] used a zonally-averaged, steady-state model of northern hemisphere macroclimate.[32,33] This model differs from the others mentioned here in that it considers a latitudinally varying but longitudinally homogeneous surface, which is described in terms of the proportions of land, water and ice along a latitude circle. The solutions of this model therefore only represent the zonal average climatic state and cannot be identified with the local climate at a point. Another difference results from the use of a vertical average rather than values of meteorological variables at different levels in the atmosphere. The model considers only northern hemisphere conditions and the authors present steady-state solutions for winter (October–March) and summer (April–September).[31]

Gates[34] uses the RAND version of the two-level Mintz–Arakawa model, which has a global grid of 4° in the latitude and 5° in longitude. The top of the model is at 200 mbar, with the atmosphere divided into two layers of equal mass. The model climatology for January has been described by Gates.[35] The simulations were started from realistic initial conditions appropriate to 1 May and the numerical solutions were carried out for 92 days corresponding to the months of May, June and July; the July climate was then determined by time-averaging the solutions over the last 31 days of the integration.

Glacial period and present-day boundary conditions

Each of these models has been used to simulate the atmospheric circulation of the present day and of the most recent glacial maximum for July and two of the models have also simulated January conditions. All climate models require certain boundary conditions to be specified at the beginning of a run. These boundary conditions describe, in the case of the models discussed here, the orography of land surfaces, the albedo of land (and in some cases, sea) surfaces and the SST distribution. For the simulation of the present-day atmospheric circulation the boundary conditions are derived from recent observations published in several sources and referred to by the different authors. For the glacial period simulations the boundary conditions must be based on geological and palaeobiological evidence for the conditions at the Earth's surface at the maximum of the last glacial period. Each glacial period simulation has used different compilations of boundary conditions.

Alyea[25] prepared the boundary conditions for a July glacial period simulation at a time when abundant published information was not available. He used maps of snow and ice extent[36,37] and information on the orography within the glaciated areas[38] which have now been improved upon. SST data for the glacial maximum were necessarily crude; the subtropical Atlantic and Caribbean were taken to be 5 °C lower than present July, the equatorial Pacific and Indian Oceans were 3 °C lower and there was a linear temperature depression from these areas northwards to the ice edge where a temperature of 4 °C was assumed.

The boundary conditions used in the glacial period simulations with the NCAR GCM are described by Williams *et al.*[27] and Barry.[39] The sources of information on ice cover and vegetation used in the glacial period boundary conditions are given in Williams *et al.*[27] Of particular interest, however, is the distribution of glacial period SSTs used in the simulations.[27] Isotherms for the North Atlantic were based on coccilith data from ocean-bed cores.[40] Ocean surface temperatures elsewhere were estimated on the basis of information given by Emiliani.[41] So, apart from the Atlantic Ocean, for which isotherms for January glacial period temperatures were given in the literature, the glacial period ocean temperatures were derived by subtracting 3–6 °C from present-day temperatures. The exact amounts and regions for which they were used have been illustrated by Barry.[39]

Saltzman & Vernekar[31] derived the percentage of each latitude band (of the northern hemisphere) covered by ocean, sea-ice, snow and land-ice at the last glacial maximum from Flint.[22] The ocean surface temperatures (zonally-averaged) for the glacial maximum simulation were derived from Williams *et al.*[27]

Gates[34] and Manabe[24] have used the most recent compilation of boundary conditions at the maximum of the latest glacial period.[23] Using data from cores of sea-floor sediments the CLIMAP group have mapped SST at 18 000 yr ago in greater detail than previously possible. The extent and thickness of continental ice sheets and the distribution of vegetation have also been mapped by CLIMAP project members and used as boundary conditions by Gates & Manabe.

There are substantial differences between the ocean surface temperatures used by Williams *et al.*[27] and Gates[34] for the glacial period simulation. Since it is known that SST anomalies can influence the simulated atmospheric circulation, [9,10,42–45] it is of interest to look at these differences. Fig. 11.2 shows the SST profile along 30° W for the present day and glacial period maximum as used by Williams *et al.* and Gates. The CLIMAP ocean surface temperature data suggest that the glacial period temperatures were the same as those for the present day in some locations and along 30° W this is seen between 20–30° N and 30–35° S (Fig. 11.2*b*). Apart from a small area south of 40° S, where present ocean surface temperatures are close to freezing, the SST values used for the glacial maximum along 30° W by Williams *et al.* were consistently lower than those for the present day. At the equator, the glacial period temperature used by Williams *et al.* was < 22° C, while that used by Gates was > 24° C. At 30° N the difference between glacial period and present-day SSTs was about 6° C in Williams *et al.* and about 0 °C in Gates. Between 50° N and 30° N the glacial period temperature change was approximately 9 °C in Williams *et al.* and 15 °C in Gates. Such differences are large and could account for equally large differences in the climatology of the glacial period simulations.

Fig. 11.3 shows the glacial period SSTs used in the above two glacial period simulations in the Pacific Ocean along 150° W. In general, the SSTs used by Williams *et al.* are 2–3 °C cooler than those used by Gates, and the gradients are quite similar. In the equatorial area, however, the SSTs of Williams *et al.* are greater than those of Gates, which have a minimum at about 5° S. The minimum marks the area indicated by CLIMAP to have been 6 °C lower than present SST.[23]

Some results of the experiments

Each model has been used to simulate the atmospheric circulation or climate of the present day; I shall describe here in detail the simulation of the glacial period maximum conditions. Alyea[25] shows how his model simulates the present-day atmosphere. Several studies have compared the results from the NCAR GCM with observed data. [27,28,30] Saltzman & Vernekar[32] evaluate the simulation of the present macroclimate with their

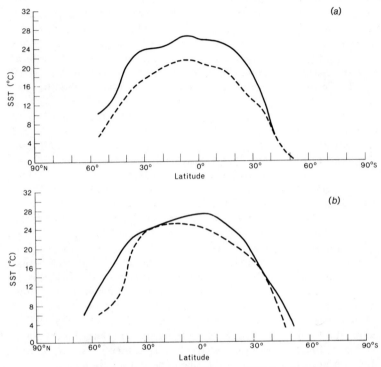

Fig. 11.2. Sea surface temperatures (SST) at 30° W for the glacial period maximum and present day for July simulations of (a) Williams et al.[27] and (b) Gates.[34] '———', present day; '– – – –', glacial period.

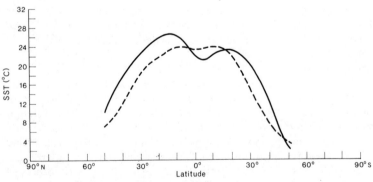

Fig. 11.3. Sea surface temperatures (SST) at 150° W for the glacial period maximum July simulations of Williams et al.[27] (– – – –) and Gates[34] (———).

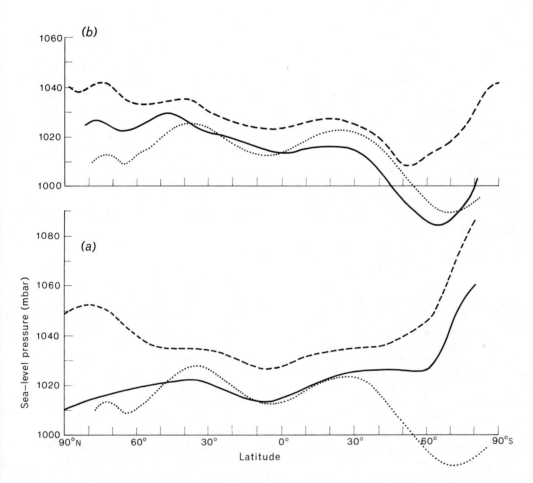

Fig. 11.4. Sea-level pressure at 30° W for the glacial period maximum and present-day July simulations of (*a*) Williams *et al.*[27] and (*b*) Gates.[34] Observed July sea-level pressure.[49] '——', control; '– – – –', glacial period; '. . . . .', observed.

model. A January and a July simulation with the RAND model have been evaluated.[46,47] Each model does reproduce reasonably well the basic features of the atmospheric circulation or macro-climate; each model is better at simulating some things than others.

In an Appendix of the report of the US Committee for GARP the results of several different

model integrations are compared in order to display the level of accuracy characteristic of present-day atmospheric GCMs. For sea-level pressure, while the models' results differ in several details, they show a useful level of accuracy. The greatest differences occur in the middle and high latitudes of the northern hemisphere, where cyclonic activity (which is hard to simulate in low-resolution models) is most frequent in the real atmosphere. Examination of the tropospheric temperature and pressure fields and the simulation of cloudiness and precipitation further confirms that the GCMs generally simulate the large-scale features of the atmospheric circulation.

The differences between simulations of the present day and of the recent glacial maximum are considered as estimates of the magnitude and direction of the differences between present-day atmospheric circulation and that which actually existed at the most recent glacial maximum. We shall emphasize results obtained for July in the northern hemisphere since only two studies[27,34] have modelled the southern hemisphere glacial period atmosphere and only two studies [27,31] have modelled the winter conditions.

The models discussed here do not all consider the same properties of the Earth's climate, so a comprehensive comparison is not possible. The following sections consider several features of the atmospheric circulation of different models and compare the different results.

Sea-level pressure and cyclone activity
For only two of the model simulations are there results in the form of sea-level pressure maps. Earlier papers have discussed the sea-level pressure fields in the experiments with the NCAR model[27,39,48] and the RAND model.[34]

Fig. 11.4 shows meridional cross sections of the sea-level pressure field along 30° W for the July control and glacial period simulations with the two models, together with the observed present-day sea-level pressure.[49]

Both present-day July experiments simulate the sea-level pressure field in the tropical latitudes reasonably well. The RAND model has the North Atlantic high pressure zone further north and stronger than observed, and while the pressures in the South Atlantic are lower than observed, the gradients are accurately simulated. The NCAR model has a weaker North Atlantic high pressure than observed, and the (fictitious) pressure over the Antarctic continent is much higher than that 'observed'. The absence of the subpolar low has been discussed.[48] Both models have greater sea-level pressure at all latitudes along 30° W in the glacial period case than in the control case. In the RAND model the South Atlantic subpolar low has moved 15° equatorward and pressure has increased most

over Greenland and Antarctica. In the NCAR model the largest changes along 30° W also occur over Greenland and Antarctica, with stronger pressure gradients between the equator and poles.

Cyclone activity in the glacial period and present-day experiments has been examined for three models.[25,27,31] Alyea found that transient activity was considerably less in the present-day July simulation than in the glacial period simulation across the North Atlantic and in south-central to eastern Europe. Williams *et al.*[27] noted the influence of the ice sheets and sea-ice in displacing the zones of cyclonic activity southward. In the July glacial period case there was a major storm track across the North Atlantic to the west of Ireland, and a continuation of this or a further major track, from eastern Europe into Asia. Similarities between this pattern and another reconstruction[50] have been noted.[28] Saltzman & Vernekar[31] found an intensification of baroclinic activity (or storminess) in their glacial period simulation.

Zonal wind
Several reconstructions of the glacial period climate made before the model reconstructions suggested that the zonal winds were stronger at the glacial maximum than at present.[50,51,37] Fig. 11.5 shows the zonally-averaged, west–east component of the wind for July in the control (present-day) and glacial period cases at 500 mbar in Alyea's experiments, at 6 km (approximately 500 mbar) in the experiments of Williams *et al.*, for the vertically averaged troposphere in the experiments of Saltzman & Vernekar and at 400 mbar in Gates' experiments.

All of the models showed an increase in the zonal wind speed in the middle latitudes of the northern hemisphere in the July glacial period case, but there are large differences in the results. The differences between the present-day and glacial period zonal winds are not very great in the experiments of Alyea and Saltzman & Vernekar. Williams *et al.* find the glacial period winds weaker between 50° N and 70° N and stronger between 50° N and 20° N. The absence of any easterly winds in either case of Williams *et al.* is noticeable.

For January, two experiments[27,31] find an increase in the strength of the zonally-averaged, west–east component of the wind.

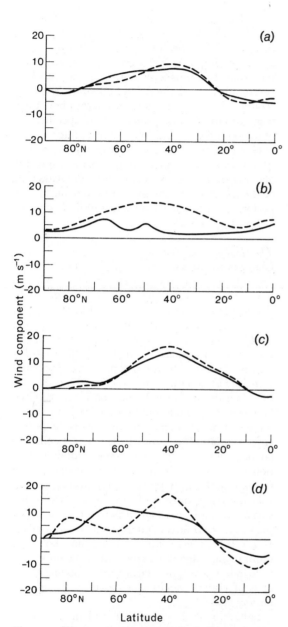

Fig. 11.5. West–east component of wind (zonal average) for July glacial period maximum (– – – –) and present-day (——) simulations of (a) Alyea[25] (500 mbar), (b) Williams et al.[27] (6 km), (c) Saltzman & Vernekar[31] (vertical average) and (d) Gates[34] (400 mbar).

Temperature

Alyea[25] finds the north–south temperature gradient between the subtropics and the mid-latitudes was increased in the glacial period case because of a large reduction of heating between 50° N and 70° N.

The differences in ground temperature between the glacial period and present-day simulations with the NCAR model have been studied.[28] The temperatures fell everywhere, with the largest changes, not surprisingly, occurring over the middle and high latitudes of the northern hemisphere. Some very large changes occur where large increases of cloudiness took place and caused unrealistically large cooling.[52] Williams[28] has also discussed tropospheric temperature changes in the NCAR experiments. In the January cases in the lower troposphere the largest changes occurred in the polar areas, while in the upper atmosphere the January glacial period case was colder at all latitudes but the temperatures were reduced more in the tropics than in middle latitudes. In the July cases the change from the present-day case to the glacial period case was greater, with the largest reductions at about 3 km poleward of 60° N. At 1.5 km the equator-to-pole temperature gradient increased in the glacial period case but at 10.5 km temperature gradients were reduced in nearly all regions. The latter result supports a suggestion regarding glacial period temperature gradients.[53]

Saltzman & Vernekar[31] found zonal mean surface and mid-tropospheric temperatures are uniformly cooler in the glacial period case, with the surface values at high latitudes in winter showing an increased temperature gradient near the edge of the ice sheet at 55° N.

Gates[34] found markedly lower surface air temperature in the July glacial period case over regions of North America and Europe; however, over scattered locations in Africa and Asia and in western Siberia the glacial period case temperatures were higher than those simulated for the present day by up to 6 °C. At higher levels in the atmosphere, Gates found that the thermal effect of the surface was reduced and the air temperature was more influenced by the large-scale atmospheric circulation.

Kinetic energy

Usually the kinetic energy of the atmosphere is resolved into the amounts associated with the zonally-averaged fields of motion (zonal kinetic energy) and the amounts associated with the eddies (eddy kinetic energy). Two studies[25,31] found an increased amount of eddy kinetic energy in the glacial period model atmosphere, while one (Williams[28]) found a decrease in the July glacial period simulation. Since eddy kinetic energy in July is generated largely by east–west circulations (monsoons) established in response to land/ocean thermal contrasts, the reduction of eddy kinetic energy in the glacial period simulation with the NCAR model represents a reduction or elimination of monsoon circulations in July. Alyea's model[25] was unable to simulate the monsoon successfully for present-day July[26] so the kinetic energy results of Alyea can not represent the change in monsoon as found by Williams.[28] The zonally-averaged model of Saltzman & Vernekar[31] is also unable to take account of monsoon changes, which might explain the absence of a decrease in eddy kinetic energy in their experiments.

In the January experiments both the atmospheric eddy kinetic energy and zonal kinetic energy were about the same in the present-day and glacial period cases.[28]

Northern hemisphere summer monsoon circulation
As described above, Williams[28] found in the atmospheric eddy kinetic energy amounts an indication of weakening or elimination of the northern hemisphere summer monsoon in the July glacial period case. Other features supported this conclusion. The tropical easterly jet, which is associated with the monsoon circulation was simulated in the July present-day case but not in the July glacial period case. Consideration of the pressure distribution at 12 km showed that in the present-day July case the large-scale circulation features, associated with the monsoon,[54,55] were simulated. In the July glacial period case the pressure distribution at 12 km shows a steady increase from low pressure at the poles to higher pressure at the equator, with no pronounced circulation features. The temperature field at 12 km also showed the absence of the monsoon circulation in the July glacial period case. Williams & Barry[52] pointed out that the inclusion of glacial period boundary conditions in the GCM caused the July glacial period case ground temperature distribution to resemble strongly that of the January present-day case in the northern hemisphere. The elimination of the normal July land/ocean heating contrasts could be a major reason for the absence of the northern hemisphere summer monsoon in the glacial period simulation.

As I have mentioned, Alyea's model failed to simulate the monsoon circulation in the present-day case. It has been suggested[26] that the intensity of heating in Alyea's model was too weak in the subtropics and thus could not maintain a blocking thermal high over the Asian highlands. Since the model only covers the northern hemisphere there were apparently also problems with specification of the boundary conditions at the equator.

Gates[34] found that the monsoonal low-pressure system was simulated in the July present-day and glacial period cases. Manabe[56] found a weakening of the monsoon in the glacial period simulation performed with the Geophysical Fluid Dynamics Laboratory (GFDL) model.

The Hadley circulation
Changes in the Hadley circulation between the present day and the glacial period maximum have been discussed, and it has been suggested[51] that the Hadley circulation would be stronger in the glacial period, and that the rising branch of the Hadley circulation would not migrate into the northern hemisphere, as it does in present-day summer.[57] Alyea[25] examined the vertical pressure velocities at 500 mbar and 1000 mbar to see how the meridional circulation patterns differed between cases and found that the Hadley cell was somewhat more intense in the July glacial period case. Williams[28] examined the vertical component of the wind and found that the Hadley circulation was stronger in the January glacial period simulation than the January present-day case. In July the rising branch of the Hadley circulation was as strong in the glacial period case as in the present day, while the sinking branch was weaker. In the July cases, Williams noted that the maximum rising velocity was located at the same latitude in both cases but the latitudinal extent of the area of rising velocity decreased in the glacial period case.

Saltzman & Vernekar[31] found in winter an intensification and southward shift of the polar direct cell along with smaller intensifications of the Ferrel and Hadley cells; they pointed out that these changes were dynamically consistent with the changes in the surface west–east wind component and with the increases of baroclinic eddy amplitudes and transports in the middle latitudes. The summer changes in meridional circulation were the same as those found by Alyea. As in the study of Williams, Saltzman & Vernekar found no change in the position of the rising branch of the Hadley circulation. Gates[34] found that the mean meridional circulation was about 30 per cent weaker in the July glacial period case than in the present-day case in tropical regions, although the positions of the strongest mean rising and sinking motions were found at 10° N and 30° S in both cases.

Atmospheric moisture content and precipitation
The earliest theories devised to explain the origin of till and erratics, deposits now known to have been left by glaciers, were based on the idea of transport by water; the deposits were believed to be the result of the biblical Noah's flood. In the middle of the nineteenth century, the idea of transport by ice was developed. From the first geological evidence it was easy to conclude that during glacial periods an increase in precipitation produced the ice sheets, while in warmer latitudes the rainfall increased. Evidence for 'pluvial' periods corresponding in time to the glacial periods seemed overwhelming. More recent evidence on various aspects of Late Pleistocene climates has suggested that the assumption of pluvial–glacial synchrony is not necessarily true.

Fairbridge[58,59] has found much evidence for glacial period aridity and suggests that the maxima of glacial periods were characterized by world-wide aridity. Geological and palynological evidence has been found increasingly in recent years to support this concept.[60–66]

Alyea[25] does not report changes in precipitation and moisture balance in his two July cases but does find that the presence of ice in the glacial period case inhibited the latent heat release through condensation processes.

Williams[28] found that the specific humidity of the atmosphere in the glacial period case was reduced in amount and vertical and latitudinal extent compared with the control cases. For the glacial period the meridional and vertical transports by the mean circulation and eddies were weaker except in the tropics in January where transports by the mean circulation were stronger than in the control case. Williams[28] also found that in general the precipitation in the glacial

period cases was less than in the control cases over the continents and greater than in the control cases over the oceans. It was suggested that the primary reason for this occurrence was that while the oceans experienced a decrease of temperature of the order of 5 °C, the continents became much colder in the glacial period cases. The extreme cold of the continents probably led to a reduced moisture flux from the surface, while air flowing out over the warmer oceans would pick up moisture, and therefore have more available for condensation and precipitation.

Saltzman & Vernekar[31] found that both evaporation and precipitation had a global decrease from the control to the glacial period cases and there was also an increase in the poleward eddy flux of water vapour. This led to an equatorward-shifted pattern of evaporation-precipitation, with increased amplitude in winter, giving a more positive surface water balance from 25° to 75° N and a less positive balance in the remaining tropical and polar latitudes.

Gates[34] found that the precipitable water amount was decreased by about 37 % from the control case to the glacial period case in the northern hemisphere and 23 % in the southern hemisphere. The glacial period case evaporation and precipitation was also reduced by 15 % in comparison with the control case values. Manabe's results[24] also showed more arid conditions in the July glacial period case with precipitation increasing over the oceans but decreasing over the continents.

Discussion
The results of several simulations using present-day and glacial period boundary conditions have been compared. The model experiments may be viewed as sensitivity experiments, used to investigate the response of the climate system to different conditions, specifically those of the period 18 000–22 000 yr ago. Each of the models produces a set of results and in several cases the responses of the models are the same – for example each model found that the atmosphere was colder and drier in the glacial period case. There are some differences between the results; for example, the Hadley circulation was found to be stronger in the glacial period simulation of Alyea and Saltzman & Vernekar, little changed in the case of Williams and weaker in the case of Gates. It is not clear, however, whether these differences

are because of differences in model formulations or in boundary conditions used, or in a combination of these. The differences between the results of Gates[34] and Williams *et al.*[27] with regard to sea-level pressure distribution must be a result of model differences and the different SST distributions used as described above. Such problems of interpretation will be eliminated by running the models with the same input conditions, specifically those compiled by the CLIMAP project. Plans are being made for such a series of experiments.

In spite of the limitations of climate models, such as those briefly described earlier and discussed in detail by Chervin (Chapter 12 of this volume), the models are an ideal tool for studying the climate system. Since they include many of the physical processes that we know are important and also include non-linear interactions, the climate models such as those described above represent the most appropriate way now available for investigating the sensitivity of the climate system to various sets of boundary conditions. As the models are improved, for instance by the addition of interactive ocean models and realistic cloud parameterizations, the results from the above experiments will be improved upon and knowledge of the climate system and past climates will be increased. Nevertheless, the results of these first experiments and the several areas of agreement are encouraging and emphasize that a new tool is now available for the palaeoclimatologist to study equilibrium climates of the past.

The use of climate models in studying climatic change

The amount of computer time required to run a global or hemispheric model is such that it is prohibitive to run for several simulated years or hundreds of simulated years to mimic the changing climate, although more parameterized models[67] can be used in such a way. The GCMs can only be used to investigate the response over a 30-day or 40-day period, in which the model has supposedly reached its equilibrium state, to changes in boundary conditions. As I mentioned at the beginning of this chapter, Barry[16] has discussed the use of these models in palaeoclimatic reconstruction – models have been used to investigate the response of the climate system to removal of Arctic Ocean ice,[68-70] for example. The use of models to investigate possible future climates has also begun. For example, the response of the

NCAR model to thermal pollution has been studied.[71] Others[8] have studied the response of the GFDL model to a doubling of the level of carbon dioxide in the atmosphere. Also, model experiments to investigate the effects of SST anomalies[9,10,43-45] and albedo changes[72] have been carried out, to give more information about feedbacks in the climate system. The question of climate predictability has been considered by Lorenz[73] who discusses which models would be most suitable for studies in this area. The GCMs are physically most acceptable but prohibitive in terms of computer resources required. The very simple models do not have the latter restrictions but, as Lorenz points out it can be very dangerous to place too much confidence in models whose behaviour depends very strongly upon details of parameterization. While the ideal model for studying the climate system is not yet available, the studies described in this chapter illustrate the best use that can be made of such models at the present time to study climatic change or past climates.

I would like to thank my colleagues at the National Center for Atmospheric Research, Boulder, Colorado, USA, at the Institute of Arctic and Alpine Research, University of Colorado, Boulder, Colorado, USA, and at the Climatic Research Unit, University of East Anglia, Norwich, UK, for their support, guidance and interest as I have worked on this topic. I also am grateful to R. M. Chervin, W. W. Kellogg and R. G. Barry for their comments on the first draft of this paper.

References

1. GARP (1975). *GARP Publ. Ser.* No. 16. WMO–International Council of Scientific Unions. 265 pp.
2. Schneider, S. H. & Dickinson, R. E. (1974). *Rev. Geophys. Space Phys.*, **12**, 447.
3. Budyko, M. I. (1969). *Tellus,* **21**, 611.
4. Manabe, S. & Wetherald, R. (1967). *J. atmos. Sci.,* **26**, 786.
5. Schneider, S. H. & Gal-Chen, T. (1973). *J. geophys. Res.,* **78**, 6182.
6. Sellers, W. D. (1969). *J. appl. Met.,* **8**, 392.
7. Sellers, W. D. (1973). *J. atmos. Sci.,* **12**, 241.
8. Manabe, S. & Wetherald, R. T. (1975). *J. atmos. Sci.,* **32**, 3.
9. Rowntree, P. R. (1972). *Q. Jl R. met. Soc.,* **98**, 290.

10. Houghton, D. D., Kutzbach, J. E., McClintock. M. & Suchman, D. (1974). *J. atmos. Sci.*, **31,** 857.
11. Washington, W. M. & Daggupaty, S. M. (1975). *Mon. Weath. Rev.*, **103,** 105.
12. Shukla, J. (1975). *J. atmos. Sci.*, **32,** 503.
13. US Committee for the Global Atmospheric Research Program (1975). National Academy of Sciences, Washington D.C. 239 pp.
14. Arakawa, A. (1975). *Rev. of Geophys. Space Phys.*, **13,** 668.
15. Smagorinsky, J. (1974). In *Weather and climate modification*, ed. W. M. Hess, pp. 633–86. Wiley, New York.
16. Barry, R. G. (1975). *Palaeogeogr. Palaeoclimatol. Palaeoecol.*, **17,** 123.
17. Manabe, S. (1969). *Mon. Weath. Rev.*, **97,** 739.
18. Bryan, K. (1969). *Mon. Weath. Rev.*, **97,** 806.
19. Wetherald, R. T. & Manabe, S. (1972). *Mon. Weath. Rev.*, **100,** 42.
20. Schneider, S. H. (1972). *J. atmos. Sci.*, **29,** 1413.
21. Haltiner, G. J. (1971). *Numerical weather prediction*. Wiley, New York.
22. Flint, R. F. (1971). *Glacial and Quaternary geology*. Wiley, New York.
23. CLIMAP Project Members (1976). *Science*, **191,** 1131.
24. Hammond, A. L. (1976). *Science*, **191,** 455.
25. Alyea, F. N. (1972). *Atmospheric Science Paper* No. 193. Colorado State University, Fort Collins, Colorado. 120 pp.
26. Krishnamurti, T. N., Daggupaty, S. M., Fein, J., Kanamitsu, M. & Lee, J. D. (1973). *Bull. Am. met. Soc.*, **54,** 1234.
27. Williams, J., Barry, R. G. & Washington, W. M. (1974). *J. appl. Met.*, **13,** 305.
28. Williams, J. (1974). *NCAR Cooperative Thesis/ Institute of Arctic and Alpine Research Occasional Paper* No. 10.
29. Kasahara, A. & Washington, W. M. (1967). *Mon. Weath. Rev.*, **95,** 389.
30. Kasahara, A. & Washington, W. M. (1971). *J. atmos. Sci.*, **28,** 657.
31. Saltzman, B. & Vernekar, A. D. (1975). *Quaternary Res.*, **5,** 307.
32. Saltzman, B. & Vernekar, A. D. (1971). *J. geophys. Res.*, **76,** 1498.
33. Saltzman, B. & Vernekar, A. D. (1972). *J. geophys. Res.*, **77,** 3936.
34. Gates, W. L. (1976). *Science*, **191,** 1138.
35. Gates, W. L. (1975). *J. atmos. Sci.*, **32,** 449.
36. Cox. A. (1968). *Met. Monogr.*, **8**(30). **112.**
37. Lamb, H. H. (1961). In *Descriptive palaeoclimatology*, ed. A. E. M. Nairn, pp. 8–44. Interscience, New York.
38. Flint, R. F. (1957). *Glacial and Pleistocene geology*. Wiley, New York.
39. Barry, R. G. (1973). *Arctic Alpine Res.*, **5,** 171.
40. McIntyre, A. (1967). *Science*, **158,** 1314.
41. Emiliani, C. (1971). In *Late Cenozoic glacial ages*, ed. K. K. Turekian, pp. 183–198. Yale University Press, New Haven.
42. Spar, J. (1973). *Mon. Weath. Rev.*, **101,** 91.
43. Spar, J. (1973). *Mon. Weath. Rev.*, **101,** 767.
44. Spar, J. (1973). *Mon. Weath. Rev.*, **101,** 554.
45. Chervin, R. M., Washington, W. M. & Schneider, S. H. (1976). *J. atmos. Sci.*, **33,** 413.
46. Gates, W. L., Batten, E. S., Nelson, A. B. & Kahle, A. B. (1971). Publication R–877–ARPA. The Rand Corporation, Santa Monica, California.
47. Gates, W. L. & Schlesinger, M. E. (1977). *J. atmos. Sci.*, **34,** 36.
48. Barry, R. G. & Williams, J. (1975). In *Quaternary Studies*, ed. R. P. Suggate & M. M. Cresswell, pp. 57–66. Royal Society of New Zealand, Wellington.
49. Data from Schutz, C. & Gates, W. L. (1972). Publication R–1029, Rand Corporation, Santa Monica, California. Compiled from Crutcher, H. L. & J. M. Meserve (1970). NAVAIR–50–1C–52. Naval Weather Service Command, Washington D.C. and Taljaard, J. J. *et al.* (1969). NAVAIR–50–1C–55. Naval Weather Service Command, Washington D.C.
50. Bryan, K. & Cady, R. C. (1934). *Am. J. Sci.*, **27,** 241.
51. Kraus, E. B. (1960). *Q. Jl R. met. Soc.*, **86,** 1.
52. Williams, J. & Barry, R. G. (1975). In *Climate of the Arctic*, ed. G. Weller & S. A. Bowling, pp. 143–9. Geophysical Institute, University of Alaska.
53. Kraus, E. B. (1973). *Nature*, **245,** 129.
54. Krishnamurti, T. N. (1971). *J. appl. Met.*, **10,** 1066.
55. Krishnamurti, T. N. (1971). *J. atmos. Sci.*, **28,** 1342.
56. Unpublished results presented at CLIMAP meeting, Lamont-Doherty Geological Observatory, November 1975.
57. Newell, R. E. (1973). *Nature*, **245,** 91.
58. Fairbridge, R. W. (1964). In *Problems in paleoclimatology*, ed. A. E. M. Nairn, pp. 356–9. Interscience, New York.
59. Fairbridge, R. W. (1970). *Rev. Geogr. phys. Geol. dyn.*, **12,** 97.
60. Galloway, R. W. (1970). *Ann. Am. Ass. Geogr.*, **60,** 245.
61. Damuth, J. E. & Fairbridge, R. W. (1970). *Bull. geol. Soc. Am.*, **81,** 189.
62. Frenzel, B. (1968). *Science*, **161,** 637.
63. Street, F. A. & Grove, A. T. (1976). *Nature*, **261,** 385.
64. Galloway, R. W. (1965). *J. Geol.*, **73,** 603.
65. Bonatti, E. (1966). *Nature*, **209,** 984.

66. Bonatti, E. & Gartner, S. (1973). *Trans. Am. geophys. Un. (EOS)*, **54,** 327.

67. Sellers, W. D. (1976). *Mon. Weath. Rev.*, **104,** 233.

68. Fletcher, J. O., Mintz, Y. Arakawa, A. & Fox, T. (1973). In *Energy fluxes over polar surfaces*, *WMO Tech. Note* No. 129. WMO, Geneva.

69. Newson, R. (1973). *Nature,* **241,** 39.

70. Warshaw, M. & Rapp, R. R. (1975). *J. appl. Met.*, **12,** 43.

71. Washington, W. M. (1972). *J. appl. Met.*, **11,** 763.

72. Charney, J., Stone, P. H. & Quirk, W. J. (1975). *Science,* **187,** 434.

73. Lorenz, E. N. (1975). In *GARP Publ. Ser.* No. 16. WMO–International Council of Scientific Unions.

12

The limitations of modelling: the question of statistical significance

ROBERT M. CHERVIN

The use of general circulation models (GCMs) to simulate the atmospheric response to ice age boundary conditions is discussed in detail in the previous chapter by Williams. The ice age experiment series is a specific example of the general category of prescribed change GCM experiments. In this context, the term 'prescribed change' denotes any modification to the model formulation (for example, altering boundary conditions and testing new parameterizations or numerical schemes) from which tests of the sensitivity of the model's climatic statistics to that prescribed change are to be made. Other examples of such experiments in which the model's response to a boundary condition change is examined (as a possible causal mechanism of real climatic change) include ocean surface temperature anomalies[1-3] and thermal pollution.[4] The relative ease with which this type of experiment can be performed often masks the difficulties involved in assessing the significance of the results. Basically, the problem is one of distinguishing, in an objective fashion, between signal (that part of the difference between the results of a prescribed change experiment and an unperturbed control case attributable to the prescribed change) and noise (some measure of inherent variability in the model) in a prescribed change response.

In this chapter, the discussion on the limitations of modelling considers only the limitations on the use of atmospheric GCMs for prescribed change experiments, with the emphasis on the question of statistical significance of the atmospheric response in prescribed change experiments designed to test hypotheses and mechanisms of climatic change. However, a brief description of the general limitations and uncertainties, common to most GCMs, which arise from model formulation is presented first. The rest of the chapter is devoted to demonstrating the necessity of considering a model's inherent variability when assessing the statistical significance of any prescribed change response. Specific results are shown for a standard GCM in use at the National Center for Atmospheric Research (NCAR) for several years, but the approaches and methods should be applicable more generally for GCMs of comparable complexity and sophistication. Most of the following is based on recent efforts of the author and his colleagues at NCAR and the interested reader is referred to Chervin, Gates & Schneider,[5] Chervin & Schneider,[6] Chervin & Schneider,[7] and Chervin, Washington & Schneider[3] for more detail.

General GCM limitations and uncertainties

An atmospheric GCM is a set of partial differential equations which are presumed to govern the dynamics and other physical processes of the atmosphere, *taken together* with the approximation techniques of solution (usually a discrete version of the set of equations); the 'model' is designed for solution on a modern, high-speed digital computer. Simplifications introduced either in the formulation of the model physics *or* in the approximation techniques could cause errors, and since these are part and parcel of the models' individual behavior, the basic equations and their discrete form together comprise the 'model'. Descriptions of many of the GCMs currently in use at various meteorological research and operational centers around the world are given in a report of the Global Atmospheric Research Program (GARP).[8]

Discretization of the equations and solution on a finite spatial grid is necessary since no analytical methods are yet available to solve the equations completely in their continuous form. Naturally, the mathematical approximations used to solve equations introduce errors into the solutions. In general, these errors may be reduced by resorting to a higher grid resolution which unfortunately necessitates a shorter time step (to avoid computational instabilities) and hence more computer time for climate simulation purposes. So limitations and uncertainties must exist because of the trade-off between numerical accuracy and computer time.

Furthermore, a model's finite grid resolution also requires the formulation of subgrid-scale statistical parametric representations ('parameterizations') of the large-scale (grid scale or greater) effects of those physical processes not explicitly resolved by the model. Often, there are adjustable parameters in the parameterizations which can be 'tuned' to produce as close a match as possible between the model's long-term mean state (that is, climate) and that of the real atmosphere. Again, compromises are struck among the number and treatment of physical processes, the fidelity of the simulations and computer time. Finally, the tuning of the model to present climate statistics (a simulation experiment) may not necessarily improve its ability to reproduce the true atmospheric *response* to altered forcing (what is called a sensitivity experiment).

The uncertainties imposed by non-linear interactions in the model and the consequent predictability limit from the growth of small random errors in initial conditions are the subject of the next section and the basis of the methods described in the rest of the chapter.

Predictability considerations

Predictability studies, including those surveyed by Kasahara,[9] show that GCMs exhibit a characteristic growth rate for small initial random errors such that, after a few weeks' model time, the detailed solutions of two runs, nearly identical at the start, are indistinguishable from any solutions selected at random from an ensemble of solutions consistent with the same external (boundary) conditions. That is, the deviations (with respect to an unperturbed case) in the evolution of the instantaneous solution of a perturbed case become comparable to the standard deviations of the instantaneous state, which are associated with the simulated day-to-day

synoptic fluctuations. After this 'predictability period' of a few weeks, the detailed behavior of the atmospheric model (the instantaneous state of the atmosphere or 'weather') becomes essentially unpredictable from an initial condition of arbitrary specificity. The models, moreover, respond in this manner to *any* perturbation, whether it is associated with the inevitable uncertainty in specification of the initial conditions, the unavoidable errors in computation due to truncation or rounding, or a supposedly significant prescribed change to model formulation or boundary conditions. For example, if a control case and a prescribed change experiment with a negligible 10^{-4} change in, say, solar input, started from *identical* initial conditions, the evolution of their states would differ in the same manner as would the evolution of the states of two identical cases differing only by a negligible change in their respective initial states. In fact, even if after only the first numerical time step the 10^{-4} change in solar input were removed, the evolution of states would still diverge. So any prescribed change response also contains a random component in addition to the signal of interest.

This inherent variability arising from day-to-day unpredictable weather fluctuations is due to the high degree of non-linearity present in the models (and in the atmosphere) which causes changes on even the smallest scales resolved to be transmitted to all other spatial scales. Although the inclusion of dissipation and smoothing in GCMs may serve to modify the rate of spread of such errors, the existence of unstable physical modes, such as large-scale baroclinic instability, serves to spread the unpredictability rapidly throughout the atmospheric system.

An example of such error growth for a version of the six-layer, 5° latitude/longitude resolution NCAR GCM (described in detail by Kasahara & Washington[10]) is shown in Fig. 12.1, adapted from Williamson & Kasahara.[11] A random error taken from a rectangular distribution with minimum -1 m s^{-1} and maximum $+1$ m s^{-1} was added to each grid point of the 13.5 km zonal wind field on day 20 of a control experiment that had started at day 0 with the atmosphere at rest and isothermal. Both the random and control cases were run in the perpetual January mode, with the Sun declination angle, ocean surface temperatures and other parameters of the external forcing system fixed at the climatological

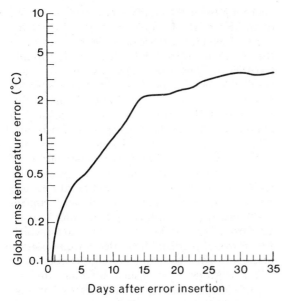

Fig. 12.1. Growth of global root mean square (rms) error in temperature (adapted from Williamson & Kasahara[11]).

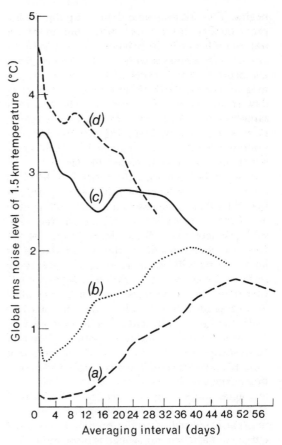

Fig. 12.2. The estimated global root mean square (rms) noise level of 1.5-km temperature as a function of time-averaging interval computed from differencing random and control cases beginning at (a), day 21, (b) day 31, (c) day 41 and (d) day 51.

values for mean January conditions, and were continued to day 80. The curve in Fig. 12.1 shows the growth of the instantaneous global root mean square (rms) 'error' (the difference between random and control cases) in temperature, averaged over the six levels in the model, as a function of the time after random error insertion at day 20. The error growth rate slows down about 15 days after insertion and only begins to level off after 30 days. Predictability curves for other fields (and other models) exhibit similar behavior.

These instantaneous error growth curves are obviously of considerable importance for short-to-medium range weather forecasting where the chief concern is the degradation of forecast quality in time (that is, an increase in the uncertainty of the detailed behavior of the atmosphere) owing to the growth of initial condition uncertainties. But, for climate studies, information is needed about the uncertainties or noise levels of the *time-averaged* or 'climatic' behavior of the simulated atmosphere after the initial errors have largely randomized the evolution of the model's detailed solutions.

Time-averaged predictability or climatic noise

The analogue of the evolution of the instantaneous global rms error for GCM climatic statistics is

the global rms time-averaged difference between a randomly perturbed and a control case as a function of time-averaging interval. This latter quantity is commonly referred to as the estimated global rms noise level.

For the same pair of random and control cases used in Fig. 12.1 and described in the previous section, global rms noise levels for several model variables were estimated for different averaging intervals. Fig. 12.2 shows curves describing global rms noise level estimates made as described above for the 1.5-km temperature as a function of averaging interval. The shape of the curves depends strongly on the time, relative to the error insertion time, at which the time-averaging begins. This starting time defines a point on the predictability curve which turns out to be an important factor in the behavior of the noise level curves. The averages were all determined from consecutive 12-h samples with curve (*a*) beginning at day 21, curve (*b*) at day 31, curve (*c*) at day 41 and curve (*d*) at day 51. The last 12-h sample used in all cases was day 80. The estimated noise levels for zero averaging interval are considerably lower for the curves in which time-averaging was begun closer to day 20 – an obvious indication that these small differences are not independent of the initial errors, and are biased by their closeness to the initial point (day 20). Clearly curves (*a*) and (*b*) include the transient predictability effects which, in spite of the time-averaging, result in the accumulation of larger errors with an increase in the averaging interval. Of course, it is possible (and expected) that these curves eventually would decline after an averaging interval sufficiently long to overcome the initial transients, but to demonstrate this would require continuing both random and control experiments for a considerable period of time and could therefore involve prohibitive computer expense. Curves (*c*) and (*d*) show that time-averaging finally begins to reduce the estimated global rms noise level only when the averaging interval is started after the predictability error curve has flattened out (which is more indicative of a situation in which the realizations differenced to create the curves are *independent* of the initial state and in a quasi-equilibrium state). The close-to-monotonically decreasing noise reduction character of curve (*d*) is a result of the relative flatness of the predictability curve for times greater than 30 days after the error insertion. This same behavior is also typical of the estimated global rms noise level of other variables (see Chervin & Schneider[6]).

The global rms noise level, however, is merely a measure of the average noise level at a single grid point for the variable of interest and is approximately an average over the probability distribution functions at different grid points as opposed to an average over independent samples of the same statistic. This rms quantity is of substantial importance for basic predictability studies; it also shows the general effects of time-averaging on the average over all grid points in a given spatial domain (global, zonal, or any arbitrary region), but does not provide much guidance for judging the significance of time-averaged quantities in GCM climatic change experiments.

GCM experiments are often designed to study the sensitivity of the model's regional climatic statistics to prescribed changes in boundary conditions such as ocean surface temperature anomalies, altered surface albedo and thermal pollution-type heat sources. Thus, it is necessary to obtain estimates of the geographical distribution of noise levels for GCM-generated climatic statistics. Crude estimates of the distribution of noise levels for January statistics are possible by taking the difference at each grid point between the previously described random and control experiments (the 'perturbation response') averaged from day 51 to day 80 – an interval for which the transient predictability effects are no longer dominant (since the error source was inserted at day 20). An example of such differences for the 1.5-km temperature field ($T_{1.5}$) is shown in Fig. 12.3. The inhomogeneous distribution of noise levels estimated from a pair of runs is clearly indicated on this two-dimensional noise level map. In these experiments no physical significance can be construed from the existence in a given area of a negative difference instead of a positive one, or vice versa. It is the absolute value of the plotted differences that represents the estimated noise level, since a different pair of runs could well reverse the sign of this particular perturbation response.

The response map for $T_{1.5}$ indicates the highest levels in the mid-latitude and polar regions. The very large estimated noise levels over the Antarctic can be attributed in part to computational difficulties connected with the special vertical differencing scheme used at grid points above high topography. The '1.5-km' level in the model is actually defined as that level midway between ground elevation and 3 km. Consequently, in regions where the average elevation for a 5° grid square is close to 3 km, the vertical increment in the differencing scheme becomes quite small, seeming to contribute to the large noise levels over mountains. (In a sense this thin layer becomes more a part of the highly variable boundary layer than a representation of a thicker atmospheric layer.)

To this point, estimation of GCM noise levels has been derived from the time average of the difference between two independent realizations. Naturally, a one-month average of the difference

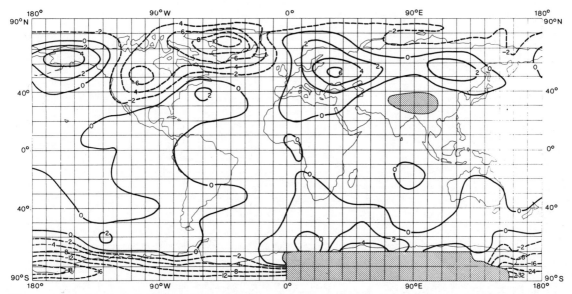

Fig. 12.3. Geographical distribution of 1.5-km temperature perturbation response (averaged over the interval from day 51 to day 80) computed from differencing random and control cases. The estimated noise levels are the absolute values of the indicated perturbation response.

between two GCM-generated time series will differ from the monthly average difference of any other such pair of independent time series because of the sampling error inherent in this limited sample. Therefore, Chervin & Schneider[6] also examined the noise levels estimated from a different independent pair of GCM time series in order to illustrate the magnitude of the uncertainties in noise statistics estimated from only one pair of realizations, a sample size of two independent monthly averages.

Other estimates of noise levels were obtained by differencing the same control run and the model run of Williamson & Washington[12] in which they investigated the effects of using lower precision computer arithmetic on a long-term simulation. In this case the noise source was the rounding-error accumulation resulting from using computer words with 24-bit mantissas instead of the standard 48-bit mantissa arithmetic. This noise is smaller in magnitude than the random 1 m s^{-1} perturbation but was applied continuously, as opposed to at only one point in time. However, it

is known from basic predictability studies that even an infinitesimal initial perturbation will also grow rapidly into large errors. As before, time averages of the differences were computed from 12-hr samples in the interval from model day 51 to day 80.

The geographical distribution of noise levels as estimated by taking the differences between lower precision and control experiments (another perturbation response) averaged from day 51 to day 80 is shown in Fig. 12.4 for $T_{1.5}$. The large perturbation response in the mid-latitudes and polar regions and small response in the tropics found previously for the random minus control case is also evidence in the lower precision minus control case. The magnitude of the largest estimated noise levels and some of their locations do, however, differ in the estimates provided by the two different noise sources.

These differences, which arise from sampling errors, suggest that a Monte Carlo approach may be necessary for a more definitive determination of the noise levels of GCM-generated statistics.

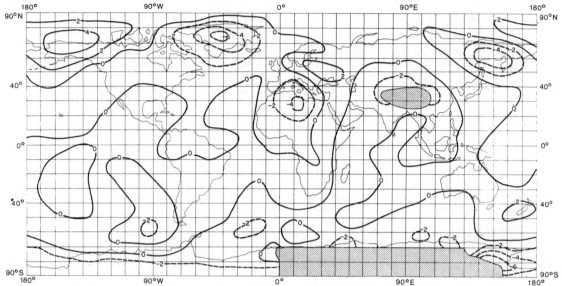

Fig. 12.4 Geographical distribution of 1.5-km temperature perturbation response (averaged over the interval from day 51 to day 80) computed from differencing lower precision and control cases. The estimated noise levels are the absolute values of the indicated perturbation response.

That is, a few differently perturbed random cases could be combined to form a better estimate of the inherent variability of the model. By increasing the sample size (number of independent sets of runs) in this way, one could avoid an erroneous conclusion of significance for a prescribed change response resulting from the chance occurrence of a low noise level in a particular area as estimated from a single pair of random and control experiments (or conversely, for chance high noise levels). A method for obtaining better estimates of inherent variability and a statistical significance-estimating procedure from Chervin & Schneider[7] is presented in the next two sections.

Sampling and estimating inherent variability

Statistically, the climate as simulated by a model may be determined from the ensemble of independent realizations or internal states generated by the model, with fixed external conditions and identical model formulations. The probability distribution associated with an infinite number of such realizations completely describes the climate

ensemble. For example, the first moment and the second central moment of the distribution are the climatic mean and variance, respectively. In the case of GCM climatic studies, estimates of the climate ensemble (and its derived climatic statistics) may be made with a finite number of realizations collected from a number of GCM simulations differing only by small random perturbations in the initial conditions. Naturally, as a consequence of traditional sampling theory, estimates of climatic statistics become closer to the 'true' ensemble quantities as the sample size of realizations increases. An obvious constraint to greatly reducing the uncertainties in such estimates is the large amount of computer time at present required to general multiple GCM simulations with only initial conditions perturbed.

Chervin & Schneider[7] assembled a set of five available Januaries simulated by the six-layer, 5° latitude/longitude resolution version of the NCAR GCM described by Kasahara & Washington[10] for the purpose of estimating the model's inherent variability. First, 30-day means, based on

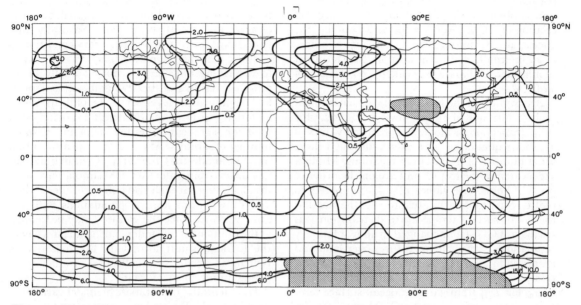

Fig. 12.5. Geographical distribution of σ_{30} estimated from five simulated Januaries for the 1.5-km temperature

the same sequence of 59 12-h samples, were computed at each grid point in the model for each simulated January. An ensemble (of January model calculations) average was then estimated by taking the average of the five 30-day means. Finally, an estimate of the standard deviations at each grid point for the model's January climate, σ_{30}, was made from the standard deviations of the five individual 30-day means about the aggregate average (where the sum of the five square-departures from the overall mean was divided by four). Statistically, the results of these calculations constitute an estimate of σ_{30} with four degrees of freedom – not an impressive number when compared to the corresponding calculations with the larger data sets available for many variables of the real atmosphere. Nevertheless, from sampling theory considerations, these calculations provide a far more reliable estimate of the model's inherent variability than the estimated noise level information based on only a single pair of randomly perturbed and unperturbed control experiments.

The validity of the above estimates is predicated on the assumptions that the time series from which the 30-day means are computed are stationary, independent and members of the same population. These conditions are not easily verifiable but some precautions are possible to minimize the likelihood that the assumptions are violated. Firstly, each of the five Januaries was run in the perpetual January mode with the solar declination angle, ocean surface temperatures and other parameters of the external forcing system fixed at the climatological values for mean January conditions. Secondly, the individual realizations differed from each other by the addition of *small* random errors to the initial conditions. Finally, the 30-day means were computed from a time interval within which the transient predictability effects are no longer dominant – that is, more than 30 days after initial conditions are perturbed.

Comparisons of such estimates of σ_{30} with the variability of monthly means computed from real atmospheric data have been cursory to date since the chief concern in this effort has been to develop a measure of model variability suitable for judging the significance of changes (relative to a control run) in monthly means, Δ_{30}, resulting from prescribed change experiments with the NCAR GCM. Furthermore, there are no assurances that the causes of inherent variability are the same in the model as in the real atmosphere. This question is obviously important in establishing the relation of GCM results to the atmosphere, but, once again, the initial purpose here was to judge, in an internally consistent fashion, the statistical significance of climate experiments *performed with the GCM*.

An example of the geographical distribution of such an estimate of σ_{30} is shown in Fig. 12.5 for $T_{1.5}$. Inspection of model vs. observed values of σ_{30} shows qualitative agreement, although there are a number of important exceptions. Minima are located in the tropics and are on the order of 0.1 °C. These low values of σ_{30}, compared to reality, can probably be attributed to the fact that the ocean surface temperature distribution was (as in most GCMs) unchanged in all five January simulations (surface temperatures over land and ice are computed from energy balance considerations, however). Thus, any possible contribution to σ_{30} for $T_{1.5}$ from variable ocean surface temperatures was missing in the GCM data. Again, the high degree of variability over the model's Antarctic is probably a consequence of the particular computational scheme used in that region of extensive orography.

Work continues at NCAR on estimating the model's inherent variability for other months and grid resolutions. It is expected that detailed comparisons of these estimates with the variability of monthly means computed from real atmospheric data may well prove to be a powerful diagnostic technique for model verification against observations. Thus, these techniques may be more general than mere statistical significance testing tools.

Estimating statistical significance of prescribed change response

With the aid of these estimates of σ_{30}, a judgment on the significance of the results of a prescribed change GCM experiment can be made by generating the geographical distribution of the ratio

$$r = |\Delta_{30}| / \sigma_{30} \qquad (12.1)$$

where Δ_{30} is the algebraic difference between 30-day means in the prescribed change experi-

ment and an unperturbed control experiment (the prescribed change response) calculated at each grid point. Traditionally, for the sake of convenience, only one control experiment has been used in calculating Δ_{30}. The advantage of using the average of several controls in Δ_{30} will be made clear at the end of this section. Naturally, one would expect $r > 1$ to be a necessary condition to consider a change significant. However, in order to determine how large r must actually be to judge the results statistically significant at a specified significance level, it is helpful to relate equation (12.1) to the Student's t statistic often used for significance testing involving a small number of samples. At this point, the assumption must be made that the samples (individual 30-day means from different GCM simulations) used in calculating r at each grid point are independent and drawn from normal populations with the same variance since the variate t may be defined as the quotient of two independent variates where the dividend is unit normal and the divisor is the root mean square of a collection of other independent unit normal variates.

(Incidentally, the estimate of model variability used by Shukla[13] based on an estimate of the process autocorrelation function from a single time series is not strictly applicable for formal significance testing because the sample distribution function of such an estimate of the variance of a time-averaged quantity is not proportional to chi squared, as required for the t test (see Jones[14]). Furthermore, the uncertainties in such variance estimates are not simply obtainable from resorting to the concept of degrees of freedom. Nevertheless, in the absence of a sufficient number of independent realizations for formal significance testing, the autocorrelation technique could be useful as an economical method to make intuitive estimates of significance.)

The following may be used as a working definition of the test variate at each grid point for the difference of the means of two independent sets of samples:

$$(12.2)$$

$$t = \left[\frac{1}{N_1} \sum_{i=1}^{N_1} x_i - \frac{1}{N_2} \sum_{i=1}^{N_2} y_i \right] \left[s^2 \left(\frac{1}{N_1} + \frac{1}{N_2} \right) \right]^{-1}$$

with the combined estimate of variance, s^2, given by

$$s^2 = \left[\sum_{i=1}^{n_1} (x_i - \bar{x})^2 + \sum_{i=1}^{n_2} (y_i - \bar{y})^2 \right] / (n + n_1 - 2)$$

$$(12.3)$$

where

$$\bar{x} = \sum_{i=1}^{n_1} x_i/n_1, \ \bar{y} = \sum_{i=1}^{n_2} y_i/n_2$$

and x_i and y_i are sample values (30-day averages) at the grid point of interest from the prescribed change and control populations (experiments), respectively. In the above, N_1 and N_2 are the sample sizes (number of experiments) used to estimate the prescribed change and control population means, respectively, and form the prescribed change response; and n_1 and n_2 are the sample sizes from the prescribed change and control populations, respectively, used to form the combined estimate of variance. The number of degrees of freedom is equal to $n_1 + n_2 - 2$. In the case of σ_{30} estimated from five Januaries and Δ_{30} computed from two samples (a typical form of prescribed response), the sample value of t equals $r/\sqrt{2}$ and there are four degrees of freedom (since $N_1 = 1, N_2 = 1, n_1 = 1$, and $n_2 = 5$.) Thus, using a table of the distribution functions of the t-distribution, such as in Smirnov,[15] Table 12.1 was con

Table 12.1. *Significance levels for a selection of values of* $r = |\Delta_{30}| / \sigma_{30}$ *corresponding to four degrees of freedom and* $t = r/\sqrt{2}$

r	Significance level (%) (two-sided t-test)	Significance level (%) (one-sided t-test)
1	51.69	25.93
2	22.86	11.65
3	10.02	5.05
4	4.69	2.37
5	2.38	1.20
6	1.31	0.66
7	0.77	0.39
8	0.47	0.24
9	0.31	0.15

structed to indicate the significance levels corresponding to four degrees of freedom and a selection of values of r for both the two-sided and the one-sided t-test. (The use of the two-sided t-test is based on the hypothesis that no specific signed difference, either positive or negative, in the population means is anticipated. If, however, some physical argument can be offered to expect a specific signed difference, then the one-sided t-test with its associated smaller numerical values of significance level would be appropriate.) The significance level is an estimation of the probabil

ity that a given value of r could be exceeded merely by chance sampling and hence is a measure of the lack of confidence associated with rejecting the null hypothesis that no significant change occurred in the model's climate due to a prescribed change in the model. A rejection of the null hypothesis at a specified significance level implies an acceptance of the alternate hypothesis that there is a statistically significant difference in sample means at that significance level. Therefore, the smaller the numerical value of the significance level (say, 1 % vs. 5 %) at which the null hypothesis is rejected, the more confident one can be that the sample means truly represent different populations and that, indeed, the prescribed change response is statistically significant.

Typically, in significance testing the null hypothesis is rejected (somewhat intuitively) at either the 5 % or 1 % significance level, depending upon the particular application. In the case of four degrees of freedom and $t = r/\sqrt{2}$, these significance levels for the two-sided t-test correspond to critical values of r (values above which the null hypothesis is rejected) equal to 3.92 and 6.48, respectively.

At this point it is instructive to compare these critical values with those associated with a significance test using the estimates of noise level based on a sample of only two simulated Januaries. This measure of inherent variability, of course, corresponds to only one degree of freedom. If r is defined as the ratio of $|\Delta_{30}|$ to the absolute value of the difference between the 30-day averages of a single randomly perturbed case and one control case, then t equals r. Here, the critical values of r are 12.71 at the 5 % significance level and 63.66 at the 1 % significance level for the two-sided t-test. Thus, for a prescribed change response to be statistically significant at the 5 % significance level, r needs to be more than three times larger than is required when the noise level is estimated from a sample of two as opposed to a sample of five. At the 1 % significance level, r needs to be almost 14 times larger than is required with the smaller sample size! The necessity and advantage of using more than two samples in estimating a model's inherent variability should now be obvious.

The methods described here for prescribed change experiment significance testing were put into practice by Chervin *et al.*[3] In this instance the prescribed change was a dipole-like North Pacific

Ocean surface temperature anomaly pattern and the prescribed change response examined was the 1.5-km temperature field both above the anomaly and downstream over the continental US. It was found that a statistically significant signal could be easily detected in the immediate vicinity of large ocean surface temperature anomalies, but that teleconnections of this signal downstream over the US could be identified only with very small statistical confidence. The physical significance of this result is, as pointed out by the authors, not yet clear without further modelling and data comparisons.

An alternative approach for calculating Δ_{30} (which should be used as much as is feasible in GCM experiments) to the one considered so far is based on the difference between the 30-day mean of the prescribed change experiment and the average of the five 30-day means. Using the revised definition in equations (12.1) and (12.2), then the sample value of t equals $r/\sqrt{1.2}$ since $\mathcal{N}_1 = 1$, $\mathcal{N}_2 = 5$. $n_1 = 1$, and $n_2 = 5$. So for a given magnitude of r, a larger value of t would result and this yields a lower numerical value of significance level for rejecting the null hypothesis, as shown in Table 12.2. When $\mathcal{N}_2 = 5$, there

Table 12.2. *Significance levels for a selection of values of* $r = |\Delta_{30}|/\sigma_{30}$ *corresponding to four degrees of freedom and* $t = r/\sqrt{1.2}$

r	Significance level (%) (two-sided t-test)	Significance level (%) (one-sided t-test)
1	41.43	20.71
2	14.12	7.06
3	5.19	2.60
4	2.16	1.08
5	1.03	0.52
6	0.54	0.27
7	0.31	0.15
8	0.19	0.09
9	0.12	0.06

would be less uncertainty in the estimate of the mean of the control climate ensemble and this would allow for the possibility of attributing more significance to a given value of prescribed change response, Δ_{30}. Even better significance would result if $\mathcal{N}_1 > 1$, but trade-offs with computer time are involved.

Conclusions and implications

The statistical significance testing methods discussed here should be of use to those contemplating or involved in analyzing GCM climate experiments. It should also be obvious that a knowledge of the geographical distributions of the estimated inherent variability of model statistics is a prerequisite for the statistical design and planning of such experiments. This approach would increase the likelihood of recovering a significant signal out of the model's inherent noise.

It must be emphasized here, however, that a judgment of significance consistent with the model's variability does not necessarily have to reflect a climate change or physical cause-and-effect in the real atmosphere. Neither does the absence of a significant response in a GCM experiment indicate that the proposed climate change mechanism does not operate in nature. The formalism presented here merely provides a method of estimation of significance that is internally consistent with the model. How well the results of this technique reflect reality depends strongly on the fidelity of the numerical model itself and the extent to which it reproduces the real climate ensemble. Consequently, model development should not be neglected in favor of fashionable climate change experiments but should be a collateral and continuing process aimed at producing a model that more exactly duplicates the properties of the real atmosphere (cf. Schneider & Dickinson[16]).

The most telling (and perhaps sobering) implications of these results, however, is that the statistical significance questions discussed here could mean that a large percentage of the total computational effort in a particular prescribed change GCM experimental series may have to be spent on generating standard deviations – unless, of course, only one prescribed change at a time is effected, or additional changes are small enough not to appreciably change the normal climatology so that the pre-existing estimates of standard deviations may be used. That is GCM modellers must face the prospect that if a new or improved parameterization (cloudiness, radiation and so on) leads to a statistically significant change in the model statistics and also improves the model's fidelity, then it is possible that this prescribed change will also alter the inherent variability. If so, a new data base of noise levels may have to be recomputed every time the standard or control model formulation is changed. Therefore, careful advanced planning of GCM climate experiments is an essential step to insure a statistically significant experiment within a realistic computer budget. It could also turn out to be expedient to perform climate experiments with a more established model where its noise statistics as well as its characteristics are well documented.

One of the most difficult questions to decide in advance is how accurately inherent variability and significance levels need be determined before proceeding with a proposed prescribed change experiment. Similarly, it must often be decided how much signal should be required to accept or reject as statistically significant the results of an already completed prescribed change experiment. In some cases, the answers to these questions will be clear: when the significance levels are either very large or very small. In those cases where an intermediate level of significance is obtained, however, the success of such experiments will depend upon the physical intuition of the experimenters and the existence of other supporting evidence to reinforce the hypothesis being tested. Therefore, a judgment on the significance of the results of a prescribed change experiment will have to be based on an evaluation of many factors, including the significance levels from a formal significance test. In any case, the evidence for claims of significance or insignificance by those reporting on prescribed change experiments with GCMs needs to be clearly stated.

References

1. Spar, J. (1973). *Mon. Weath. Rev.,* **101,** 91.
2. Houghton, D. D., Kutzbach, J. E., McClintock, M. & Suchman, D. (1974). *J. atmos. Sci.,* **31,** 857.
3. Chervin, R. M., Washington, W. W. & Schneider, S. H. (1976). *J. atmos. Sci.,* **33,** 413.
4. Washington, W. W. (1972). *J. appl. Met.,* **11,** 768.
5. Chervin, R. M., Gates, W. L. & Schneider, S. H. (1974). *J. atmos. Sci.,* **31,** 2216.
6. Chervin, R. M. & Schneider, S. H. (1976). *J. atmos. Sci.,* **33,** 391.
7. Chervin, R. M. & Schneider, S. H. (1976). *J. atmos. Sci.,* **33,** 405.
8. *Modelling for the First GARP Global Experiment, GARP Publ. Ser.* No. 14. WMO, Geneva. 261 pp.
9. Kasahara, A. (1972). *Bull. Am. met. Soc.,* **53,** 252.
10. Kasahara, A. & Washington, W. M. (1971). *J. atmos. Sci.,* **28,** 657.
11. Williamson, D. & Kasahara, A. (1971). *J. atmos. Sci.,* **28,** 1313.
12. Williamson, D. & Washington, W. W. (1973). *J. appl. Met.,* **12,** 1254.
13. Shukla, J. (1975). *J. atmos. Sci.,* **32,** 503.
14. Jones, R. H. (1976). *J. appl. Met.,* **15,** 514.
15. Smirnov, N. V. (ed.) (1961). *Tables for the distribution and density functions of 't'-distribution.* Pergamon Press, London. 129 pp. (English translation.)
16. Schneider, S. H. & Dickinson, R. E. (1974). *Rev. Geophys. Space Phys.,* **12,** 447.

SECTION 5

Climate and Man

13

Global influences of mankind on the climate

WILLIAM W. KELLOGG

Whether mankind has the *capability* of altering the climate of the Earth is no longer a debatable topic. The inadvertent changes we have already made on the face of the land and the composition of the atmosphere are of such a magnitude that they should now be taken into account in any equation describing the global 'balance of nature'; and, if we were to apply the kind of human resources we currently spend on armaments to changing the climate purposefully, there is no doubt we could succeed though we would not necessarily improve matters (Kellogg & Schneider, 1974).

As with so many aspects of the extremely complex system that we call 'the environment', the difficulty that we face in assessing mankind's present and potential influences on the climate arises from that very complexity. In other parts of this book the many non-linear interactions that govern the system that determines climate are discussed, and I will not belabor the details here.

For this reason, because of the many uncertainties involved with any quantitative assessment of the climate system, there has been an understandable reluctance on the part of many scientists who are studying climatic change to apply their shaky knowledge to drawing a scenario of the future directions that the climate might take under the growing pressure of civilization. Since I will attempt to do just that, it is pertinent to say at the outset why I believe, first, that one *can* sketch a rough scenario of the future, and, secondly, that we have a certain *obligation* to do so.

While we cannot pretend that we understand all the natural causes of the many climatic changes that have occurred in the past, nor the myriads of interactions of 'feedback loops' within the climate system itself, a variety of climate models and atmospheric general circulation models (such as they are) have been developed to the point where we know how to include most of the dominant factors that govern the energy balance and energy transport processes. They are crude, and do not adequately include ocean circulations and the effects of cloudiness, but they seem to display the main features of the atmosphere as we know it – specifically its surface temperature distribution, and to a lesser degree the patterns of precipitation (Gates, 1975; NAS, 1975, Appendix B). Because these models simulate the present climate quite well, it is reasonable to believe that they will simulate the response of the system to a small change. Our confidence in them is reinforced when several different models show approximately the same response to a given change.

Turning now to my contention that it is not only possible but necessary to attempt to draw a climatic scenario of the future, we must first make a few fairly obvious observations – at least, they have become obvious recently. To begin with, the fact that mankind can change its environment, often for the worse, has become so evident that there has been a growing sense of responsibility for preserving the environment, a kind of tribal guilt that finds its expression in 'the environmental ethic'. Considering the consternation expressed by environmentalists over each inroad into a wilderness area by a road or pipeline, the resistance to building new power plants, and the very real progress made in some countries to curb urban air pollution (at considerable expense), it is curious that there is not a more widespread awareness and concern over the potential for altering the planet's environment as a whole.

In the summer of 1972 there was an historic UN

Conference on the human environment, held in Stockholm, and one of the reports that was required reading for many of the delegations was the *Study of Man's impact on climate* (SMIC Report, 1971), written by an international group of distinguished scientists the year before. This UN Conference, with the SMIC Report in hand to make its members fully aware that an international problem of unprecedented scope was being faced, ended its deliberations with a flurry of sensible and resounding resolutions concerning the need to protect or conserve the global environment of our planet, and enunciated the following Principle 24: 'International matters concerning the protection and improvement of the environment should be handled in a cooperative spirit by all countries, big or small, on an equal footing...'

Subsequent to that 1972 meeting the United Nations Environmental Program (UNEP) was established, with headquarters in Nairobi, Kenya. It has labored to get the countries of the world to protect and improve the environment 'in a cooperative spirit', but so far there has been little indication that any nation is really willing to relinquish its way of doing things for the good of its neighbors, and there is no serious move among the community of nations to establish an international mechanism with any authority. Nor has any case yet been brought to the World Court in the Hague involving litigation over the effects of one country's activities on the environment or climate of another.*

As will be shown presently, the scale of human interventions, both purposeful and inadvertent, on the system that determines our global climate is bound to increase, and it seems very likely that the effects will become noticeable in the next few decades. We should start now to determine to the best of our scientific ability what the extent of these physical changes will be for a given future course of action – how much overall warming will take place? where will it be greatest? what will be the changes in rainfall? how will the desert and 'bread baskets' shift? and so forth. These are *physical* questions that can be approached by our scientific knowledge and our computer-based

*The closest approach to an environmental dispute to be brought before the World Court was the proposed case against France for testing nuclear devices in the South Pacific, but Australia and New Zealand were unsuccessful in their attempts to get it on the docit of the Court.

models, though we cannot yet claim any great precision in our answers.

The next set of questions are even more difficult, since their answers involve *value judgments* and generally revolve around what changes are 'acceptable': how much are we willing to sacrifice by a reduction in some pleasurable activity (such as driving our cars or eating meat)? to what extent can long-established societal patterns be changed (such as moving our agricultural areas or our coastal cities)? how will we allocate finite resources between developed and less developed countries? and so forth.

While it is tempting for a physical scientist to indulge in such value judgments, he should be careful to make it clear when he is doing so. In these matters he is just another 'man in the street', somewhat better informed than the average, perhaps, but not necessarily a better judge of such intangibles.

Can we forecast climatic change?

There is no reason to keep the reader in suspense, awaiting the answer to that question. The answer as I see it, is that we cannot yet usefully predict *natural* climatic change (nor even short-term seasonal fluctuations of average weather conditions beyond a month or two), but that we have probably achieved enough insight into the way the climate system works to predict the *effects of human activities* on some of the factors in this system. In other words, nature could have some surprises in store for us, and we must be satisfied for the most part with past climatic statistics to tell us their probabilities; but, to the extent that mankind's escalating activities can be foreseen, we can estimate some of the future climatic implications of these activities (Schneider, & Kellogg, 1973; Kellogg, 1974, 1975a,b; Mitchell, 1971a, 1972, 1975).

Not every scientist working in this field would agree with what has just been said. Some would argue optimistically that we can even predict some of the natural climatic changes in store for us – relying, for example, on sunspot or solar activity cycles (Willett, 1949; Roberts & Olson, 1973; King, 1975; Wilcox, 1975), or slow changes, in the coupled ocean–atmosphere system (Namias, 1974; Newell, 1974; Quinn, 1974; Adam, 1975). Others would be so pessimistic as to doubt seriously that we can make any useful predictions, even in the face of mankind's massive interventions (Smagorinsky, 1974). Thus I will briefly explain my stand.

So far as we can tell, the Sun, the Moon, and the Earth itself have been behaving pretty much the same way for the past several billion* years, during which time life had a chance to evolve on Earth. While there have been drifts and re-arrangements of entire continents, alternations between ice ages and long periods when there was no permanent ice and snow in the polar regions, periods of greater and less volcanic activity and so forth, the fact that *we are here* is adequate testimony to the remarkable stability of the system that has governed our planetary environment. It has, apparently, never gone too far 'out of bounds' – though there are startling hints of several climatic 'crises' in the distant past that we would very much like to understand better. (I will return to that matter later.)

The history of past climates has been reconstructed well enough for us to apply statistical tests to the record, and we see what seem to be random fluctuations on short time-scales (years to decades) superimposed on some rather striking periodic changes, with periods ranging from a century or less to 100 000 yr or more. There is evidence for some other shorter periodicities in the climate, and the ones corresponding to the solar cycle of 11 and 22 yr show up rather persuasively in some regional climatic records (King, 1975).

The Panel on Climatic Variations of the US Committee of the Global Atmospheric Research Program (GARP) (NAS, 1975) has summarized such climatic statistics, based on a number of different paleoclimatic records (for example, Shackleton & Opdyke, 1973; Kutzbach & Bryson, 1974; CLIMAP, 1976) covering the past 700 000 yr, and has described the history of the mean surface temperature in terms of a superposition of five periodic functions with different amplitudes, as shown in Fig. 13.1. No single paleoclimatic record looks just like this reconstruction of the history of mean surface temperature back to 10 000 yr ago, but it does display most of the important changes of the past, such as the cooling trend since about 1940, the warming trend before that, and the remarkable decrease in temperature about 4000 yr ago from the warm period known as the Altithermal. It will be noted that we have purposely blurred the curve

*Thousand million.

as it goes back in time, taking the shorter-period oscillations (whose period and phase we do not know at all well) to be a kind of 'noise' superimposed on the longer-period fluctuations. This blurring also helps to convey our uncertainty about the correct mean temperature values to assign as we go further back in time.

This reconstruction does not agree in detail with any of the individual temperature records deduced from deep-sea cores, lake beds, peat bogs, tree rings, or the Greenland ice sheet. In fact, we do not know how to draw a reliable global mean temperature record. One record that applies to an extensive region, however, has been deduced by Nichols (1974, 1975) from the study of ancient pollens in peat deposits at six locations across northern Canada (and is shown in the upper part of Fig. 13.1). From the pollen counts one can derive the relative abundances of various kinds of shrubs, grasses, lichens, and trees, and this in turn indicates the mean summertime temperature or length of growing season (probably closely related). It does not, however, tell much about winter temperatures. In this record for northern Canada the Altithermal period shows up quite strikingly, and there are apparent shorter-term fluctuations that should be noted. For example, the abrupt cooling that occurred about 4800 yr ago killed off the spruce forests of northern Canada and forced the line separating forest from Arctic tundra southward almost to its present position. The tree line slowly moved back northward in the succeeding century or two, but then there was another abrupt cooling around 3500 yr ago, accompanied by an increase in forest fires in summer, that forced the tree line far south again.

I have gone into these matters in some detail here in order to provide some background on the natural climatic changes that have occurred, long before mankind can have had any significant influence on the surface temperature. It also provides a useful gauge of the point at which human influences can begin to be considered as 'significant', and it is evident from Fig. 13.1 that the *rate of change* must be a major criterion. For example, we should expect one or two degrees of change over a period of several thousand years from the natural fluctuations, whereas the expected natural rate of change of the next few decades is at most about 0.15 °C per decade. (However, these statistics do not take account of the occasional abrupt cooling events that appear

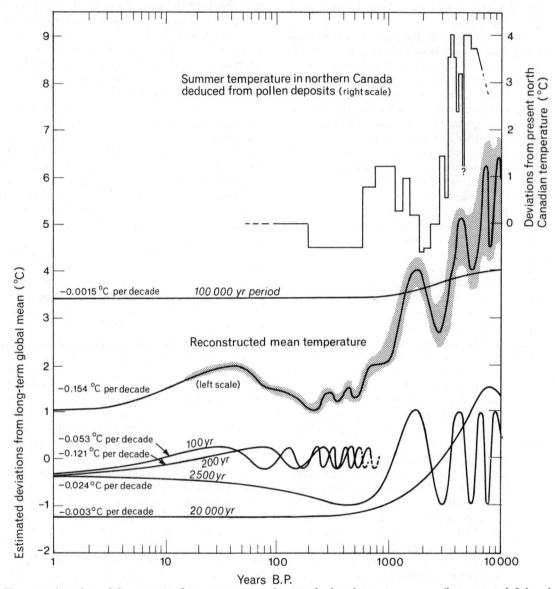

Fig. 13.1. A review of the mean surface temperature changes during the past 10 000 yr (lower part, left hand scale). This is intended to show the general features of the changes; the five periodic functions (with periods from 100 to 100 000 yr) from which the mean temperature was reconstructed were derived from a wide variety of paleoclimatic records (NAS, 1975), no one of which can be considered as entirely representative. It will be noted that the shorter-period fluctuations largely account for the rate of change (noted at the left end of each curve), while the longest-period fluctuation has the largest amplitude and largely accounts for the major alternations between ice ages and interglacials (such as the present one). The temperature record for northern Canada (upper part, right hand scale), obtained by Nichols (1974, 1975) from studies of pollen in lake sediments, is roughly representative of records obtained elsewhere at middle-to-high latitudes. Dating of such records is generally done by ^{14}C analysis.

in the paleoclimatic record from many places and which have no generally accepted explanation, such as the cooling in northern Canada 4800 yr ago. There does not seem to be a good way of assigning a probability to their occurrence until we understand them better.)

Turning now to the matter of mankind's present and future influences on the climate, I will be dealing with rather well-formulated questions involving the specific effects of specific changes in the boundary conditions or 'forcing functions' of the climate system. These boundary conditions, such as the radiational input at the surface or the heat added to the air from our release of stored energy, are explicitly specified in our climate models, so we can perform 'experiments' with our models to see how the models respond – all other boundary conditions being kept the same.

Such experiments are not easy to interpret, however, and there are several reasons for being doubtful that the result of an experiment with a climate model will necessarily be the same for 'the real thing'. If we consider a hierarchy of such models, from simple one-dimensional (vertical), globally-averaged models to three-dimensional and time-dependent models (Schneider & Dickinson, 1974), and if we experiment with them and see that a given change of boundary condition influences the surface temperature (for example) in all the models in approximately the same way, then one can have some confidence that the real system would do approximately the same under the same change of boundary condition.

If we have been blind to a dominant factor in *all* the models used in such an experiment, then this feeling of confidence could be false. The most frequently cited 'blind spot' in current climate system modelling is the lack of consideration of interactions between large-scale ocean circulations and the atmosphere – an omission due to the difficulty of modelling the oceans themselves. Another blind spot is the proper treatment of global cloudiness, since clouds act as a kind of Venetian blind to change the solar radiation reflected by the Earth. We do not yet have a very good way of assessing just how cloudiness on a large scale responds to changes in mean temperature or circulation, though the influence of cloudiness on such a change appears to be fairly small (Cess, 1977). In a later section I will discuss, in the specific context of a predicted climatic

change due to human activities, how these two blind spots might affect the prediction.

Thus, we will proceed to make our best guesses about the future effects of human activities on the climate, mindful of the possible deficiencies in our physical models, and equally mindful of the uncertainties of mankind's future. We have, as we have said, an obligation to try. We cannot wait until we know all about the climate system and have developed a 'perfect model' – which will probably never be. The dialogue concerning our long-term future on the Earth is underway, and decisions will have to be made soon in the face of all the uncertainty (Gribbin, 1974; Schneider, 1976). We can hope that in the next few years we will narrow this degree of uncertainty.

Leverage points of the climate system

To affect the climate, mankind must succeed in having an appreciable impact on some important parts of the climate system, or 'leverage points'. Most of the effects that I will describe have to do with the energy balance, and hence the temperature distribution. But there is also the definite possibility of affecting the ability of the atmosphere to rain by control of the nuclei required for rain to form from tiny droplets, by changing the stability of the lower atmosphere, or by control of the electric field that is involved in the precipitation process.

A description of a new heat balance, mean temperature distribution, or mean rate of precipitation is not enough to define the corresponding new climate, for the environment in which we live is shaped by the large-scale atmospheric circulation patterns that girdle the Earth and determine the ever-changing weather. It is the general circulation patterns that bring rain to some regions in a given season and deny rain to desert areas; and these shifting patterns of wind may vary from year to year as well, bringing famine or plenty to hundreds of millions of people (Lamb, 1972; Schneider, 1976). Meteorologists have names for these large circulation patterns: the monsoons, the trade winds, the mid-latitude westerlies, the subtropical pressure areas and so forth. In the next section I will discuss some of the more general implications for changes in these patterns.

Levers involving the heat balance of the atmosphere
Patterns of land use. There are many ways by which mankind can influence the heat balance of the

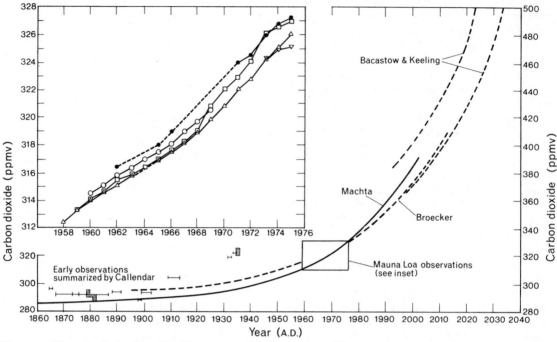

Fig. 13.2. The record of carbon dioxide concentration (parts per million by volume, ppmv) from 1860 to 1975, measured at several locations, and some estimates of future trends. The early data were critically reviewed by Callendar (1958) and subsequently re-evaluated by Barrett (1975). The current series of observations for Mauna Loa (□–□) are those reported by Keeling *et al.* (1976a) and C. D. Keeling (personal communication), for the South Pole (△–△) by Keeling *et al.* (1976b) and Keeling (personal communication), for American Samoa (▽–▽) and Point Barrow (●–●) by NOAA/GMCC (1975) and T. Harris (personal communication) and for the Swedish aircraft observations (○–○) by Bolin & Bischof (1970). Note that the carbon dioxide concentrations are given in terms of the 'adjusted index values' (for the sake of continuity with the earlier data); it may be necessary to adjust these values upward by about 3 to 4 ppmv, according to Keeling *et al.* (1976a), to obtain the correct mole-fraction, but this would not affect the slopes of the curves. The model calculations predicting future carbon dioxide increases by Machta (1973), Broecker (1975) and Bacastow & Keeling (1973) all take account of the take-up of anthropogenic carbon dioxide by oceans and the biomass (but in somewhat different ways), and assume a quasi-exponential increase in the rate of burning of fossil fuels (notably coal) in the next half-century or more. It is expected that in this time period about half of the new carbon dioxide released will remain in the atmosphere; and due to the slow mixing of deep ocean waters with the upper layers the decay time of the carbon dioxide, were we to stop producing it, would be on the order of 1000 yr.

Earth, and the one that he has been working at longest is the alteration of patterns of vegetation. When a forest is cleared for a pasture or wheat field the result is an area that generally reflects more sunlight, since crops and grassland are usually less absorbing than trees. The same is true when nomadic tribesmen allow their cattle or (especially) goats to over-graze on marginal land, since the destruction of vegetation markedly increases the reflectivity of the surface (Bryson & Baerreis, 1967; SMIC Report, 1971; Eckholm, 1975; Glantz & Parton, 1975).

Such changes in the solar radiation absorbed by the surface must certainly have an effect on the heat balance and climate of a region, and influence its precipitation as well as its mean temperature (Charney, Stone & Quirk, 1975; Otterman, 1974). A secondary effect is very likely to be the increase in wind-blown soil and sand, which also affects the radiation at the surface and the stability of the atmosphere above it (Bryson & Baerreis, 1967).

So far as I know, there has never been a world-wide inventory of this kind of effect, though we do know that some regional surface changes have been and will be very extensive (Newell, 1971; SMIC Report, 1971; Bryson, 1974). Unfortunately we do not know how much of a cumulative effect all these changes have had on our climate, but it is suspected that it has not been as extensive as some of the other effects to be described shortly.

Carbon dioxide. Since the beginning of the Industrial Revolution more than a century ago we have been taking carbon out of the Earth in the form of coal, petroleum and natural gas, and burning it to make carbon dioxide and water vapor – plus heat, which is of course our main reason for doing it. Of the carbon dioxide that has emerged from countless chimneys and exhaust pipes, about half is still in the atmosphere and the other half has been dissolved in the oceans or has gone into the Earth's biomass – mostly the forests.

The carbon dioxide in the atmosphere has risen from an estimated 280 to 290 parts per million by volume (ppmv) to the present 320-plus ppmv, and it is estimated that it will reach some 380–390 ppmv by 2000 A.D. (Bacastow & Keeling, 1973; Ekdahl & Keeling, 1973; Machta & Telegadas, 1974; Broecker, 1975), and may double by the next mid-century (Bacastow & Keeling, 1973).

Fig. 13.2 depicts the past history of the carbon dioxide concentration and how it is expected to increase in the future. The records at Point Barrow, Mauna Loa, and the South Pole show that there is a couple of years' lag between the northern hemisphere (where most of the carbon dioxide is released) and the southern hemisphere, as would be expected because of the slow exchange of air between hemispheres. Also, the slopes of the curves are not exactly constant, there being a slackening in the mid-1960s followed by an acceleration in the 1968–71 period. Since world-wide release of carbon dioxide cannot have changed much from its steady rise of about 5.7 % yr^{-1} (SCEP Report, 1970), the explanation for these changes in carbon dioxide rate of increase probably lies in changes of the rate of uptake by the oceans (Bacastow, 1976).

The chief concern that we have with this changing component of the atmosphere is its effect on the heat balance, since carbon dioxide is virtually transparent to solar radiation but absorbs outgoing terrestrial infra-red radiation in several infra-red bands, radiation that would otherwise escape to space and result in a loss of heat from the lower atmosphere. The result, therefore, of an increase in carbon dioxide is a cooling of the stratosphere and an increase in surface temperature (Schneider & Kellogg, 1973).

There have been several model calculations to show the influence of increased carbon dioxide on the surface temperatures, some globally-averaged one-dimensional models such as that of Manabe & Wetherald (1967), some latitude-dependent, with the oceans taken into account crudely (Sellers, 1974; Manabe & Wetherald, 1975). These various results have been reviewed most recently by Schneider (1975) and Budyko & Vinnikov (1976). A representative set of estimates is as shown in Table 13.1.

Table 13.1. *Effects of adding carbon dioxide to the atmosphere*

Factor of change of carbon dioxide (%)	Expected time for change to occur (years A.D.)	Mean surface temperature increase (°C)
+25	2000	0.5–1
+100	2050	1.5–3

These surface changes, it should be emphasized, refer to the weighted average (by surface area) for the globe. Both observed changes of climate and climate models indicate that in the polar regions, above about 50° latitude, any climatic change would be expected to be from three to five times larger (perhaps even more in winter) than an average change, such as those shown in the table (SMIC Report, 1971; Sellers, 1974; Manabe & Wetherald, 1975; van Loon & Williams, 1976).

Chlorofluoromethanes. A contaminant added to the atmosphere in the past few decades by mankind in large quantities, one that has been most notorious for its possible effect on the ozone layer in the stratosphere, is the chlorofluoromethanes, referred to as FC-11 and FC-12. (They are sometimes also referred to as 'freons', but that is a trade name.) These gases, used both as refrigerants and as aerosol propellants, are extremely stable, non-toxic and persist in the troposphere for very long periods of time (about 40 yr mean residence time for FC-11, 70 yr for FC-12), and the observed build-up of the FCs in the lower atmosphere suggests that virtually all of the gas released to date can still be found resident in the troposphere. There are probably small sinks at the surface and in the troposphere, and there is a long-term sink in the stratosphere, since those molecules that diffuse upward into the stratosphere are broken down by the ultra-violet radiation there (Crutzen, 1974; Rowland & Molina, 1975; Wofsy, McElroy & Sze, 1975).

While I will not comment here on the effect of these compounds on the ozone layer (though this is, in a broad sense, a change of the environment), it turns out that they have a direct effect on the temperature balance of the atmosphere that has only recently been identified (Ramanathan, 1975). Like carbon dioxide, the FCs have absorption bands in the part of the infra-red 'window' between 8 and 15 μm, where there is relatively little water vapor absorption. Thus, the FCs prevent some of the infra-red terrestrial radiation from the surface that would otherwise escape to space from passing through the lower atmosphere. The result is an increase in the surface temperature and a corresponding decrease in the stratospheric temperature.

The present mean tropospheric concentration of total FCs is about 0.2 parts per billion by volume (ppbv), and calculations by P. J. Crutzen (personal communication) and others indicate that if FCs continued to be produced at the 1974 production rate FC-11 would reach about 0.6 ppbv and FC-12 about 1 ppbv by 2000 A.D., and would level off in the middle of the next century at about 0.8 and 2 ppbv, respectively. This quasi-steady state would result in a decrease of outgoing infra-red radiation of some 0.3 % and an associated mean surface temperature increase of 0.5 °C, based on Ramanathan's model calculations (which I take to be correct to within a factor of two). If, on the other hand, the worldwide production rate of FCs continues to increase, as it has in the past (Howard & Hanchett, 1975), and the total concentration of FCs were to increase to 4 ppbv, then the temperature rise would be about 1 °C. We cannot say exactly when this would be (since it would depend on the actual production rate), but it is not inconceivable that it could occur as early as 2000 A.D.

Projections of the future use of FCs will depend very much on the passage of legislation in various countries limiting the use of FCs as propellants in spray cans. At present, roughly one half of the production is in the US, where such legislation is being seriously considered. In any case, it is likely that the FCs will continue to be used extensively as refrigerants, for which they are probably ideally suited, and it would be difficult (if not unnecessary) to prevent the continued escape of these FCs into the atmosphere.

Nitrous oxide and other infra-red absorbing gases. Marked increases in surface temperature can be produced by large-scale releases of carbon dioxide and the chlorofluorocarbons, the effect being due to their ability to absorb infra-red radiation in the atmospheric 'window', and their long persistence. This suggests that we should be alert to the build-up of any other trace gases that have similar properties, and there are a great many of them.

One such trace gas is nitrous oxide, which is mainly maintained at its present concentration of about 0.28 ppmv by biological decay and conversion processes taking place in soil and in the oceans – processes referred to as 'de-nitrification'. It has been suggested that the increasing use of nitrate fertilizers by mankind may accelerate the biological production of nitrous oxide and raise its atmospheric concentration (Crutzen, 1976; McElroy, Elkins, Wofsy & Yung, 1976), with implications for both surface temperature increase and stratospheric ozone concentration decrease. The amount of this increase in nitrous oxide concentration is still uncertain, since estimates vary from a trivial increase to as much as a factor of two in the early part of the next century. The latter would produce a warming in the order of 0.5 °C (Yung, Wang & Lacis, 1976), but this may be considered as an estimate on the high side until we understand the global nitrogen cycle better, and specifically the relative productions by ocean and land (soil) biota.

Aerosols. Another product of human activity is the particles that are produced by industry, power generation, automobiles, space heating, slash-and-burn agricultural practices and so forth. These particles, commonly known as aerosols, are obvious additions to the atmosphere of the large cities of the world, where they are largely produced by a combination of coal burning and the creation of particles from unburned hydrocarbons in the atmosphere by the action of ultra-violet radiation. Such secondary particles tend to be somewhat smaller in size than the directly produced smoke or soot particles, though after they have existed for a time in the air they attach themselves to the larger particles, forming particles that are a combination of both (NSF, 1976).

There is little doubt that since the turn of the century there has been an increase in the rate at which aerosols have been produced by mankind, particularly in the more industrialized countries (Cobb & Wells, 1970; Pivovarova, 1970; SMIC Report, 1971; Budyko & Vinnikov, 1973; Bryson, 1974; Dyer, 1974; Machta & Telegadas, 1974; Mitchell, 1974; Ellsaesser, 1975), and many non-urban stations (but definitely not all) have recorded some long-term upward trends in the total aerosol content. If this is so, then one must ask how extensive the aerosols really are, and what their effect will be on the regional or global radiation balance if the upward trend were to continue.

Aerosol particles can both scatter and absorb sunlight, and they also absorb and re-emit infrared radiation to a limited extent. When a particle scatters solar radiation some of the scattered radiation will be directed upward as well as downward, and the upward component will be lost to space. This results in less sunlight reaching the ground and an increase in the net albedo, or reflectivity, of the atmosphere–Earth system, which would cause a net cooling. However, when a particle absorbs some of the solar radiation it heats the particle and the air around it, and the effect of this is to reduce the net albedo. Theory tells us that in order to decide whether aerosols cause an increase in the net albedo (cooling) or a decrease (warming) we must take into account the ratio of the particle absorption to its backscatter, which I will call a/b, and also the albedo of the underlying surface (Mitchell, 1971a,b; Chýlek & Coakley, 1974; Weare, Temkin & Snell, 1974; Schneider & Kellogg, 1973; Coakley & Chýlek 1975). When aerosols of a given a/b are over a dark surface, such as the ocean, they are more likely to increase the net albedo than when they are over a light surface, such as a snow-field or a low cloud deck – or over land generally. This relationship is summarized in Fig. 13.3, calculated by Chýlek & Coakley (1974).

There has been a widely shared belief that such aerosols generally cause a cooling, the argument being that when spread evenly around the Earth

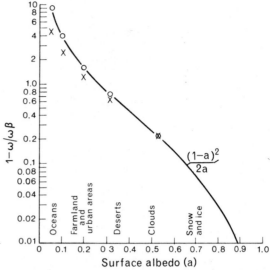

Fig. 13.3. Critical ratio of absorption to average upward-scattering cross sections $(1 - \omega/\beta\omega$ or $a/b)$ as a function of surface albedo (a). The curve with circles represents results of the radiation model of Chýlek & Coakley (1974), which takes account of solar radiation only; above this curve there will be a decrease in the net earth-atmosphere albedo as a result of the aerosols, and consequently a warming. The '×' symbols represent a typical case, calculated by J. A. Coakley (personal communication), in which both solar and infra-red effects are combined, showing that the infra-red effects tend to enhance the warming influence of aerosols.

their effect over the dark oceans is to increase the albedo and thereby prevent some of the sunlight from being absorbed by the Earth-atmosphere system (Rasool & Schneider, 1971; Yamamoto & Tanaka, 1972; Budyko & Vinnikov, 1973; Bryson, 1974; Bolin & Charlson, 1976). Recently, however, it has been pointed out that most of these anthropogenic aerosols exist over the land, where they are formed, and that they are sufficiently absorbing to reduce the albedo rather than increase it (Eiden & Eschelbach, 1973; R. E. Weiss, R. J. Charlson, D. L. Thorsell and D. D. Duncan, personal communication; Kellogg, Coakley & Grams, 1975).

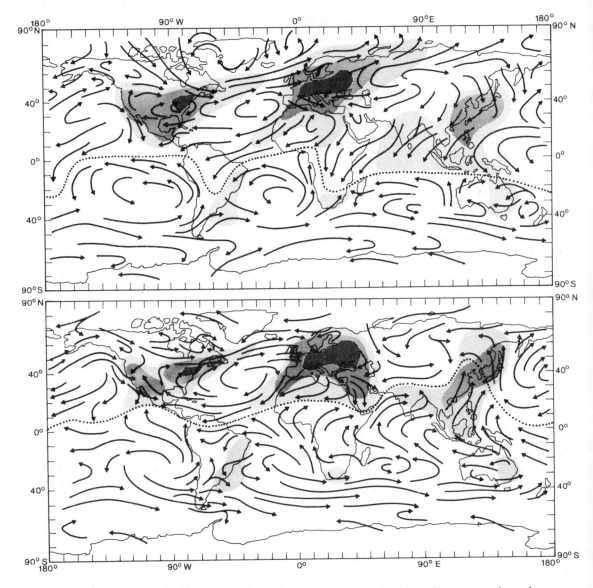

Fig. 13.4. Estimated global distributions of anthropogenic aerosols, based on the assumptions that production rate is proportional to the gross national product of each country, there is a mean residence time of five days (Martel & Moore, 1974), and surface winds (taken from Lamb, 1972) for (*a*) January, (*b*) July.

In Fig. 13.4 a theoretical global distribution of mankind's aerosols is shown, assuming that the production in each country is proportional to gross national product, and that the aerosols drift with the surface winds and remain in the atmosphere with a mean residence time of five days before they are rained out or washed out or directly deposited at the surface (Martel & Moore, 1974; Kellogg *et al.*, 1975). Their distribution is very uneven, being mostly confined to the industrialized regions of the northern hemisphere, though a certain portion does drift out over the Atlantic and Pacific Oceans, and a considerable part of Europe's 'gross national pollution' drifts over North Africa, particularly in the wintertime. These aerosols absorb more solar radiation than natural aerosols, and their a/b values are generally large enough to cause a lowering of the albedo of the atmosphere–Earth system *over the land*, and thereby cause a warming. However, largely due to our lack of quantitative knowledge of the optical characteristics of these aerosols and their distribution, we cannot yet assign any number to this warming effect.

There are other effects that aerosols may have on the climate of a region, especially its rainfall. In a later section I will discuss their role as condensation and freezing nuclei, which may be significant; another effect that may be significant is their influence on the stability of the lower layers of the atmosphere. Since, as has been pointed out, they absorb a certain amount of solar radiation, the upper part of a low-lying aerosol layer will be warmed, and the absorption and scattering processes will both cause a decrease in the solar radiation reaching the ground. The result is a warming of the upper part of the aerosol layer (in the daytime) and a decrease in the rate of warming at the ground, and this causes the stability of the atmosphere near the ground to be larger than it would be in the absence of the aerosol particles. Bryson & Baerreis (1967) have suggested that the radiational effect of aerosols may decrease convective-type precipitation, especially in subtropical places such as northwest India, and Wang & Domoto (1974) and Atwater (1975) have investigated the effect theoretically. Unfortunately, again we do not have enough information about the optical properties of aerosols to make a quantitative evaluation of this influence on rainfall.

Before leaving the effects of anthropogenic aerosols on the radiation balance, there is one more point that may be important but has often been overlooked. Clouds have a fairly high albedo or reflectivity, as is obvious, but theoretical calculations involving the scattering and absorption of plain water droplets indicate that they ought to be more reflective than they are in fact (Twomey, 1972; Liou, 1976). The difference is thought to be due to the presence of absorbing aerosol particles, and the decrease in reflectivity will occur whether the particles are included within the cloud droplets or floating between them. Since the apparent reduction of reflectivity is quite marked (10 to 20 %), it is clear that any increase in absorbing aerosols will cause additional absorp-

tion of solar radiation by the clouds, and another source of heating if anthropogenic aerosols are added to the lower atmosphere.

Aerosols, probably because they are so very obvious to the eyes of the population in large cities, have been the subject of vigorous attempts to control them. The result is that in many cities of the world the aerosol content, *particularly of larger particles*, has shown a definite decrease (Ellsaesser, 1975). The same cannot in general be said for the *total aerosol content* of the atmosphere observed in Europe and the eastern US, where secondary aerosol production of smaller submicron particles, especially sulfates from the sulfur dioxide produced by burning high sulfur fuels, have become a dominant factor in *regional* air pollution (Kellogg *et al.*, 1972). It should be noted that the practical problems posed in these regions by the ecological and health effects of increasing quantities of sulfate particles (e.g. Bolin *et al.*, 1971) probably far outweigh their influence on the regional climate, but that is beyond the scope of this chapter.

Thermal pollution. When we are considering the heat balance of the atmosphere it is clear that the direct addition of heat in any form should be taken into account. In some of the large cities of the world, especially those at high latitudes where there is relatively less sunlight, the amount of heat released per square meter is equal to or even greater than the average flux of sunlight absorbed at the surface during the year. On a regional basis, however, this is at present rarely more than a few percent, and on a global basis the total amount of heat released by all of mankind's activities is roughly 0.01 % of the solar energy absorbed at the surface (SMIC Report, 1971; Kellogg, 1974; 1975*a,b*). Such a small fraction would have a negligible effect on the total heat balance of the Earth.

Several projections of the future world population and energy production have been made by 'futurists', and most have concluded that a population of 20 billion people could be supported, given high technology (and avoidance of nuclear war), and that it is not inconceivable that the per capita energy consumed by this 'post-industrial society' could be as much as four times that of the US in 1976 (Weinberg & Hammond, 1970; Häfele, 1974; Kahn, Brown & Martel, 1976). This would mean 20 billion people each

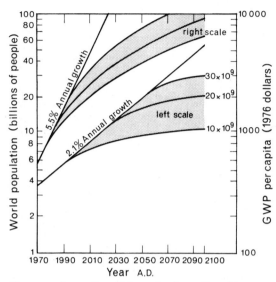

Fig. 13.5. A set of possible projections of world population and gross world product (GWP) per capita. The GWP per capita follows approximately the scenario described by Kahn, Brown & Martel (1976, Fig. 5, p. 56). It will be noted that these curves tend toward a levelling-off or steady state, which is obviously more realistic than any continued exponential growth. Nevertheless, they should not be taken as 'predictions', but rather as a rough indication of the time-scale involved in any such evolutionary process.

requiring 40 kW,* and the total rate of heat production and release into the atmosphere would then be 8×10^5 Gigawatts (GW). The total amount of solar energy absorbed at the surface of the *Earth* is about 8×10^7 GW, so this would amount to 1 % of the total.

This large amount of heat would presumably be released over the continents where the people are, and that would give an uneven distribution of heating as seen on a global scale and produce marked regional effects (Washington, 1972;

* In the US the per capita energy consumption is about 10 kW, which corresponds to the total raw energy before it is converted to 'useful' energy. The energy released, whether it goes up a smoke-stack or comes out of an electric outlet, ends up as heat added to the atmosphere – a minor exception being the radiation *directly into space* of light, radiowaves and so forth.

Llewellyn & Washington, 1977). We can assume that this heat will end up by being more or less evenly distributed in a given hemisphere and then use our climate models to estimate the effect that this would have on mean surface temperature. The additional heat can be considered in these model experiments as if it were an increase in the total amount of solar radiation reaching the surface. There is not universal agreement among the various climate models (Schneider & Dennett, 1975; Gal-Chen & Schneider, 1976) but the present set of models seem to converge on the answer that a 1 % increase in the heat available to the system would result in about a 2 °C increase in the mean surface temperature, within a factor of two. As mentioned earlier, any mean surface temperature change would correspond to a very much larger temperature change in the polar regions, perhaps three to five times greater (Wetherald & Manabe, 1975).

Again we must ask what kind of time-scale is involved in such a major increase, if it should occur. Figure 13.5 shows a simplified but reasonable ('surprise-free') scenario in terms of population or per capita gross world product (related to per capita energy), in which the exponential increase of the present gradually gives way to a steady-state condition of the future; and the *earliest* time at which this kind of levelling off could occur is about 2100 A.D. There is no assurance that the population will even grow to 20 billion – and many, including the author, hope that it never will, while others firmly believe that it cannot (Heilbroner, 1974; Holdren & Ehrlich, 1974). I use this scenario merely to convey a kind of upper limit to the possible global thermal pollution from a post-industrial society, and the emphasis here is on the probable time-scale involved.

Summary of the heating effects. I have mentioned several anthropogenic causes for climatic change in terms of their effects in the mean temperature of the surface, and Table 13.2 summarizes these effects. To a first approximation they can probably be considered as additive, because they represent small fractional changes (and therefore follow more or less linear relationships, though for larger changes some of the effects must definitely become non-linear).

The rate of change of the temperature is probably the most important factor to consider,

Table 13.2. *Summary of anthropogenic influences on the global mean surface temperature*

Effect of mankind	Time period for the effect to occur	Influence on surface temperature (°C)	Rate of change toward the end of the time period (°C per decade)
Raising the carbon dioxide content of the atmosphere	+25% by 2000 A.D. +100% by 2050 A.D.	+0.5 to 1[a] +1.5 to 3[a]	0.2 to 0.4 0.3 to 0.6
Adding chlorofluorocarbons to the troposphere	0.9 ppbv by 2000 A.D.[b] 2.7 ppbv by 2050 A.D.[b] 3.5 ppbv by 2000 A.D.[c]	0.1 to 0.4[d] 0.2 to 1[d] 0.4 to 1.5[d]	0.04 to 0.2 0.02 to 0.1 0.2 to 0.8
Adding aerosols to lower troposphere	?	Heating[e]	?
Direct addition of heat	100-fold increase by 2100 A.D.	+1 to 4	0.1 to 0.4

a See Table 13.1.

b Assuming continued chlorofluorocarbon (FC) production at the 1973 level (NAS, 1976).

c Assuming 10% yr^{-1} increase in FC production rate (NAS, 1976).

d Estimate by Ramanathan (1975), revised by NAS (1976).

e It is not clear whether the upward trend in anthropogenic aerosols will continue – it will depend to a large extent on control of sulfur-dioxide emissions.

f Estimated under the assumption that energy production would continue to grow as the product under the two central curves in Fig. 13.5. In this case the *total* effect (2°C) would be more significant than the *rate of change* because it builds up over a fairly long period.

since it can be compared to the statistics of natural climatic change summarized in Fig. 13.1, and this comparison gives a measure of the 'significance' of the effects. After all, we are dealing with a natural system that would fluctuate and change regardless of what we did to it, and our influence could only be noted if it turned out to be significantly larger than the 'background noise'. Currently, and for the next decade or so, the *natural* rate of change of mean surface temperatures can be estimated to be about -0.15 °C per decade (Fig. 13.1), and before 2000 A.D. it could turn up again at a comparable rate (Broecker, 1975). In any case, such natural trends are not likely to exceed ±0.25 °C per decade (Kutzbach & Bryson, 1974; NAS, 1975).

The conclusion to be drawn from this is that the anthropogenic influences on global mean surface temperature, especially the effects of carbon dioxide and FCs (unless the use of the latter is severely restricted), *will probably begin to dominate over natural processes of climatic change before the turn of the century*, and will result in a decided warming trend that will accelerate in the decades after that. It is quite likely, in fact, that this warming trend has already begun (Damon & Kunen, 1976; Borzenkova, Vinnikov, Spirina & Stekhnovskiy, 1976).

Again we must take care to remember that when we attempt to estimate the effect of a change in the 'boundary conditions' of the climate system using our models we must recognize that these

models are still fairly primitive, and do not necessarily simulate the 'real thing' exactly. Judging from the fact that a variety of models have been intercompared, and that they agree remarkably well, we can probably have faith that they represent the correct answer to within a factor of two or better. While there are a number of factors that we recognize as not having been adequately taken into account, such as the ocean circulations and the effects of cloudiness, there is convincing evidence that when these are properly taken into account they will not greatly influence the answer (Cess, 1977).

Changes effecting the precipitation process

Condensation and freezing nuclei. Many of the aerosols produced by industry have the property of acting as condensation nuclei or freezing nuclei – that is, they can initiate the formation of cloud droplets or hasten the freezing of cloud droplets at temperatures below 0 °C. Notable among freezing nuclei sources are steel mills and lead compounds from automotive exhausts. Also, the most common kind of aerosol produced by burning coal and fuel oil, sulfates, are very good condensation nuclei.

While the effects of these condensation and freezing nuclei on the precipitation process are bound to be significant regionally, and while it has been clearly demonstrated that precipitation has indeed increased in and down-wind from certain cities such as Saint Louis, Chicago and

Paris (Dettweiler & Changnon, 1976), it is difficult to assess quantitatively the effect of these activities even on a regional scale. We must merely, for the time being, recognize this as a potentially important effect (Hobbs, Harrison & Robinson, 1974).

Krypton-85 from nuclear power generation. Several radioactive gases are released into the atmosphere from nuclear power plants and from the plants that reprocess nuclear fuel. Notable among these are tritium, with a half-life of 12.5 yr, and krypton-85, with a similar half-life of 10.7 yr. Krypton-85 is a noble gas that remains more or less permanently in the atmosphere without undergoing any chemical combinations, so it builds up in the atmosphere, subject only to its slow radioactive decay.

When a krypton-85 atom disintegrates it produces an energetic electron that ionizes the air in its vicinity. There are other sources of ionization in the lower atmosphere, such as cosmic rays and the radioactive products of uranium, notably radon and its decay products. As the concentration of krypton-85 builds up in the troposphere, assuming a continued increase of the use of nuclear power in the world, the ionization from this source will begin to compete with all the other natural radioactive sources. One estimate has been made of this effect, and the prediction is that there will be a 10 to 15 % increase in the total ionization or conductivity of the lower atmosphere in about 50 yr (Boeck, Shaw & Vonnegut, 1975; Boeck, 1976).

Such a change in the ionization of the atmosphere would have little or no direct effect on living things that we can identify (the level of krypton-85 discussed by Boeck is 100 times less than the maximum permissible airborne concentration in unrestricted areas), but if the conductivity of the lower atmosphere is increased one may expect that there will be an effect on the fair-weather electric field, which is maintained by all the thunderstorms of the world acting together as a direct current generating mechanism. This electrical system is in effect a global spherical condenser, with a positive charge in the upper atmosphere (the outer shell, which is a good conductor) separated from a negative charge on the ground (the inner shell) by the relatively non-conducting lower atmosphere. The lower atmosphere is not a perfect insulator, however, and a

steady leakage of current takes place from upper atmosphere to the ground that must be just balanced by the upward countercurrents produced in the thunderstorms. If the conductivity of the lower atmosphere were increased due to krypton-85 ionization, as suggested by Boeck, then the leakage between the two regions would be increased and (as when a condenser is discharged) the electric field would be decreased – unless the thunderstorm generators worked correspondingly harder. Actually, there is good reason to believe that the efficiency of the thunderstorm charge separation process depends on the fair-weather electric field (Sartor, 1967, 1969), so a decrease in the electric field would decrease the rate at which the global generating mechanism worked to maintain it – a positive feedback.

It is generally believed that the process of rain formation, especially in thunderstorms, depends on the existence of strong electric fields in the clouds, and these electric fields in the clouds (closely related to the recharging processes just discussed) are initiated by the fair-weather electric field that was there before the cloud formed (Sartor, 1969). Thus, it has been hypothesized that a change in the fair-weather electric field would decrease the rate of electric field generation in clouds, and that this in turn would result in a decrease in the rate of formation of precipitation and perhaps a weakening of the cloud dynamical processes as well (Ney, 1959; Vonnegut, 1963; Kellogg, 1975c; Markson, 1975).

Unfortunately, these processes are not understood quantitatively, and it is impossible at this time to assign any value to the effect of an increase in conductivity or a decrease in the fair-weather electric field on precipitation – though I would guess that it would be a *negative* effect. Any such change on a global scale would also affect the heat balance, since thunderstorms account for a major part of the vertical exchange of heat and momentum at low and middle latitudes (Palmén & Newton, 1969).

What would a warmer Earth be like?

I have shown that most of the things that mankind does to affect the heat balance of the climate system result in a warming. Mankind also affects precipitation processes and the distribution of heat and moisture. The next obvious question to be asked is whether such a warmer Earth, with its new distributions of temperature and precipitation, would be a 'better' or 'worse' place to live in, and in this section I will attempt to throw some light on this point.

Simulating and reconstructing the climate of a warmer Earth

It is the changing large-scale circulation patterns of the atmosphere that are largely responsible for the climate of a region: its temperature and rainfall or snowfall, the march of the seasons and the variability that we associate with the changing weather. There must be a relationship between these circulation patterns and the large-scale heat balance (or mean global surface temperature), of course, since they are both measures of the activity of the atmospheric heat engine. The first is, in general terms, a measure of the kinetic energy of the system, and the other is a measure of the thermal energy available to run it. We would like to know more about this relationship.

Our most elaborate models are called 'general circulation models' (or GCMs for short), and these involve a detailed integration on a computer of the time-dependent equations of motion and state that govern the atmosphere, and the result is what amounts to a moving map of the global circulation system that looks remarkably like the real pattern. There have been several experiments with GCMs (already referred to), in which changes in the heating applied to the system by the Sun have been introduced, and the resulting change in the circulation pattern noted (e.g. Wetherald & Manabe, 1975). In a different category, a number of other experiments have been done in which the surface boundary conditions of the most recent ice age, roughly 18 000 yr ago, have been introduced into the computation (e.g. Williams, Barry & Washington, 1974). Unfortunately, we can no longer go back to that period and verify how well the model has reproduced the ice age climate.

Such experiments have been most instructive, but we must recognize that there are limits to the ability of a GCM to simulate reality, particularly where the subtle variations of seasonal precipitation patterns are concerned (Gates, 1975; Manabe & Holloway, 1975). In a very real sense, it is these precipitation patterns that determine where the deserts, marginal lands and 'food baskets' will be, and that is what should concern us in a world where the climate may be changing.

Our GCM experiments have shown dramatically that when there is a change in the heat input to the system the model atmosphere responds in a most complex way. For example, with an increase in the total heat supplied to the system there is an overall warming of mean surface temperature, but some regions will warm very much more than others, and there may even be a cooling in some places. The real atmosphere behaves the same way (van Loon & Williams, 1976). The same complex response undoubtedly refers to the patterns of precipitation, and we would expect that there will be places where the precipitation will increase and others where it will decrease in the course of any marked climatic change.

Another way to find out what a warmer Earth might be like is to study a time when the Earth itself was warmer than it is now. Such a time actually existed roughly 4000 to 8000 yr ago, during the period known as the 'Altithermal' (also known as the Hypsithermal, Atlantic, or Climatic Optimum – *optimum* for whom?), and paleoclimatologists are beginning to piece together the strikingly complex picture of the conditions that existed then, at the dawn of civilization. This warming is shown clearly in Fig. 13.1.

Evidence for the conditions at that time is derived from the distribution of fossil organisms in ocean sediments and of pollens in lake sediments, the history of the amounts of water in lakes, the extent of mountain glaciers, the distribution of trees and other vegetation in swamps, widths of tree rings, the location of ancient sand dunes and so forth (Lamb, 1972). Out of many such investigations the picture of the conditions during the Altithermal period can be pieced together, as shown in Fig. 13.6, referring to the precipitation relative to the present (Kellogg, 1977). It will be noted, for example, that North Africa was generally more favorable for agriculture than it is now, that Europe was wetter, Scandinavia drier, and a belt of grass lands (sometimes called 'the Prairie Peninsula') extended across North America in what subsequently became forested land.

We must not accept this as a literal representation of what might occur if the Earth becomes warm again, since the causes and the nature of the warming 4000 to 8000 yr ago could have been quite different from the nature of society's future effects. While we do not really know what caused

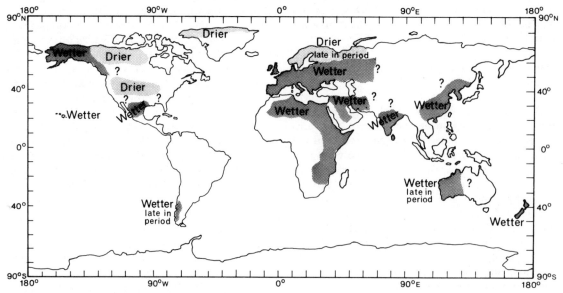

Fig. 13.6. A somewhat schematic map of the distribution of rainfall, predominantly during the summer, during the Altithermal period of 4000 to 8000 yr ago when the world was generally several degrees warmer than now. The terms 'wetter' and 'drier' are relative to the present. Blank areas are not necessarily regions of no rainfall change – the information is still far from complete, though I have collected more than 100 references on the subject.

that high level of mean temperature to be maintained during the Altithermal, one likely cause is an increase in the total output from the Sun, and another possibility is the distribution of sunlight between the northern hemisphere and the southern hemisphere as the Earth's elliptical orbit around the Sun changed. It is even possible that there was more carbon dioxide then, though we have no good evidence for this. In any case, each of these mechanisms to account for the higher mean temperature would presumably result in a different distribution of that heat energy, and therefore the patterns of the general circulation and precipitation would also depend on the mechanism involved.

In spite of these reservations, it seems reasonable to study the way the world was when it was warmer than it is now, and to note that this at least represents a possible pattern for the future warmer Earth. Using the real Earth as our model is at least as good as, and probably better than, the theoretical numerical models that we currently run on our computers.

Looking at the patterns shown in Fig. 13.6 one can conclude that a warmer Earth would by

no means be a less desirable place to live. In general, the patterns of precipitation seem to be more favorable to more people than the present ones, but there are some disturbing exceptions, such as the Prairie Peninsula in North America. In addition to the generally augmented precipitation patterns, one should keep in mind that a 1 °C increase in mean summertime temperature corresponds to roughly 10 days greater average growing season at middle and high latitudes. Thus, on the whole, the prospect for a warmer Earth is a favorable one, though a few places may suffer.

The fate of the ice masses

I have pointed out above that, while the most probable change in the mean surface temperature is a warming, the greatest changes will be in the polar regions, above 50° or 60° latitude, and this would certainly have an effect on the extent of ice and snow.

There are four distinct regimes of ice and snow: the winter snow cover on the land that melts in the summer; floating sea-ice, or 'pack-ice', some of which now survives through the summer in both polar regions; mountain glaciers, that can occur at any latitude; and the great ice sheets of Greenland and the Antarctic that have remained more or less intact for many millions of years. Each of these regimes of ice and snow must be considered separately when we estimate their response to a change in the mean temperature at high latitudes.

As seen on a geological time-scale of tens of millions of years, we now find our planet in the unusual condition of having some permanent ice and snow in both polar regions. For roughly 90 % of the time in the past 500 Myr the poles have been virtually ice-free, at least in summer. However, on a time-scale of hundreds of thousands of years we are in an unusually warm period, and there are no longer massive ice sheets covering northern North America and northern Europe, as there were 18 000 or 20 000 yr ago – the vestiges of which were apparently still in retreat only 8000 yr ago. Thus, the degree of polar glaciation, or the extent of ice at high latitudes that survives the summer, has varied greatly in the past.

These changes of the quasi-permanent ice masses must have been the result of natural processes governing the heat balance of the Earth, processes connected with the ocean circulation, the Earth's orbit, the positions of the continents, the Sun's output and so forth. Now we are about to change that heat balance again.

Of the four regimes of ice on Earth, the two that are probably most important to consider in our scenario are the floating sea-ice, and the great ice sheets. The floating sea-ice in the Antarctic appears and nearly disappears each year, while in the Arctic Ocean there is always a substantial area of sea-ice the year round. The contrast between the two polar regions can be illustrated by the fact that the area of pack-ice frozen each winter (and melted each summer) around the Antarctic Continent is larger than the area of the entire Arctic Ocean.

Referring to the Arctic Ocean specifically, the major question is whether a large warming can result in removing the pack-ice completely, and whether such a complete removal will mean that it will remain open and not freeze over again in winter. There are several reasons for arguing that it probably would tend to remain open once the ice pack had been melted, barring a major change in sea-level (Ewing & Donn, 1956;

Fletcher, 1965; Donn & Ewing, 1966; SMIC Report, 1971; Kellogg, 1975*b*). For one thing, the Arctic Ocean would present a dark surface in summer compared to the highly reflecting ice pack that exists now, so, even with low clouds covering the area, a great deal more solar energy would be absorbed by the system. Another rather compelling reason for thinking that the Arctic Ocean would be harder to freeze over once the ice pack had been removed is that there is now a layer of relatively low-salinity water floating under the ice pack (to a depth of 10 to 30 m), and, since this relatively fresh water has a lower density than the normal salt water of the ocean, it produces a stable layer that inhibits mixing and exchange of heat between the surface layers and the warmer waters below (Aagaard & Coachman, 1975). With the ice pack removed, wave action and surface currents would almost surely eliminate this thin stable upper layer. For both of these reasons, it is expected that if and when the Arctic Ocean ice pack is removed the open, freely mixing ocean will not freeze over again until another very major cooling of the northern hemisphere occurs.

So far there is no adequate combined atmosphere–ocean–sea-ice model that can be used to estimate the response of Arctic sea-ice to a global warming, though there have been some notable advances in this area (Maykut & Untersteiner, 1971; Rothrock, 1975; Washington, Semtner, Parkinson & Morrison, 1976). It must require a considerable warming to remove the ice, however, since evidence from Arctic Ocean sediments suggests that it has never been ice-free for the past million years or more. For example, Budyko (1974) estimates, based on a pack-ice model, that at least a 4 °C warming in summer would be required to eliminate it.

An open Arctic Ocean would of course allow a great deal more evaporation than the frozen Arctic Ocean, and this would presumably result in more rain in summer and snow in winter around its shores. What this would do to the mean snow cover on land, or to the size of the Greenland ice sheet, is still a matter of speculation, but it would certainly represent a major difference in the patterns of temperature and rainfall that exist now. Experiments with the two-layer, Mintz–Arakawa GCM have been performed at the Rand Corporation to determine the effect on temperature and precipitation of an open Arctic Ocean (Fletcher, Mintz, Arakawa & Fox, 1973), and the result was a warmer temperature at the edge of the ocean by 10 °C, and an even larger increase in the central Arctic. These results are for a wintertime situation. The change in precipitation in the model simulation experiment does not seem to have been very significant, however, which is surprising.

Turning to the ice sheets of Greenland and the Antarctic, their total volume is determined over a long time period by a balance between the snowfall on the tops and the melting or break-off at their edges. Also, the effect of 'surges' of an ice sheet must be considered (Hughes, 1973; Flohn, 1975 and Appendix to this volume) since these are dynamic systems. Thus, it is not evident that a warming will necessarily result in a decrease in the size of these ice sheets, since a warmer atmosphere can also hold more moisture, and this in turn can result in more snowfall on their tops. There is apparently evidence that the East Antarctic ice sheet (see below) shrank during the period of the last glaciation in the northern hemisphere and then enlarged during the warming period (Denton, Armstrong & Stuiver, 1971; Flohn, 1973), out of phase with the continental ice sheets of North America and Europe.

Each of these ice sheets should be considered separately in such a discussion, since their characteristics are very different. Greenland, with a total volume that corresponds to about 7 m of ocean depth, is influenced by the Arctic Ocean and the other sources of moisture in the northern hemisphere. It receives more snow than the Antarctic ice sheets, and its southern end extends well below the Arctic Circle. The East Antarctic ice sheet is by far the most massive in the world, with a volume eight to ten times greater than that of the Greenland ice sheet, and its highest point is not far from the South Pole. The West Antarctic ice sheet, with a slightly smaller volume than that of the Greenland ice sheet, has less snowfall to replenish it, and unlike the other ice sheets its edges are partly grounded below sea-level. There are already some signs of a retreat of this ice sheet (Denton *et al.*, 1971), and if there is a major warming and it retreats so that the Antarctic Ocean water can flow under it it would presumably begin to melt quite fast (Mercer, 1968; Hughes, 1973). On a geological time-scale it is this ice sheet that we should watch with some concern, but obviously one should not expect much action in the time-scale of human affairs – that is, for the next few centuries at least.

It must be clear, however, that considering the immense volumes of these three ice sheets, even a relatively small and unmeasurable fractional change of their volumes would affect mean sea-level. Since the turn of the century the sea-level has risen about 20 cm (SMIC Report, 1971), but this rate of rise has slowed since 1940 (Hicks & Crosby, 1975). Can this have been due to some melting of the ice sheets?

In the glaciological literature there is a type of event that has attracted much attention, known as a glacier 'surge'. It is well known that mountain glaciers under certain circumstances can move very rapidly for a period of a few months or years, and then more slowly again. The explanation is that melting at the bottom of a glacier allows it to slide with less friction over the underlying rock, and the greater motion (once a surge starts) helps to generate heat at the interface, and that in turn maintains the motion until a new equilibrium distribution of mass is attained. The same could, in principle, happen to the ice sheets of the Antarctic (Wilson, 1969; Hughes, 1970) with very pronounced effects on the climate of the world as these great blocks of ice were carried to other latitudes by the ocean currents (Flohn, 1975).

When conditions are warmer there is more likely to be water on the underside of an ordinary glacier, so an ice sheet might also be expected to respond to a warming by moving faster. Actually, in the case of the Antarctic ice sheets, changes in the air temperature *on the time-scale in which we have been dealing* would probably not be felt at the bottom, since conductivity in these ice sheets is poor and they are extremely massive. Therefore, it is highly unlikely, regardless of whether such Antarctic ice surges could occur or have occurred in the distant past, that they would be a part of our scenario of warming under the influence of mankind in the next century or so, but we cannot exclude it completely as an unlikely event that might take place anyway.

Epilogue to the Scenario

I have referred to the future of the world's climate in terms of a scenario, based on a set of 'best guesses' about the course mankind would take in the decades ahead, how its future activities would pull on certain leverage points in the climate system, and how the climate system as a whole would respond to these pulls. I have tried at each step in the development of the scenario to indicate my reasoning and the uncertainties involved.

Many scientists familiar with these subjects will probably wish to emphasize the uncertainties and the need for more research – and I would not disagree with them. Others, with an eye for the dramatic, will be tempted to elaborate on this tenuous scenario with a full orchestration – but I hope they will resist the temptation.

The conclusion that can be drawn from this long and complex story is, after all, a fairly simple one. It is that, barring a major natural perturbation to the climate, a perturbation larger and more rapid than anything we have witnessed in the past 10 000 yr of the climate record, mankind's activities will in all likelihood warm the Earth in the next few decades. Our estimate, based on the best theory and best models of the climate system that we know how to construct, is that this rate of warming will be appreciably larger than any change of mean surface temperature we have seen in the past 1000 yr, and could roll back the clock to 4000 to 8000 yr ago when the Earth was warmer than now.

It is not for me to pass judgment on whether this change would be 'good' or 'bad' – that is a value judgment involving the future well-being of people and of nations. It is my unscientific opinion, as one who visualizes his children and grandchildren involved in this scenario, that some will fare better, some worse, but on the whole the Earth will have a climate more favorable for feeding the increasing population.

That is some slight consolation to a society that is obviously going to have to face a difficult transition from a period of exponential growth to some kind of steady state in terms of population, food, energy, and resources. Perhaps the best way we can summarize the outlook is to say that, in the priority list of problems to be faced by society during this transition, long-term change of the average climate probably does not belong at the top. More important, I believe, will be next year's climate, and the year after that. The famines ahead, wherever and whenever they occur, will assure that millions in the poorer, less developed countries will not survive to witness a 'warmer Earth'.

References

Aagaard, K. & Coachman, L. K. (1975). Toward an ice-free Arctic Ocean. *Trans. Am. geophys. Un. (EOS)*, **56,** 484–6.

Adam, D. P. (1975). Ice ages and the thermal equilibrium of the earth, II. *Quarternary Res.*, **5,** 161–71.

Atwater, M. A. (1975). Thermal changes induced by urbanization and pollutants. *J. appl. Met.*, **14,** 1061–71.

Bacastow, R. B. (1976). Modulation of atmospheric carbon dioxide by the southern oscillation. *Nature*, **261,** 116–18.

Bacastow, R. & Keeling, C. D. (1973). Atmospheric carbon dioxide and radiocarbon in the natural carbon cycle. I. Changes from A.D. 1700 to 2070 as deduced from a geochemical model. In *Carbon and the biosphere*, ed. G. M. Woodwell & E. V. Pecan, pp. 86–135. US Atomic Energy Commn CONF–720510 (available from Natl Tech. Info. Service).

Barrett, E. W. (1975). Inadvertent weather and climate modification. *CRC Crit. Revs. Environ. Contr.*, (*Chem. Rubb. Co.*), **6,** 15–90.

Boeck, W. L. (1976). Meteorological consequences of atmospheric krypton-85. *Science*, **193,** 195–8.

Boeck, W. T., Shaw, D. T. & Vonnegut, B. (1975). Possible consequences of global dispersion of krypton-85. *Bull. Am. met. Soc.*, **56,** 527.

Bolin, B. & Bischof, W. (1970). Variations in the carbon dioxide content of the atmosphere of the northern hemisphere. *Tellus*, **22,** 431–42.

Bolin, B. & Charlson, R. J. (1976). On the role of the tropospheric sulfur cycle in the shortwave radiative climate of the Earth. *Ambio*, **5,** 47–54.

Bolin, B., Granat, L., Ingelstam, L., Johannesson, M., Mattsson, E., Oden, S., Rodhe, H. & Tamm, C. O. (1971). *Air pollution across national boundaries: the impact on the environment of sulfur in air and precipitation.* Royal Ministry for Foreign Affairs and Royal Ministry of Agriculture, Stockholm, Sweden.

Borzenkova, I. I., Vinnikov, K. Ya, Spirina, L. P. & Stekhnovskiy, D. I. (1976). Change in the air temperature of the northern hemisphere for the period 1881–1975. *Meteorologiya i Gidrologiya*, **7,** 27–35.

Broecker, W. S. (1975). Climatic change: are we on the brink of a pronounced global warming? *Science*, **189,** 460–3.

Bryson, R. A. (1974). A perspective on climatic change. *Science*, **184,** 753–60.

Bryson, R. A. & Baerreis, D. A. (1967). Possibilities of major climatic modification and their implications: northwest India, a case for study. *Bull. Am. met. Soc.*, **48,** 136–42.

Budyko, M. I. (1974). *Climate and life,* English edn ed. D. H. Miller, *Intl Geophys. Ser.,* vol. 18. Academic Press, New York & London. 508 pp.

Budyko, M. I. & Vinnikov, K. Ya. (1973). Recent climatic changes. *Meteorologia Hydrol.*, **9**, 3–13. (In Russian.)

Budyko, M. I. & Vinnikov, K. Ya. (1976). Global warming. *Meteorologiya i Gidrologiya*, **7**, 16–26. (NOAA Tranl. UDC 551, 583.)

Callendar, G. S. (1958). On the amount of carbon dioxide in the atmosphere. *Tellus*, **10**, 243–8.

Cess, R. D. (1977). Climate change: an appraisal of atmospheric feedback mechanisms employing zonal climatology. *J. atmos. Sci.*, **33**, 1831–43.

Charney, J., Stone, P. H. & Quirk, W. J. (1975). Drought in the Sahara: a biophysical feedback mechanism. *Science*, **187**, 343–435.

Chýlek, P. & Coakley, J. A. (1974). Aerosols and climate. *Science*, **183**, 75–7.

CLIMAP Project Members (1976). The surface of the ice-age Earth. *Science*, **191**, 1131–7.

Coakley, J. A., Jr. & Chýlek, P. (1975). The two-stream approximation in radiative transfer: including the angle of the incident radiation. *J. atmos. Sci.*, **32**, 409–18.

Cobb, W. E. & Wells, H. J. (1970). The electrical conductivity of oceanic air and its correlation to global atmospheric pollution. *J. atmos. Sci.*, **27**, 814–22.

Crutzen, P. J. (1974). Estimates of possible future ozone reductions from continued use of chlorofluoromethanes (CF_2Cl_2, $CFCl_3$). *Geophys. Res. Lett.*, **1**, 205–9.

Crutzen, P. J. (1976). Upper limits on atmospheric ozone reductions following increased application of fixed nitrogen to the soil. *Geophys. Res. Lett.*, **3**, 169–72.

Damon, P. E. & Kunen, S. M. (1976). Global cooling? *Science*, **193**, 447–53.

Denton, G. H., Armstrong, R. L. & Stuiver, M. (1971). The late Cenozoic glacial history of Antarctica. In *Late Cenozoic glacial ages*, ed. K. K. Turekian, pp. 267–306. Yale University Press, New Haven.

Dettweiler, J. & Changnon, S. A., Jr (1976). Possible urban effects on maximum daily rainfall rates at Paris, St. Louis, and Chicago. *J. appl. Met.*, **15**, 517–19.

Donn, W. L. & Ewing, M. (1966). A theory of ice ages, 3. *Science*, **152**, 1706–12.

Dyer, A. J. (1974). The effect of volcanic eruptions on global turbidity, and an attempt to detect long-term trends due to Man. *Q. Jl R. met. Soc.*, **100**, 563–71.

Eckholm, E. P. (1975). Desertification: a world problem. *Ambio*, **4**, 137–45.

Eiden, R. & Eschelbach, G. (1973). Atmospheric aerosol and its influence on the energy budget of the atmosphere. *Z. Geophys.*, **39**, 189–288. (EPA Transl. 457–74.)

Ekdahl, C. A. & Keeling, C. D. (1973). Atmospheric carbon dioxide and radiocarbon in the natural carbon cycle. 1. Quantitative deductions from records at Mauna Loa Observatory and at the South Pole. In *Carbon and the biosphere*, ed. G. M. Woodwell & E. V. Pecan, pp. 51–85. US Atomic Energy Commn CONF–720510 (available from Natl Tech. Info. Service).

Ellsaesser, H. W. (1975). The upward trend in air-borne particles that isn't. In *The changing global environment*, ed. S. F. Singer, pp. 235–72. D. Reidel, Dordrecht (Holland) and Boston (USA).

Ewing, M. & Donn, W. L. (1956). A theory of ice ages, 1. *Science*, **123**, 1061–6.

Fletcher, J. O. (1965). *The heat budget of the Arctic Basin and its relation to climate*, Rand Corporation Rep. R–444–PR. Santa Monica, California. 179 pp.

Fletcher, J. O., Mintz, Y., Arakawa, A. & Fox, T. (1973). Numerical simulation of the influence of Arctic sea ice on climate. In *Energy fluxes over polar surfaces*, WMO Tech. Note No. 129, pp. 181–218. WMO, Geneva.

Flohn, H. (1963). Zur meteorologischen Interpretation der Pleistozä; Klimaschwankungen. *Eiszeit und Gegenwart*, **14**, 153–60.

Flohn, H. (1975). Background of a geophysical model of the initiation of the next glaciation. In *Climate of the Arctic*, ed. G. Weller & S. A. Bowling, pp. 98–110. Geophysical Institute, University of Alaska, Fairbanks.

Gal-Chen, T. & Schneider, S. H. (1976). Energy balance climate modeling: comparison of radiative and dynamic feedback mechanisms. *Tellus*, **28**, 108–21.

Gates, W. L. (1975). Numerical modeling of climate change: a review of problems and prospects. *Proceedings of the WMO/IAMAP symposium on long-term climatic fluctuations*, WMO Tech. Note No. 421, pp. 343–54. Geneva, Switzerland.

Glantz, M. H. & Parton, W. (1976). Weather and climate modification and the future of the Sahara. In *Politics of natural disaster: case of the Sahel drought*, ed. M. H. Glantz, pp. 303–22. Praeger, New York.

Gribbin, J. (1974). Climate and the world's food. *New Scient.*, **64**, 643–5.

Häfele, W. (1974). A systems approach to energy. *Am. Scient.*, **62**, 438–57.

Heilbroner, R. L. (1974). *An inquiry into the human prospect*. Norton, New York. 150 pp.

Hicks, S. D. & Crosby, J. E. (1975). *An average, long-period, sea-level series for the United States*, NOAA Tech. Memo. NOS 15. Rockville, Maryland.

Hobbs, P. V., Harrison, H. & Robinson, E. (1974). Atmospheric effects of pollutants. *Science*, **183**, 909–15.

Holdren, J. P. & Ehrlich, P. R. (1974). Human

population and the global environment. *Am. Scient.*, **62**, 282–92.

Howard, P. H. & Hanchett, A. (1975). Chlorofluoro-carbon sources of environmental contamination. *Science*, **189**, 217–19.

Hughes, T. (1970). Convection in the Antarctic ice sheet leading to a surge of the ice sheet and possibly to a new ice age. *Science*, **170**, 630–3.

Hughes, T. (1973). Is the West Antarctic ice sheet disintegrating? *J. geophys. Res.*, **78**, 7884–910.

Kahn, H., Brown, W. & Martel, L. (1976). *The next 200 years: a scenario for America and the world.* William Morrow and Co., New York. 241 pp.

Keeling, C. D., Bacastow, R. B., Bainbridge, A. E., Ekdahl, C. A., Jr, Guenther, P. R. & Waterman, L. S. (1976a). Atmospheric carbon dioxide variations at Mauna Loa Observatory, Hawaii. *Tellus*, **28**, 538–51.

Keeling, C. D., Adams, J. A., Jr, Ekdahl, C. A., Jr, & Guenther, P. R. (1976b). Atmospheric carbon dioxide variations at the South Pole. *Tellus*, **28**, 552–64.

Kellogg, W. W. (1974). Mankind as a factor in climate change. In *The energy question*, ed. E. W. Erickson & L. Waverman, pp. 241–58. University of Toronto Press, Toronto.

Kellogg, W. W. (1975a). Climate change and the influence of Man's activities on the global environment. In *The changing global environment*, ed. S. F. Singer, pp. 13–23. D. Reidel, Dordrecht (Holland) and Boston (USA).

Kellogg, W. W. (1975b). Climatic non-limits to growth. *Proceedings of the conference on atmospheric and climatic change*, pp. 76–89. University of North Carolina, Chapel Hill, North Carolina.

Kellogg, W. W. (1975c). Correlations and linkages between the Sun and the Earth's atmosphere: needed measurements and observations. *Proceedings of the symposium on possible relationships between solar activity and meteorological phenomena*, ed. by W. R. Bandeen & S. P. Maran, *NASA Spec. Publ.* 366, pp. 365–87. NASA, Washington D.C.

Kellogg, W. W. (1977). Global precipitation patterns during the Altithermal period compared with the present. (In preparation.)

Kellogg, W. W., Cadle, R. D., Allen, E. R., Lazrus, A. L. & Martell, E. A. (1972). The sulfur cycle. *Science*, **175**, 587–96.

Kellogg, W. W., Coakley, J. A. & Grams, G. W. (1975). Effect of anthropogenic aerosols on the global climate. *Proceedings of the WMO/I AMAP symposium on long-term climatic fluctuations*, WMO Tech. Note No. 421, pp. 323–330. WMO, Geneva.

Kellogg, W. W. & Schneider, S. H. (1974). Climate stabilization: for better or for worse? *Science*, **186**, 1163–72.

King, J. W. (1975). Sun–weather relationships. *Aeronautics and Astronautics*, **13**, 10–19.

Kutzbach, J. E. & Bryson, R. A. (1974). Variance spectrum of Holocene climatic fluctuations in the North Atlantic sector. *J. atmos. Sci.*, **31**, 1958–63.

Lamb, H. H. (1972). *Climate: present, past and future*, vol. 1. Methuen, London. 613 pp.

Liou, K.-N. (1976). On the absorption, reflection, and transmission of solar radiation in cloudy atmospheres. *J. atmos. Sci.*, **33**, 798–805.

Llewellyn, R. A. & Washington, W. M. (1977). Energy and climate: outer limits to growth – regional and global aspects. US National Academy of Sciences, Washington, D.C. (In press.)

Machta, L. (1973). Prediction of CO_2 in the atmosphere. In *Carbon and the biosphere*, ed. G. M. Woodwell & E. V. Pecan, pp. 21–31. US Atomic Energy Commn CONF–720510 (available from Natl Tech. Info. Service).

Machta, L. & Telegadas, K. (1974). Inadvertent large-scale weather modification. In *Weather and climate modification*, ed. W. N. Ness, pp. 687–726. Wiley, New York.

Manabe, S. & Holloway, J. L., Jr (1975). The seasonal variation of the hydrologic cycle as simulated by a global model of the atmosphere. *J. geophys. Res.*, **80**, 1617–49.

Manabe, S. & Wetherald, R. T. (1967). Thermal equilibrium of the atmosphere with a given distribution of relative humidity. *J. atmos. Sci.*, **24**, 241–59.

Manabe, S. & Wetherald, R. T. (1975). The effects of doubling the CO_2 concentration on the climate of a general circulation model. *J. atmos. Sci.*, **32**, 3–15.

Markson, R. (1975). Solar modulation of atmospheric electrification through variation of conductivity over thunderstorms. *Proceedings of the symposium on possible relationships between solar activity and meteorological phenomena*, ed. W. R. Bandeen & S. P. Maran, *NASA Spec. Publ.*, **366**, pp. 171–8. NASA, Washington D.C.

Martell, E. A. & Moore, H. E. (1974). Tropospheric aerosol residence times: a critical review. *J. Rech. atmos.*, **8**, 803–910.

Maykut, G. A. & Untersteiner, N. (1971). Some results from a time dependent, thermodynamic model of sea ice. *J. geophys. Res.*, **76**, 1550–75.

McElroy, M. B., Elkins, J. W., Wofsy, S. C. & Yung, Y. L. (1976). Sources and sinks for atmospheric N_2O. *Rev. Geophys. Space Phys.*, **14**, 143–50.

Mercer, J. H. (1968). Antarctic ice and Sangamon sea level. *Int. Ass. scient. Hydrol., Publ.* No. **79**, 217–25.

Mitchell, J. M., Jr (1971a). Summary of the problems of air pollution effects on the climate. In *Man's impact on the climate*, ed. W. H. Mathews, W. W.

Kellogg, & G. D. Robinson, pp. 167–75 MIT Press, Cambridge, Massachusetts.

Mitchell, J. M. (1971*b*). The effect of atmospheric aerosols on climate with special reference to temperature near the Earth's surface. *J. appl. Met.*, **11**, 651–7.

Mitchell, J. M., Jr (1972). The natural breakdown of the present interglacial and its possible intervention by human activities. *Quaternary Res.*, **2**, 436–45.

Mitchell, J. M., Jr (1974). The global cooling effect of increasing atmospheric aerosols: fact or fiction? *Proceedings of the IAMAP/WMO symposium on physical and dynamic climatology, WMO Tech. Note* No. 347, pp. 304–19. WMO, Geneva.

Mitchell, J. M., Jr (1975). A reassessment of atmospheric pollution as a cause of long-term changes of global temperature. In *Global Effects of Environmental Pollution*, ed. S. F. Singer, pp. 149–74. D. Reidel, Dordrecht (Holland) and Springer-Verlag, New York.

Namias, J. (1974). Longevity of a coupled air–sea–continent system. *Mon. Weath. Rev.*, **102**, 638–48.

NAS (1975). *Understanding climatic change: a program for action.* US Committee for GARP, National Academy of Sciences, Washington D.C.

NAS (1976). Halocarbons: environmental effects of chlorofluoromethane release. Comm. on impacts of stratospheric change, National Academy of Sciences, Washington D.C.

Newell, R. E. (1971). The Amazon forest and atmospheric general circulation. In *Man's impact on the climate*, ed. W. H. Mathews, W. W. Kellogg, & G. D. Robinson, pp. 457–60. MIT Press, Cambridge, Massachusetts.

Newell, R. E. (1974). Changes in the poleward energy flux by the atmosphere and ocean as a possible cause for ice ages. *Quaternary Res.*, **4**, 117–27.

Ney, E. P. (1959). Cosmic radiation and the weather. *Nature*, **183**, 451–2.

Nichols, H. (1974). Arctic North American paleoecology: the recent history of vegetation and climate deduced from pollen analysis. In *Arctic and alpine environments,* ed. J. D. Ives & R. G. Barry, pp. 637–67. Methuen, London.

Nichols, H. (1975). *Palynological and paleoclimatic study of the Late Quaternary displacement of the boreal forest-tundra ecotone in Keewatin and Mackenzie, NWT, Canada. Institute of Arctic and Alpine Research, Occasional Paper* No. 15, Boulder, Colorado. 87 pp.

NOAA (1975). *Geophysical monitoring for climatic change,* No. 3, Summary Rep.–1974, ed. J. M. Miller. Environmental Research Laboratories, National Oceanic and Atmospheric Administration, Boulder, Colorado.

NSF (1976). More than meets the eye. *Mosaic*, **7**,

21–7. National Science Foundation, Washington D.C.

Otterman, J. (1974). Baring high-albedo soils by overgrazing: a hypothesized desertification mechanism. *Science*, **186**, 531–3.

Palmén, E. & Newton, C. W. (1969). *Atmospheri: circulation systems.* Academic Press, New York.

Pivovarova, Z. I. (1970). Study of the regime of atmospheric transparency. *Proceedings of the WMO/IUGG symposium, WMO Tech. Note* No. 104, pp. 181–5. WMO, Geneva.

Quinn, W. H. (1974). Monitoring and predicting El Nino Invasions. *J. appl. Met.*, **13**, 825–30.

Ramanathan, V. (1975). Greenhouse effect due to chlorofluorocarbons: climatic implications. *Science*, **190**, 50–2.

Rasool, S. I. & Schneider, S. H. (1971). Atmospheric carbon dioxide and aerosols: effects of large increases on global climate. *Science*, **173**, 138–41.

Roberts, W. O. & Olson, R. H. (1973). New evidence for effects of variable solar corpuscular emission on the weather. *Rev. Geophys. Space Phys.*, **11**, 731–40.

Rothrock, D. A. (1975). The steady drift of an incompressible arctic ice cover. *J. geophys. Res.*, **80**, 387–97.

Rowland, F. S. & Molina, M. J. (1975). Chlorofluoromethanes in the environment. *Rev. Geophys. Space Phys.*, **13**, 1–36.

Sartor, J. D. (1967). The role of particle interactions in the distribution of electricity in thunderstorms. *J. atmos. Sci.*, **24**, 601–15.

Sartor, J. D. (1969). On the role of the atmosphere's fair-weather electric field in the development of thunderstorm electricity. In *Planetary electrodynamics,* ed. S. C. Coroniti & J. Hughes, pp. 161–6. Gordon & Breach Scientific Publications.

SCEP Report (1970). *Man's impact on the global environment: study of critical environmental problems.* MIT Press, Cambridge, Massachusetts.

Schneider, S. H. (1975). On the carbon dioxide-climate confusion. *J. atmos. Sci.*, **32**, 2060–6.

Schneider, S. H. (with L. E. Mesirow) (1976). *The Genesis strategy: climate and global survival.* Plenum, New York & London. 419 pp.

Schneider, S. H. & Dennett, R. D. (1975). Climatic barriers to long-term energy growth. *Ambio,* **4,** 65–74.

Schneider, S. H. & Dickinson, R. E. (1974). Climate modeling. *Rev. Geophys. Space Phys.,* **12,** 447–93.

Schneider, S. H. & Kellogg, W. W. (1973). The chemical basis for climate change. In *Chemistry of the lower atmosphere*, ed. by S. I. Rasool, pp. 203–50. Plenum, New York.

Sellers, W. D. (1974). A reassessment of the effect of CO_2 variations on a simple global climatic model. *J. appl. Met.,* **13,** 831–3.

Shackleton, N. J. & Opdyke, N. D. (1973). Oxygen

isotope and paleomagnetic stratigraphy of equatorial Pacific core V28–238: oxygen isotope temperatures and ice volumes on a 10^5 and 10^6 year scale. *Quaternary Res.*, **3**, 39–55.

Smagorinsky, J. (1974). Global atmospheric modeling and numerical simulation of climate. In *Weather and climate modification,* ed. W. N. Hess, pp. 633–86. Wiley, New York.

SMIC Report (1971). *Inadvertent climate modification; report of the study of Man's impact on climate.* MIT Press, Cambridge, Massachusetts.

Twomey, S. (1972). The effect of cloud scattering on the absorption of solar radiation by atmospheric dust. *J. atmos. Sci.*, **29**, 1156–9.

van Loon, H. & Williams, J. (1976). The connection between trends of mean temperature and circulation at the surface: Part 1. Winter. *Mon. Weath. Rev.*, **104**, 365–80.

Vonnegut, B. (1963). Some facts and speculations concerning the origin and role of thunderstorm electricity. *Met. Monogr.*, **5**, 224–41.

Wang, W.-C. & Domoto, G. A. (1974). The radiative effect of aerosols in the Earth's atmosphere. *J. appl. Met.*, **13**, 521–34.

Washington, W. M. (1972). Numerical climatic-change experiments: the effect of Man's production of thermal energy. *J. appl. Met.*, **11**, 768–72.

Washington, W. M., Semtner, A. J. Jr, Parkinson, C. & Morrison, L. (1976). On the development of a seasonal change sea ice model. *J. Phys. Oceanogr.*, **6**, 679–85.

Weare, B. C., Temkin, R. L. & Snell, F. M. (1974). Aerosol and climate: some further considerations. *Science*, **186**, 827–8.

Weinberg, A. M. & Hammond, R. D. (1970). Limits to the use of energy. *Am. Scient.*, **58**, 412–18.

Wetherald, R. T. & Manabe, S. (1975). The effects of changing the solar constant on the climate of a general circulation model. *J. atmos. Sci.*, **32**, 2044–59.

Wilcox, J. M. (1975). Solar activity and the weather. *J. atmos. terr. Phys.*, **37**, 237–56.

Willett, H. C. (1949). Long period fluctuations of the general circulation of the atmosphere. *J. Met.*, **6**, 34–50.

Williams, J., Barry, R. G. & Washington, W. M. (1974). Simulation of the atmospheric circulation using the NCAR global circulation model with ice age boundary conditions. *J. appl. Met.*, **13**, 305–17.

Wilson, A. T. (1969). The climatic effects of large-scale surges of ice sheets. *Can. J. Earth Sci.*, **6**, 811–918.

Wofsy, S. C., McElroy, M. B. & Sze, N. D. (1975). Freon consumption: implications for atmospheric ozone. *Science*, **187**, 535–7.

Yamamoto, G. & Tanaka, M. (1972). Increase of global albedo due to air pollution. *J. atmos. Sci.*, **29**, 1405–12.

Yung, Y. L., Wang, W. C. & Lacis, A. A. (1976). Greenhouse effect due to atmospheric nitrous oxide. *Geophys. Res. Lett.*, **3**, 619–21.

14

Climatic changes and human affairs

STEPHEN H. SCHNEIDER &
RICHARD L. TEMKIN

Variations in climate have been the rule rather than the exception throughout climate history. Climatic changes and climatic variability* have been inferred indirectly from many types of 'proxy' records and, for more recent periods, directly from observational records of meteorological instruments. These data suggest that some of the variations in geographic patterns of temperature, precipitation and other climatic variables have been major (for example, advances and retreats of the polar ice-caps and continental glaciers). Smaller fluctuations, in the order of several tenths of a degree Celsius averaged over the northern hemisphere and occurring on decadal time-scales, have also been evidenced, as have intermediate-scale climatic variations such as the Little Ice Age (about the sixteenth to the nineteenth centuries), in which much of the mid-latitudes of the northern hemisphere experienced temperature reductions of about 1–2 °C.[2] Clearly, climatic variation has existed throughout geological time.

During the past four million years or so of this continual climatic variation, *Homo sapiens* has been present on Earth. Quite possibly we are approaching a stage of history in which human inputs may produce climatic fluctuations comparable to those of the past; until recently, however, the influence seems to have been unidirectional. That is, human affairs have been influenced by climatic variability (or climatic change) with little human feedback on the

climate. But, as discussed in earlier chapters, that situation is changing.

Regardless of whether particular climatic changes are based on anthropogenic or non-anthropogenic causes, human affairs are affected by climate in many ways. In the past, climate has been a factor (often the controlling one) in determining where and when people should hunt, where they should live, what they should wear, where they should practice agriculture, what they should grow and so on. Even though today human dependence on climate is modulated by highly energy-intensive technological amenities, climatic variability still remains a strong influence in our affairs. Especially now, as problems arise over approaching 'world energy and material shortages' in the face of a rapidly growing world population, climate could again come to exert an acknowledged and significant influence on civilizations.

For instance, climatic records based in part on glacial ice core samples in Greenland have shed light on the development and decline of the Norse Atlantic colonies in the ninth century.[3] During a warming period Icelandic settlers had fared rather well and had even sent some of their colonists west to Greenland. However, this Greenland colony began to collapse as climatic cooling set in. Further cooling almost destroyed the main Norse colony in Iceland itself. Thus, a climatic change had severe effects on a civilization that was living in a 'climatically marginal' area, a theme we return to later. Many other examples of the effects of climate on human affairs can be cited.[4] About 2000 yr ago, for instance, houses in Britain were generally built with a southern exposure presumably because of a warmer and drier climate. But by 1500 A.D.

---

* Climatic change and climatic variability will not be rigorously defined here, a difficult problem in itself, but will be used loosely to refer to 'long-' and 'short-term' variations in climate, respectively. For an extended discussion, see the Global Atmospheric Research Program (GARP)[1].

houses more often were constructed on northern slopes, facing either east or north probably to avoid the strong southerly or westerly winds.

In addition, famines resulting from cold or drought were not at all infrequent through recorded history; quite logically, these often led to increased death rates, particularly among the poor (who generally live a marginal existence in which an increase in stress is often fatal). Furthermore, since historical examples of implied climatic variation exist, it might be expected that climatic change would also affect many regions of the world in this century. In fact, repeated famines in climatically marginal parts of Asia and Africa have been directly implicated in the starvation of millions during the twentieth century.[5]

In the next section we will discuss several specific ways in which climatic variations may have an impact on today's world. These examples provide a framework from which some important generalizations are drawn; they exhibit recurring themes which, on the one hand, offer a foreboding picture of what society might have to face in the future, but also, on the other hand, illustrate common dangers that can be reduced by appropriate policy decisions. It will be suggested that climatic fluctuations and changes could cause severe strains on existing institutions, social mores and even governments – assuming that proper precautions are not implemented.

Implications of climatic variability in specific societal sectors
Food

In the production of food a changed or variable climate could have immediate regional and global implications for humanity. It is clear that severe weather can have a drastic impact on crops (hail storms, wind storms, early frosts and so on), although the weather event itself may occur on a very short time-scale. Seasonal droughts or heat waves, which could be considered climatic fluctuations rather than weather events, may also have harmful effects on food production. Certain technological factors, such as the appropriate choice of particular crop strains or intensive irrigation, often can ameliorate the consequences of short-term climatic anomalies. However, if an anomaly is present over a longer period (that is, if it perhaps represents a climatic change), a society may not be able to sustain all its members at a

given level of 'adequate' nutrition. It is not at all unlikely that an extended drought could occur in a major grain-growing region of the world. Paleoclimatic reconstruction suggests that a drought of some 200 yr duration gripped the region of the state of Iowa in the US beginning about the year 900 A.D.; [6] It is quite possible that a similar situation might recur at some future time. It is important, then, to inquire as to the likelihood of drought occurrence *and* to examine the sensitivity of food production centers to historically precedented climatic fluctuations.

Since the North American granary is at present the only major exporter of grains (Table 14.1), it

Table 14.1. *The changing pattern of World grain trade (from Brown[7])*

Region	Grain exports (+) and imports (−) (million metric tons)				
	1934–38	1948–52	1960	1970	1976[a]
North America	+ 5	+23	+39	+56	+94
Latin America	+ 9	+ 1	0	+ 4	− 3
Western Europe	−24	−22	−25	−30	−17
Eastern Europe and USSR	+ 5	—	0	+ 1	−25
Africa	+ 1	0	− 2	− 5	−10
Asia	+ 2	− 6	−17	−34	−47
Australia and New Zealand	+ 3	+ 3	+ 6	+12	+ 8

[a] Preliminary estimates.

is urgent to discuss the effects of climatic variability in this region. The global implications of an unfavorable yield of grain in North America are plain. About a third to a half of the grain grown in North America is exported, while imports of that grain by regions outside of North America amount to 5 to 10 % of the world's total grain crop.[7,8] A 10–20 % reduction in North American productivity (comparable to previous historical weather-induced crop losses) would probably not lead to food deficits in North America. Implications for many other countries dependent on these imports would, however, be far more serious: the price instability that follows reduced food productivity at times of depleted foods stocks would threaten the food purchasing power of chronically malnourished nations that must compete for shrinking supplies in the 'free world' market. But of course, North America would not remain unaffected since the global shortage would drive food prices up as well – and the poorest citizens would suffer disproportionately to the rest of the

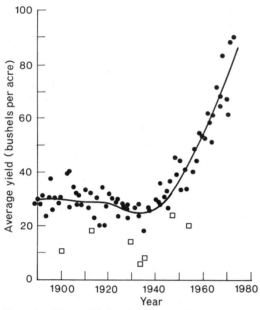

Fig. 14.1. The solid line is the trend up to 1973 in corn yield per acre (1 acre = 0.4 ha) in Missouri (after Decker[9]). The circles represent individual yearly yields for years without major drought and the squares are for years with drought. Note that since 1956 not only have yields increased significantly, but variability in yields from one year to the next has been reduced.

population. Moreover, because of the complexities of international trade arrangements, the growth of resource cartels (such as OPEC), and the threat of terrorist activities by desperate people, a major cut in North American food exports – particularly if there were simultaneous difficulties elsewhere – could well create sufficient political instability around the world that even the best-fed North Americans would not escape its effects. In any case, the issue of societal vulnerability to climatic fluctuations is not limited strictly to climate/crop relationships, but is extended by economic and political interdependencies to the social and political spheres as well.

Returning to the subject of crop yields and climate variability, we see in Fig. 14.1 the average annual corn yields in Missouri from about 1900

to 1973, as well as the long-term trend (solid line is yield per acre (0.4 ha)). Several points should be emphasized here. First, although some of these fluctuations in the *amount* of yield are strongly associated with climatic fluctuations (1930s droughts), others are associated with different factors (e.g. the 1970 corn blight). Secondly, variability may be somewhat reduced by aggregating over larger areas, but it certainly can remain high enough to fall into the realm of a 10–30 % reduction from the expected yield based upon 'normal' weather. A discussion of the concept of 'normal' weather will be given shortly.

Fig. 14.1 also shows that since about 1945 the yield has increased dramatically, with what appears to be a marked decrease in yield variability. The increased trend in yield is unquestionably a result of technological innovation, such as hybrid crops and fertilizers. Could it be that the improved technology has not only increased yields but has also significantly decreased yield variability? We will return to this question later.

But what of the world situation? The widespread introduction of the 'Green Revolution' in developing countries was largely responsible for a 3 % annual growth in world food production during the late 1960s and early 1970s, seemingly comfortably greater than the 2 % annual growth rate in world population. Had technology finally insulated societies from the vagaries of climate and, as many proclaimed, banned the spectre of Malthus from the earth? Throughout 1971 this may have appeared so – at least until the unfavorable weather (or climatic anomalies) of 1972 and 1974. The year 1972 was in fact somewhat of a climatic anomaly world-wide. For instance, in the Soviet Union a combination of insufficient snow the previous winter and severe cold in January 1972 caused the destruction of almost a third of the Soviet winter wheat crops. The hot and dry spring that followed (in the European part of the USSR) further reduced the winter grain yield and affected the spring crop as well. Other 'abnormal' climatic events of 1972 included a delay in the onset of the India monsoon, continuing droughts in central Africa, and floods in Pakistan. A change in the Peruvian coastal currents (a reduction in the cold upwelling currents), drastically reduced the anchovy yield – with the problem exacerbated by overfishing. Damaging floods also occurred in the midwest US and elsewhere. Kukla & Kukla[10] interpreted a change in the

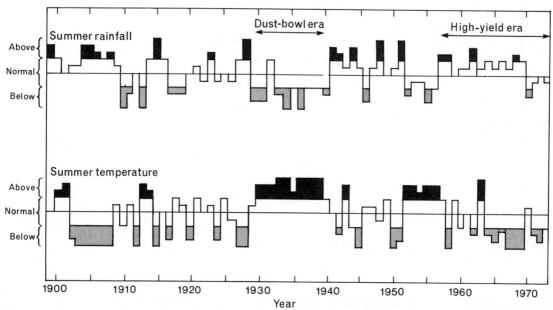

Fig. 14.2. An eight-decade record of summer average temperature and rainfall in the five major wheat-producing states (Oklahoma, Kansas, Nebraska, South Dakota and North Dakota) of the US, compiled by
Donald Gilman[13] and showing the 10-yr drought period (with above-average temperatures and
below-average rainfall) of the 'dust-bowl era' and the 17-yr recent 'high-yield era' (a period of above-
average rainfall and below-average temperature).

brightness of the earth (observed from a satellite)
as a sharp increase in snow cover and floating ice
in the northern hemisphere, and noted that this
too coincided with the 1972 climatic anomalies
(see also Chapter 6).

In 1972, global grain production, strongly
influenced by the climatic anomalies, dropped by
'only' 1 %, yet food prices skyrocketed and food
reserves plummeted from roughly two month's
supply to about half that much. The decline in
food stocks was also related to the deliberate
policy of the major food exporting countries to
liquidate government-held food reserves on the
grounds that these reserves tend to stabilize prices
and reduce the incentive for farmers to expand
production – an argument that has been strongly
challenged.[7,11]

Aside from the continuation of the devastating
Sahelian drought, 1973 was a record year for
world food production. But again in 1974 weather
was more highly variable than it had been during

much of the previous 15-yr period. In North
America, late, heavy spring rains interfered with
seeding. This, followed by a hot, dry July and
early autumn frosts, reduced crop yields sharply.
Late monsoon rains in India hampered grain
production there and the world food conference
in Rome once again focused attention on the most
vulnerable group: the chronically malnourished
billion who live primarily in developing countries.

These recent weather events obviously affected
world food production. (It should be kept in
mind that crop yields may also be strongly
influenced by other factors, such as water availability, seed stock, genetic variability of crops,
fertilizer, productivity of soils, pesticides, herbi-
cides and pest-control services of natural ecosystems.) But was the world adequately prepared
for this type of climatic variability and its impact
on grain production – even with the new technology? A document[12] issued by J. McQuigg, L.
Thompson and other agricultural climatologists

231

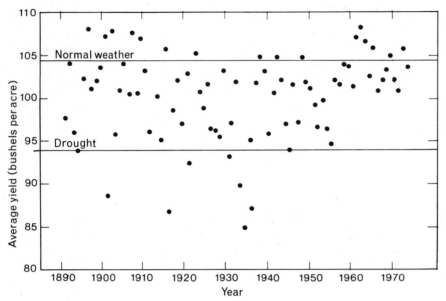

Fig. 14.3 Simulation with Louis Thompson's crop model of corn yields (bushels per acre, 1 acre = 0.4 ha) in the five major corn-producing states (Ohio, Indiana, Illinois, Iowa and Missouri) under the assumptions that technology is fixed at 1973 levels and the only influence on corn yield variability is the weather patterns that prevailed from 1890 to 1973 (after McQuigg et al.[12]).

and agronomists in 1973 said no! These researchers indicated that future crop yields in the US might be considerably more variable than those of the previous 15 yr. Evidence for their views can be seen in Fig. 14.2. This graph, from Gilman,[13] contains a 75-yr record of summer average temperature and rainfall in the five major wheat-producing states of the US. Darkened parts of this figure indicate above- or below-normal values of the temperature and rainfall. As shown, the drought of the 1930s was characterized by above-normal temperatures and below-normal rainfall. But note that the 15-yr period after 1956 was marked by generally below-normal temperatures and above-normal rainfall; unusually favorable conditions for high yields in the plains. This fortuitous weather (coupled with technological improvements) certainly helped reduce yield variability in this region.

The work by McQuigg et al.[12] considered corn yields in the 'corn belt' region of the US as one example. Employing the crop yield model of Thompson, they used as a set of independent variables (for their statistical regression) selected monthly averages of precipitation and temperature. In attempting to account for technology, however, they assumed that its influence was fixed (that is, they fixed the 'technology parameter' in their regression equation at the 1973

level). In essence, their empirical model isolated the effects of weather on crop yield variability. Although the model explained more than 80 % of the variance in corn yields, it obviously did not explain the total variance; for example, blights and early- or late-season frosts were not in their particular choice of variables. Even in an approximate sense, technology still was not strictly accounted for in the model; variability in crop yield was not properly considered as a function of technology because of the fixed technology assumption and because technology did not influence the weather terms in the model. The results of such an empirical model are likely to miss the interaction between technology and weather that may have mitigated (or enhanced) the weather effects of earlier years. But the general question asked by these researchers was appropriate to the assumptions of the model: if technology had been steady (at the 1973 level) in every year since 1890, what kind of variation in crop variability might have been expected from weather alone? Fig. 14.3 shows some of their results. Variations in yields in the model are now entirely a function of weather variability (technology is fixed); the drought line corresponds to a 10 % reduction in yield. The 'normal' line (the line corresponding to the prevailing concept of 'normal' weather) was, in essence, equivalent to optimum yield conditions. Thus, McQuigg *et al.* argued, true, average yields should in the long run be less than yields determined from what recent experience has called 'normal' weather. Furthermore, this figure suggests that crop variability caused by weather for the last 17-yr period (high-yield era) was in fact considerably less than that of the previous years. Similar plots also exist for soybeans and wheat, although weather variability effects are much less dramatic for wheat. Thus, the report concluded that the high-yield era was not normal with respect to weather variability, and that crop yield variability could be expected to increase. A reduction of more than 20 % in US corn yields in 1974 caused by the weather further reinforced this warning – unfortunately, after the fact. More implications of weather or climate variability on crop yields will be discussed later in a broader context.

Water

Consider now the relationship of climatic variability to water supplies. Since water resources are used for irrigation, for urban consumption, for many energy processes, and for recreational purposes, implications of climatic variability or climatic change as reflected in variations of precipitation or river run-off are obvious. Increasing demand from a growing, or more affluent, population leads to competition for water among various users; such demands could also place water resources at the mercy of climatic fluctuations. Hydrologists, aware of this fact, generally design water supply systems to account for weather uncertainty and thus build in some margin of safety. However, spatial and temporal variations in precipitation and hence river run-offs, as well as a lack of knowledge of future water demands, suggest that problems may still arise.

As an example, one can view the case involving the Colorado River Basin. A Colorado River Compact was drawn up some 40 yrs ago[14] which prescribed the allotment of water between those states in the upper basin and those in the lower basin. However, the two factors mentioned above were not taken fully into account. At the time of the compact, water needs in the upper basin were not very high. But, with a rapidly growing population in Colorado and prospects for energy resource mining and development, demand for water has grown dramatically. Furthermore, it

appears now that the allocation of river resources was based upon a period of time when the annual flow of the Colorado River was abnormally high.[15]

Using tree-ring data, Stockton[16] has reconstructed a 450-yr record of annual run-off for several sub-basins within the upper Colorado River Basin. His reconstructed hydrograph for the upper Colorado River Basin (Fig. 14.4) indicates that during the early part of the twentieth century, the river flow remained anomalously high for perhaps the longest sustained interval of the entire 450-yr period. Thus, since at the time the Colorado River Compact was entered only short-term records were used, the estimates of mean annual flow and its variance were not truly representative of the longer record. In fact, this has been borne out in the past few decades. Since the river basin is just as likely to have a period of low flow as another period of high flow (presumably associated with a climatic variation), water availability may not meet the legal requirements of the Colorado River Compact and thus could also have significant effects on human affairs. The situation is already one of stress. Again, the influence of climatic variability on society is clear in this example, as is, conversely, the means to reduce that influence by prudent planning.

Energy

Society is heavily dependent today, and may be more so in the future, on energy-intensive goods and services. Some of these direct energy demands are obviously affected by climate, such as 'comfortably' heated buildings in cold weather and 'comfortably' air-conditioned buildings in hot weather. In light of this growing demand for controlled interior environments, coupled with the approaching situation of limited fuel resources, it can be seen that exterior weather variations (or climatic fluctuations if they persist) will increase the problems of those responsible for delivering adequate power in many countries. Generally, the suppliers need a long lead time to prepare for a drastic fluctuation in climate, since it takes many years to construct power plants and obtain fuels. For instance, shortages in natural gas, some petroleum products and coal were only exacerbated by the colder winters of 1972–73 and 1973–74 in the US.[17] If we were to experience another OPEC petroleum boycott as in 1973, bad weather

233

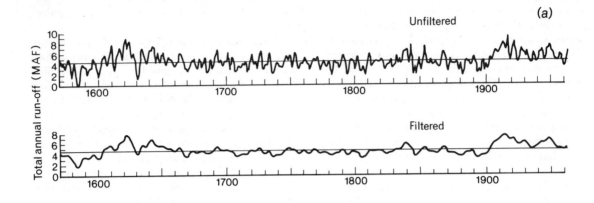

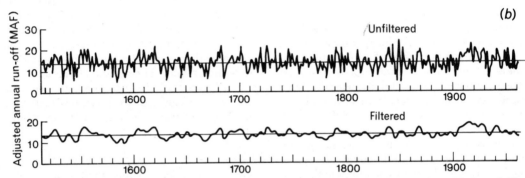

Fig. 14.4. Reconstructed hydrographs for the total annual run-off for (*a*) the Green River, Utah and (*b*) the Colorado River at Lee Ferry, Arizona. In each case the lower graph is the same time series, but with frequency components greater than 0.1 yr^{-1} removed (after Stockton[16]). MAF stands for millions of acre feet.

could proverbially be 'the straw that would break the camel's back'.

Short-term and longer-term weather fluctuations influence the electric producers directly in determining the most efficient (actually most economical) means for delivering energy to consumers. Energy producers do build a certain amount of 'spinning reserve capacity'[17] into the system (so that some operating units carry no load and are readily available in case of a malfunction somewhere). Diurnal variations in temperature, insolation, wind, humidity and so forth help determine the necessary 'spinning reserve capacity'. However, the consumption of fuel to produce electricity also requires advance planning on a

longer time-scale. Since electrical industries often operate at rather low inventories, unanticipated long-term weather changes may result in higher costs to consumers – if not curtailed service.[17] Long-term planners should thus consider not only questions of consumer demand but also weather- and climate-related input to that demand.

Weather not only influences energy usage but may also affect energy delivery systems, portions of which are very sensitive to severe weather anomalies. High winds accompanying a storm or an ice storm, could easily disrupt delivery service. Also, since atmospheric events are important in dispersing the increasing concentration of pollutants given off during some forms of energy production, long periods of stagnant air have an indirect effect on the health and well-being of humans, other animals and plants. Another indirect effect of climatic variability or energy demand relates to the use of agricultural fertilizers. At present, approximately 5 % of all crude oil used in the US is processed for fertilizer production.[17] Bad weather would not only influence food production directly as discussed earlier but could also affect it indirectly by placing demands on the energy system through the use of fertilizers. Precipitation has a significant effect on fertilizer application; heavy rains might easily wash away much of the fertilizers, perhaps decreasing food production or decreasing energy availability for other uses (if more oil is diverted for fertilizer production). Irrigation obviously requires energy input too. Clearly, the vital resources of food, water and energy are all coupled to each other as well as to climatic variability. Thus, it is reasonable to expect that anomalous weather or climatic variability might affect the food, water and energy systems simultaneously, which in turn, would affect other elements of human affairs.

Probably the most important climate–energy issue is that of the potential existence of *climatic barriers to long-term energy growth*, a problem considered in depth elsewhere.[11,18,19] For our purposes here, it is sufficient to point out that continued growth in global energy consumption could lead to inadvertent climatic change on an unprecedented scale as early as the end of this century. Considerable uncertainty remains as to the reliability of present estimates of energy-induced climatic change (climatic effects may be underestimated as well as overestimated). However, the assumptions of those who count indefinitely on increased energy consumption as a means for improving the human condition must be strongly challenged because of the possibility of significant climatic disruption if present trends of energy use were to persist.

General issues of societal vulnerability

Many imponderable questions are raised by the previous examples, even though the discussion has been limited to climate-related implications. For instance, is society any more vulnerable now to climatic change than it has been in the past? What sectors of society are really most vulnerable? Can anything be done to reduce risks associated with climatic variability? How does one

estimate these risks, especially in comparison to other types of risks (assuming a risk/benefit comparison can be made?) Can society cope with climatic variability or climatic change? These are but a few of the issues that can be considered; we will elaborate on some of them below. The preceding examples of the implications of climatic change or climatic variability on certain segments of society should be kept in mind throughout the following discussion, for it is from these examples that some of the most general inferences will be drawn.

Almost certainly, climatic change in the near future will not lead to 'the end of the world'. It is unlikely that the next major glacial advance will occur within a century, or that the polar ice-caps will soon melt, causing major flooding in coastal cities of the world. Although these extreme possibilities of climatic change cannot be entirely eliminated (especially in the light of growing anthropogenic insults), even if they occurred the capacity of the Earth to support *some* humans should not be directly threatened. What may well be most important are the seemingly small, precedented climatic fluctuations, which should be urgently considered by planners and policy makers. Before technology changed the character of human vulnerability to variability in the elements, unusual weather generally had major consequences only in the region of its occurrence, primarily affecting food, which usually was produced and consumed locally. Of course, this situation itself was often serious to the indigenous population, as evidenced by the elimination of the Norse colonies in Greenland or by the large number of deaths associated with various recorded regional droughts.

Today, societies are much more specialized; hundreds of millions of people are dependent upon energy, water and food provided by others over whom they have little control. For example, many people now depend for their food upon the continuous productivity of only a few world granaries; these food production and distribution systems are, in turn, sensitive to relatively small variations in regional climate. In essence, societal vulnerability to climatic variations has been transformed from the pre-technological condition of high-frequency, low-amplitude risk to the present situation of low-frequency, high-amplitude risk. That is, most local adverse weather situations can be smoothed out by food transfers;

but the less probable *simultaneous disruption* of growing conditions in only three or four major granaries could threaten the survival of perhaps half a billion people.

Moreover, those people who first feel the brunt of any food, water, or energy shortages are the poor of the world – 'marginal people' in the sense that they exist at marginal nutrition, marginal income, or marginal water availability in order barely to survive. Many also live in marginal climatic regions where a small change in climate would have a major local effect (such as the inhabitants of the drought-prone regions that border deserts or exist at the extremity of monsoon rain belts). It is even quite likely that climatic variations elsewhere, which lower the productivity of the food-exporting regions, would also have a major impact (through world trade) upon these 'marginal people', even if their local harvests were normal.

Another point to be stressed here is that a seemingly small change, say, in a zonally- or hemispherically-averaged sea-level temperature could cause a much larger modification in a local region which is near the boundary of different circulation regimes. A shift of a few hundred kilometers in the position of the Indian monsoon region would not signal a global catastrophe. Or if the US grain belt were to move northward by a few hundred kilometers as a result of global warming it would represent only a small perturbation, globally, to world food production systems and the Earth's carrying capacity. However, in the present world situation where most people are no longer nomadic and are locked into national boundaries, such a marginal shift could be serious. Since current cropping patterns are frequently based upon pre-existing climatic conditions and expectation of their continuance, slow and gradual changes in climate may be anticipated, allowing appropriate strategies for crop planting and so on, thus avoiding potential crises. But if changes were to occur quickly it is possible that a catastrophe might result – unless preparation for sufficient reserve capacity *and* distribution systems has been made.

The high vulnerability of marginal people to climatic variation is not primarily a consequence of a rapidly deteriorating climate, but more likely a manifestation of the inability of the global community to anticipate and hedge against the problems associated with repeated, seemingly small-scale variations in climate that are already well documented in climatic history. Global society has apparently failed to build in sufficient resilience, particularly in terms of flexibility and reserve capacities of basic resources such as food, to insure the survival of all its members. Table 14.2 provides an assessment of the evolving vul-

Table 14.2. *Index of world food security, 1961–76 (from Brown*[7])

Year	Reserve stocks of grain	Grain equivalent of idled US cropland	Total reserves	Reserves as days of annual grain consumption
	(million metric tons)			
1961	163	68	231	105
1962	176	81	257	105
1963	149	70	219	95
1964	153	70	223	87
1965	147	71	218	91
1966	151	78	229	84
1967	115	51	166	59
1968	144	61	205	71
1969	159	73	232	85
1970	188	71	259	89
1971	168	41	209	71
1972	130	78	208	69
1973	148	24	172	55
1974	108	0	108	33
1975	111	0	111	35
1976[a]	100	0	100	31

[a] Preliminary estimates.

nerability of the world to fluctuations in food production. The total world grain reserves in millions of metric tons and the corresponding number of days the reserves could feed the world population at the annual consumption rate (the 'world food security') are shown. The trend of world food security has been downward since the early 1960s, and because of the global population growth during this period, even a constant level of reserves provides less security. Furthermore, climatic variability in, say, the North American granary, could significantly lower these reserves further. Of course, one cannot yet predict the magnitude or timing of specific future climatic variation so the question of what reserve amount is 'adequate' is a difficult judgment. In addition to food reserves, water and energy reserves (or a number of energy supply options) would help to create margins of safety against climatic variability or other unforeseen factors.

The idea of building a safety factor such as food

reserves into the food producing and distributing system is certainly not new. It has been around at least since biblical times and has been espoused by many in recent history. For instance, in the Book of Genesis from the Old Testament, Joseph issued a long-range climate 'forecast' based upon his interpretation of one of Pharaoh's dreams with a skill that far exceeds the capabilities of today's forecasters. He warned that seven 'fat years' would be followed by seven 'lean years' and urged the Egyptian Pharaoh to store grain (an early food reserve). By following this prudent principle, the Pharaoh successfully built a margin of safety into his society, thereby insuring its ability to cope with the adversity of poor weather. This principle, which has been dubbed *The Genesis Strategy*,[11] could and should be applied to today's world to provide a capacity to sustain some adversity without catastrophic consequences – whether that adversity stems from climatic fluctuations, pests or pathogens, or political actions such as strikes or oil embargoes.

Reserves, of course, are not the only issue. Flexibility may also be built into a system through interconnecting grids in the energy sector or even the water supply sector (coupled, in turn, to reserves). Moreover, a diversity of growing regions and of crops (rather than the present tendency toward monoculture) may help achieve increased flexibility in the food area. Cautious implementation of new technologies (such as nuclear electricity) until their adverse consequences are better understood – or at least keeping other options open, even at an economic premium – is another means of providing flexibility. The premature commitment to large-scale unproved innovations generally creates a dependency on those systems, a dependency that could be catastrophic if unanticipated difficulties were to arise later in the absence of back-up systems. Many argue that solutions will require fundamental changes in habits. However, changes in life styles of the presently affluent may not in themselves make a major contribution to reducing societal vulnerabilities, but certainly such a 'conservationist' attitude might influence others who aspire to greater dependence on large-scale technologies to re-define their concept of necessity. In any case, regardless of their merits, these issues have been raised frequently and must be considered as one of the major options for building long-term flexibility.[20] Therefore, reserves and other means of flexibility against a constellation of uncertainties should be considered globally at a policy level in order to increase the chances for stability and security in the future.

Comparative risks

Not all climatic change or variability would necessarily be detrimental, although potential risks do exist along with the potential benefits. For example, if some now-dry regions where few crops are grown and where perhaps little animal grazing occurs experienced modest increases in rainfall, this would almost certainly have a beneficial effect. But beneficial changes on a regional scale would likely occur at the expense of other regions, perhaps producing detrimental effects somewhere else associated with decreased precipitation. In the absence of certainty as to long-range (or even medium-range) prospects for climatic variability or climatic changes the decision to hedge against potentially ill effects certainly is a major option. On the other hand, further insights into the prospects for climatic variability or climatic change and their impact on global or national food, water, or energy systems may well be appropriate before society considers implementing expensive and possibly unnecessary strategies. But whether public policy should be to hedge, study, or do both is fundamentally a *value judgment* as to how society wants to take risks – a judgment that should be made in the political arena.

Assessment of risks (and benefits) is indeed a difficult endeavor,[21] but one where expertise can be brought to bear. A comparison of risks – weighing, for example, the risk of climatic change through increasing energy use against a societal risk such as the potential unemployment costs of zero energy growth – certainly might influence how many and what kind of alternative options should be kept available. By no means are the answers to these questions easily obtained. Many risks are not yet quantifiable, or even if some aspects are quantifiable, the comparative risks are often measured in different units, again leading to the old problem of 'comparing apples and oranges'. As another example, how can we weigh the risks of climatic variability when reserves are low against the possible risks associated with implementing a reserves build-up, such as a loss of income in the grain-producing region (which might then create economic pressures that reduce

237

food productivity)? How many risks (and of what kind) can society afford to take in a given situation? Are the benefits worth the risks? And, *who benefits* and *who is at risk*? This value judgment as to what constitutes an 'acceptable risk'[22] should not be answered only by so-called experts, for any judgment of this sort is clearly *value-laden* and requires input from many quarters. Scientific input (such as statistics about climatic variability in the Colorado River Basin which relates to the river run-off volume) is necessary for informed judgments on public policy issues. However, a conscious effort must be made by those experts who do speak on the technical details of an issue to assure that, their *assessments* of risk are not implicitly confused with their value choices about the *acceptability* of various risks. This is especially important since policy matters with a scientific component are often discussed in a public forum in the context of a debate on risk acceptability.

Another point associated with climatic risks is their potential amplification with world population growth,[23] as seen in the previous examples on food, water and energy. Population pressure often tends to exacerbate environmental problems or societal vulnerabilities, or at least reduces the resources available to provide an adequate margin of safety. Dwindling food reserves, for example, now feed a smaller percentage of the world's people than an equivalent amount would have fed 10 yr ago. And in the Rocky Mountain States of the US a growing population (mostly from migration) has placed increasing demands on the variable resources of water in that region. Moreover, the expanding US population elsewhere, with its growing desire and need for energy and food, has indirectly placed further demands on the water supplies in the Rocky Mountain region. Irrigation water for the marginal croplands, water used in the mining of fuels and minerals, water proposed for some fuel delivery systems (coal and water slurry in a pipeline), and water needed for new and proposed mining developments all derive in part from this increased national population pressure. A universal demand for higher living standards also contributes.

A global population of which a significant fraction faces marginal per capita food, water and energy consumption is thus increasingly vulnerable to climatic disruption as the population continues to grow. The situation may be aggravated through non-linear interactions, which enhance each other in a 'positive feedback'. For example, mining for fuels in order to produce more energy takes more water, which may require more energy – perhaps to divert a river, to dig more wells, or even to attempt advertent weather modification. But, of course, mining for further energy resources to provide more water requires more water itself.

Population growth has traditionally been an example of exponential growth. Although food production growth might be expected to be linear it has still managed (up to 1972 at least) to stay somewhat ahead of population growth through a combination of factors over the past 15 yr. These include an ample water supply provided in part by irrigation, energy-intensive fertilizers and, not to be minimized, favorable weather. However, these factors may not remain as stable as in the recent past, and the balance between population and food production may be more difficult to maintain. Another important factor concerning population growth is timing. Since over a third of the global population is under the age of 15, the birth rate could equal the replacement rate and yet the population would still not level off for 50 to 70 yr. [[23]] Thus, such 'population momentum' creates a time lag in which population pressure on the food, water, energy and other global systems could be expected to remain high even if there were a major decline in birth rates. It follows that some sort of world strategy is necessary to consider the interactions among population, per capita consumption, per capita pollution and societal vulnerability to environmental fluctuations, whether they be of non-anthropogenic or anthropogenic origin. In particular, special attention needs to be exercised to minimize the possibility of any irreversible change in climate produced by an anthropogenic influence which could affect the global life support systems. Perhaps, the most difficult obstacle is achieving the high levels of international cooperation needed to plan and manage global resources globally.

Some related scientific issues of climatic change – predictability, latency and detectability

Although natural climatic variability can be expected to continue with a high degree of certainty and a low degree of predictability (based upon historical analogues), it still may be possible for climatologists to provide useful input in the planning process. For example, an actuarial accounting of the likelihood of climatic fluctuations of natural origin can be attempted by statistical analyses of past records. Furthermore, use of mathematical models of climate can provide estimates of the order-of-magnitude response of the climatic system to pollutants such as carbon dioxide.

But two issues, latency and detectability, need to be clarified before estimates of potential pollution effects on climate can be very useful. These issues are associated with the identification of a climatic 'signal' from, say, a pollutant, in the face of climatic 'noise' (including inherent weather variability – which is unpredictable beyond several weeks). This then leads to the broader issues of climatic predictability – can future climates in fact be predicted?

Prediction of the climate may be subdivided into two definitions.[24] 'Climate predictability of the first kind' is concerned with predicting 'climate' in a sequence. This differs from predicting a change in the long-term averaged climate in response to a change in some boundary condition (such as increased carbon dioxide in the atmosphere, or a major fluctuation in the solar insolation). This latter situation is called 'climate predictability of the second kind'. An example of the former might be the ability to foresee the seasonal rainfall for several consecutive seasons in the future. An example of the latter might be a prediction of global warming from the 'greenhouse effect' of carbon dioxide if the concentration of atmospheric carbon dioxide were doubled. The prediction of the second kind relates nothing about the sequence of climates during the change to a new long-term climatic state.

Considerable uncertainty exists as to the extent of predictability of either the first or second kind. Some preliminary evidence[25] does suggest that much of the inter-annual climatic variability of sea-level pressure for a period of about 70 yr can be attributed to unpredictable fluctuations (the weather) in the mid-latitude regions. However, there remains some optimism from this pioneering analysis for regional climatic prediction of the first kind in the tropical and polar regions. Further work along this line is proceeding with the study of other climatic variables, although it is certainly hampered by the lack of a sufficient number of long, continuous data records. Some dramatic empirical evidence does exist, however, for predictability of the second kind associated with

major external forcings. That is, a reasonably predictable climatic response occurs each year: the seasonal cycle. Thus, we can expect that as models of the climatic system improve in their capability to reproduce a seasonal cycle, they then will be more able to simulate the sensitivity of the climatic system to other large external forcings. The magnitude of an anthropogenic (or other external) perturbation to a boundary condition of the system might also be expected to govern the magnitude of the climatic response. Unfortunately, scientists face another problem since not all variations in boundary conditions or other external conditions are themselves predictable (for example, no one can say at present when the next geophysical event, such as a volcanic eruption that spews dust into the atmosphere, will occur). Uncertainty in other boundary conditions along with 'climatic noise' caused by unpredictable weather fluctuations can often mask any correlation between the response of the climatic system and a variation in a particular boundary condition that might be under study.

It is also important to note that because of the problem of identifying climatic signals, certain climatic changes may occur and may even be predictable, although not for an extremely long lag time. This problem, referred to earlier as 'latency' may arise for two reasons. First, geophysical processes have certain inherent time constants associated with them, which may be as short as a fraction of a second (for example, molecular relaxation times) or as long as millenia (for example, glacial size changes). For instance, chlorofluoromethanes released near the Earth's surface as aerosol spray propellants may take years to diffuse gradually to the stratosphere, where they will be degraded through the process of photolysis, thereby producing chlorine atoms believed to be involved in ozone gas destruction. This delay in the actual climatic effect (relative to the time of effluence) is governed by the timescales of atmospheric transport and chemistry; thus the ozone destruction may be delayed for decades after the initial release of the chlorofluoromethanes. Similarly, a glacial advance or retreat might be delayed long after a global cooling or warming event.

A second problem for the scientist wishing to establish the sensitivity of the climatic system to an external perturbation relates to the time delay involved in establishing the detectability of the perturbation. Because of the inherent variability, an important signal of climatic change may not be evident for a long time because of the noise in the climatic system – again implying a latent period before which the signal strength relative to the noise is sufficiently large as to be judged 'detectable'. For a constant-strength signal, statistical techniques (including ordinary time-averaging) do exist for increasing the signal-to-noise ratio, but sufficient data from observations or model experiments are still required. A case in which the signal-to-noise ratio would be expected to increase might be the gradual amplification of, say, a global temperature increase from carbon dioxide which is monotonically accumulating in the atmosphere from fossil fuel combustion (if this actually dominates other potential anthropogenic or non-anthropogenic changes). Even if this is the case, the ratio may not be judged statistically significant until the effect has reached what may be called a 'threshold of detectability'. (Perhaps 'threshold' is too vague a term here and the appropriate concept should be expressed as a probabilistic confidence limit.) It is this 'threshold of detectability', which occurs at some chosen value of the signal-to-noise ratio, above which the possibility of climatic change is highly probable. The magnitude of the threshold would certainly depend on such factors as the climatic time period under study, the inherent variability of the climatic system, and the accuracy and spatial coverage of the instrumental observations of climatic variables. One may compare this type of threshold to an 'effects threshold' (a concept often used in toxicology), below which no effect is present as a result of the particular agent or insult. An effects threshold may exist for some anthropogenic insults or changed boundary conditions that perturb the climatic system sufficiently to excite some atmospheric or climatic instability (such as cloud formation). However, it is more likely that any large-scale climatic changes caused by a continuously varying boundary condition or a continuously increasing insult (such as an anthropogenic emission) will be characterized by a continuously increasing climatic change (with perhaps some latency delay). Thus, the concept of a threshold of detectability becomes more useful for the purposes of determining a climatic change than the more classical concept of an effects threshold.

In view of these issues of latency and detect-

ability, it is problematical that the tentative estimates of state-of-the-art models often cannot be verified from actual data until a large, perhaps irreversible climatic change has *already* occurred. Does this mean that we should hedge against not only the uncertainties of natural climatic variability, but also those of anthropogenic climatic changes that present models may be able to predict? Existing models do enable predictions of the second kind to be made of major climatic effects, although at present they cannot incorporate all the potentially important physical and chemical interactions.[26] Society may need to plan for climatic events that can now be deemed only probable. In some cases, scientifically specified confidence limits can be obtained, although in others the range of uncertainty surrounding present estimates cannot even be quantified – and the best that can be done is to assess the intuition of the experts. Hope still remains, however, for a narrowing of these uncertainties. Meanwhile, a difficult question to consider is how long society should wait before implementation of hedging policies.

Technological fixes for climatic change or variability

It might be argued that if the climate fluctuates or changes, perhaps some advertent strategy by humans could at least balance the 'expected' variation. Often such suggestions emanate from those who do not fully appreciate the complexities of the climatic system, who may have a special interest in some climate modification scheme, or who may be driven simply to consider such drastic measures because of some perceived (real or imaginary) detrimental climatic effect on one of their major climate-related survival systems (such as food, water, or energy). Climate control has been contemplated, and actually attempted in some situations, even though the climatic system interactions are not fully understood and the potential effects (beneficial and harmful) are at present difficult, if not impossible, to validate. Whatever the actual effects may be any *perceived* harmful effects of a climate modification scheme could easily make the situation worse, since there is often a sharper human reaction to damage caused by (or assumed to be caused by) deliberate interventions than to that traceable to natural or inadvertent human acts.

It would not be unexpected for weather modi-

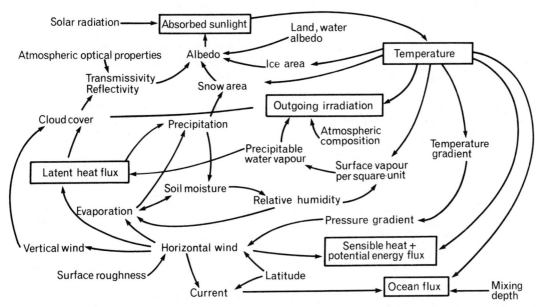

Fig. 14.5. Schematic illustration of many of the potential climatic feedback interactions that need to be considered in a climatic model.[27]

fication (such as cloud seeding) to be attempted when a drought is occurring in a food-producing region, when additional rainfall might be able to salvage a harvest. Especially today, when world reserves are low and food demand is high, any scheme could be looked on as favorable if it held out hope for increased productivity. A similar form of modification is the attempt to enlarge snowpack in the mountains during the winter season, thus leading to increased run-off or availability of water. Such attempts might occur as state or national decisions, and yet have national or even international ramifications.[27]

The following brief description of the climate system illustrates some of the complex interactions involved. The climate system (which consists of interacting subcomponents – atmosphere, oceans, land surface, ice- and snow-covered regions, and biota) is driven by energy reaching the Earth from the Sun. Energy absorption occurs both at the surface (most of the incident solar energy) and in the atmosphere (in a very non-uniform manner dependent upon the types and distribution of various atmospheric constituents). Some of the energy absorbed at the Earth's surface is then reradiated upward, evaporates water or is directly transported to the atmosphere

through heat convection. Solar incident energy itself is non-uniformly distributed over the Earth because of the Earth's spherical shape, its rotation and its axial inclination as it travels an elliptical path about the Sun. All of these factors lead to very uneven differential heating, manifested as thermal gradients which in turn drive a very complex non-linear circulation system (nature's attempt to equalize energy distribution). Everyone familiar with the weather experiences the results of the meteorological or climatic processes: winds, clouds, rain, snow, etc. These processes are in turn coupled to each other through positive or negative feedbacks (Fig. 14.5). Scientists have long appreciated these complexities but are only beginning to understand some of the details of how these couplings occur both spatially and temporally. Because of the many potential feedback mechanisms, any disruption of a natural energy flow may be expected to affect other energy flows related to weather and climatic manifestations. Clear-cut scientific proof of any advertent scheme's intended effects (either potential benefit or harm) would be difficult to obtain since cause-and-effect linkages remain cloudy. However, any climatic changes caused, or thought to be caused, elsewhere on Earth, by advertent modification could well lead to international conflicts. In consideration of these issues, it is worthwhile now to look at a few examples where advertent climatic change has been attempted or proposed.

Since 1970, a cloud-seeding project of the US Bureau of Reclamation has been underway in the San Juan Mountains of southwestern Colorado.[28] The object is to determine whether seeding clouds in winter might increase snowfall in the high mountains. These orographic (mountain) clouds are considered by some scientists to be potentially good candidates for successful cloud seeding. Augmented snowfall and run-off would perhaps be the most valuable and desired economic result of this experiment, although evaluation of this result is extremely difficult in view of previously described natural fluctuations. Other economic effects (good or bad) would also be possible. The additional water might provide a supplementary source of hydro-electric power, and would undoubtedly be useful for agricultural, municipal and industrial needs. However, increased flooding and longer periods of snowpack coverage might severely handicap mining operations. While ski areas could benefit, hunting, fishing and other summer sport seasons could be shortened. Avalanches and floods might increase, causing destruction of mountain dwellings and highways.

There is a long list of other possible side effects from this type of seeding, considered either as benefits or detriments, depending upon one's inherent values. Perhaps the most profound effects would be on ecological systems and on the environment in general. Because of the many interdependencies in an ecosystem, a change in water availability that affects one species (either plant or animal) might lead to changes in other species of vegetation, herbivorous animals, birds, insects and carnivorous animals. Increased mountain run-offs from the additional snow would certainly affect aquatic ecosystems as well. Furthermore, as difficult as it is to assess risk and benefit questions of climatic modification, it is, as pointed out repeatedly earlier, even more difficult to establish which of these are 'acceptable', since this judgment is not only value-laden but must be based on significant uncertainties in scientific assessment.

The case of the flash flood in Rapid City, South Dakota, illustrates some of the legal ramifications of weather modification in a national context. On 9 June 1972, the South Dakota School of Mines and Technology, under a contract with the US Department of Interior, participated in cloud-seeding operations. An intense downpour followed, with severe flooding that caused millions of dollars in damages; subsequently a class action suit was filed against the federal government, arguing that the seeding operation was dangerous. For some time after the event meteorologists continued to debate whether this cloud seeding actually influenced the flood, and doubts remain even today. The mode for a lawsuit in this case is the existing court system; however, for a case in which the climatic effects of a modification scheme appear to cross national boundaries, no such accepted arbitration system exists and national sovereignty reigns.

The possibility of 'geo-engineering' – large-scale climatic modification as a strategy to offset an 'expected' major global climatic change – may not be that far off in the future. Control strategies for counteracting the theorized warming effect of atmospheric carbon dioxide produced from the combustion of fossil fuels have already been proposed: C. Marchetti has suggested, for example, that the carbon dioxide be removed at the energy source and placed in the deep oceans, which are undersaturated with carbon dioxide.[29] Another proposed solution to the carbon dioxide–climate problem is to place a layer of dust in the upper atmosphere which would then reflect more solar energy and theoretically cool the Earth, balancing the warming expected from the carbon dioxide. However, the risks of these climatic control measures must be considered along with any possible benefits that might be derived. First, in the face of the present theoretical uncertainties, any specific operation to compensate for an 'expected' climatic effect could conceivably produce an over-response in the system. It is quite possible with today's rather crude estimates that the magnitude of the 'expected' climatic perturbation could be overestimated and the magnitude of the deliberate intervention could be underestimated. Also, as stressed earlier, if climatic deterioration were to follow climatic control operations, some nations could, in the absence of a definitive theory of climate, *perceive* that such climatic 'corrective' measures directly caused the deterioration. Thereby, these nations could demand restitution from the climate controllers, leading to a scenario in which the political – and possibly military – implications might be difficult to foresee.[27]

The overall situation

There is no doubt that climate influences human affairs in many ways, primarily through its influence on food, water and energy resources; through effects, in the long run perhaps, on the level of the sea and on coastal geography; and even through many immeasurable psychological factors. These influences have existed in the past, and there is no reason to believe they will diminish in the future. Humans build homes to protect themselves from the climate, the manner of construction being affected not only by the available materials and culture, but also by the natural elements of the environment. When humans first began to cultivate crops their choice of growing regions also was strongly determined by the expected patterns of the weather. Although with the help of technology, humans have increased their tolerance to some environmental stresses, considerable vulnerability remains. Furthermore, many doubts still remain about the capabilities of society to influence the climate or weather, and perhaps even more doubts about the desirability of such interventions.

Since climate itself is not fixed, the patterns of its variations have also influenced human development. Weather obviously changes day to day, and a society learns to anticipate a certain amount of variability. However, climatic variability of a larger-than-usual magnitude or variability on a longer time-scale, which we unrigorously have called climatic change, is more difficult to anticipate with the present knowledge of the climate system. About the only thing that can be stated with a high degree of certainty is that, based upon historical analogues, precedented climatic variations could well recur (possibly superimposed on any potential long-term anthropogenic influences). Society must now consider building into the system sufficient resilience to cope with the known climatic irregularities, and must augment this actuarial approach with studies based upon the analysis of possible climatic scenarios[30] (for instance, by using analogous climates of the recent past in combination with present-day estimates of the effects of changing external factors like carbon dioxide).[31]

As a rapidly expanding world population leads to increased demand for food, water and energy, society will be faced with many complex national and international decisions, where often climatic (or other environmental) risks attendant with increased consumption patterns need to be weighed against social and economic risks of slowed development, particularly in the poorer countries. Because of the low level of world grain reserves, climatic variability or change could lead to significant short-range stress on food security which, in turn, might have 'unforeseen' international consequences. Preparation for this highly precedented climate-related event is essential. Some argue for a change of life style away from high growth rates in the consumption of resources toward a more conservation-oriented society with less waste in the more affluent countries. Yet this is very unlikely to occur in the near future without significant changes in political orientation – or several convincing *visible* catastrophic resource shortages or environmental disasters. On the other hand, reduction of population growth rates, particularly in poorer countries with the highest birth rates, needs to be achieved as soon as possible if the likelihood of a smooth transition to a sustainable world order is to be realized. Again, some significant social and political changes may be a prerequisite. The concept of 'planetary bargaining' may be useful in this context.[32] In any case global population growth is likely to continue for some time, with the population momentum continuing even after the population replacement rate is reached. Thus, even if it proves difficult to reduce the population and affluence growth pressures on the global environment quickly, at the least then we should build into our major survival systems a higher degree of flexibility and reserve capacity than is present today. Of course, this raises serious questions of institutional and political realities beyond the scope of this chapter. A note of optimism could be sounded here, however, for no one has proved that society cannot, if it wanted, cope with most of these problems. But the most important problem may well be the pressure of time, since exponential growth makes the search for societal resiliency in the face of possible climatic changes extremely urgent.

As for the matter of predicting climatic change or climatic variability, decision making must still proceed even with present lack of certainty. A decision only to study but not to act *is* a decision. It should be made consciously and not implicitly by default in the face of bewildering uncertainty. Moreover, perhaps one of the greatest threats to future global security lies in the common misinterpretation by many citizens and decision makers of the frequent and confusing debate among 'experts' on the technical component of many public policy issues pertinent to future survival. For example, two still-irreconcilable views among atmospheric scientists relate to the issue of whether the globe is cooling or warming, or whether pollution will cause future cooling or warming. Faced with such scientific uncertainty, many scientists, and often the public as well, feel that they should merely study the questions more before any policy is implemented, rather than acting on tentative scientific input.

The often conflicting testimony of experts leaves doubts about whose 'expert' opinions to trust on the bewildering issues of climatic change, technology and human survival. Unfortunately, a choice often made is to trust no one, which thus conveniently avoids the issues and translates into the maintenance of the status quo. It certainly is a temptation to rationalize a postponement of action on many of these perplexing issues until the scientists, economists, moralists, politicians, or others involved in the debates are in agreement, or at least are much more certain about what they espouse, especially since many measures that might add resilience to global systems may also be expensive. All too often, we are preoccupied with more immediate concerns, such as inflation, employment, or taxes. Most elected governments generally respond only to short-term concerns – usually concentrating on issues with time-scales shorter than the period between elections. Since many issues of food, water, population, energy and climate are important over time-scales of generations, mechanisms must be considered immediately to encourage government officials to place higher priority on such long-term matters as the effect of climatic change on human survival. Changes in the political consciousness of the voters, which ultimately bring about changes in the political process, seem essential.

From the scientific point of view, issues of climatic change and public policy present a particularly disturbing situation because of the inability of present climatic theory and observations to yield more than order-of-magnitude esti-mates of either future climatic variations or their impacts on society. Moreover, even the uncertainties often cannot be quantified. At a minimum, it thus seems prudent that more research be aimed at reducing the wide range of uncertainty surrounding both the estimates of possible climatic futures and the impact of these on society. Such improved assessments would help decision makers clarify their options and would thereby aid them in choosing appropriate policies to hedge against possible disruptions while minimizing wasteful or risky developments. With regard to making the water or food systems more resilient, actuarial studies of past fluctuations in climate should be accelerated; attempts should be made to relate these fluctuations to, say, fluctuations in food production, water resources and energy demand. Because of the lack of reliable forecasting skill, this approach may be the most feasible for some time in estimating potential consequences of inherent climatic variability – but it must be recognized that both inherent observational and theoretical difficulties remain for improving forecasting skill. The potential value of actual (not actuarial) forecasts justifies, in our opinion, additional efforts to determine the feasibility of prediction; then, if feasible, researchers must search for those factors that might permit predictions of the first kind. In this connection, studies of the *social value* of forecasts can be revealing,[33] and more effort is needed in this area as well.

Climatic prediction of the second kind, which produces a forecast of changes in long-term average climate caused by external forcing variations, is a task that relies heavily on mathematical models of the climate, since some changes in boundary conditions do not have known historical analogues (such as carbon dioxide or halocarbon releases). Verification of these models, however, often occurs with simulation of known climatic variations (such as a reproduction of the seasons) and with comparisons to observational data (such as inter-annual changes in radiation balance via satellite measurements). Unfortunately, such data are often sparse and a large degree of uncertainty in model predictions will remain an unhappy reality for some time to come.

It cannot be irrefutably stated that climatic research will immediately – or even ever – narrow the present range of uncertainties surrounding all important climatic futures. But in our opinion *it is absolutely necessary*, in view of the significant human dependence on climatic stability, that greater consideration both of alternative future climates and of their potential impact on critical global systems be given a high priority. Whether research *alone* is a sufficient hedge against future catastrophes, however, is a policy issue needing careful review on a case-by-case basis. But without other preparations by global communities, preventable crises could be induced by climatic variability. On the other hand, valuable resources needed for development could be wasted in hedging unnecessarily against the predictions of tentative climatic theories. Schneider & Mesirow,[11] groping with this predicament, remarked that 'This dilemma rests, metaphorically, in our need to gaze into a very dirty crystal ball; but the tough judgment to be made here is precisely how long we should clean the glass before acting on what we believe we see inside.'

The authors thank Elmer Armstrong for editing the manuscript and Eileen Boettner for typing it.

References

1. GARP (1975). The physical basis of climate and climate modelling. *GARP Publ. Ser.* No. 16. WMO, Geneva. 265 pp.
2. *NAS* (1975). *Understanding climatic change: a program for action.* US Committee for GARP, National Academy of Sciences, Washington D.C. 239 pp.
3. Gribbin, J. (1976). Climatic change and food production. *Food Policy,* **1,** 301–12.
4. Lamb, H. H. (1966). *The changing climate.* Methuen, London; and Ladurie, E. L. (1971). *Times of feast, times of famine,* translated by Barbara Bray. Doubleday, New York.
5. An account of a number of terrible famines is given by P. R. Ehrlich and A. H. Ehrlich (1972). *Population resources environment, second edn.* Freeman, San Francisco.
6. Bryson, R. A. (1974). *World climate and world food systems* III, *The lessons of climatic history,* Report 27. Institute for Environmental Studies, University of Wisconsin.
7. Brown, L. R. (1975). The world food prospect. *Science,* **190,** 1053–9.
8. Economic Research Service (1975). *World agricultural situation.* US Department of Agriculture, Washington D.C. 47 pp.
9. Decker, W. (1974). *The climatic impact of variability in world production.* Prepared for the 1973 Annual Meeting of the American Associa-

tion for the Advancement of Science, San Francisco 27 February 1974. Reprinted in the *American Biology Teacher,* **36,** 543–40.
10. Kukla, G. J. & Kukla, H. J. (1974). Increased surface albedo in the northern hemisphere. *Science,* **183,** 709–14, and Wiesnet D. R. & Matson, M. (1976). A possible forecasting technique for winter snow cover in the northern hemisphere and Eurasia. *Mon. Weath. Rev.,* **104,** 828–35. The latter paper shows that the 1972 albedo increase appears to be a fluctuation.
11. Schneider, S. H. (with L. E. Mesirow) (1976). *The Genesis strategy: climate and global survival.* Plenum, New York & London. 419 pp.
12. McQuigg, J. D., Thompson, L., LeDuc, S., Lockard, M. & McKay, G. (1973). *The influence of weather and climate on United States grain yields: bumper crops or drought.* A Report to The Associate Administrator for Environmental Monitoring and Prediction, National Oceanic and Atmospheric Administration, US Department of Commerce, December 14.
13. Gilman, D. (1974). Paper presented at the 140th meeting of the American Association for the Advancement of Science, San Francisco, 27 February 1974.
14. Committee on Water (1968). *Water and choice in the Colorado Basin.* National Academy of Sciences, Washington D.C.
15. Wallis, J. R. (1977). Climate, climatic change and water supply. In *Climatic variability and the impact of possible future climatic changes on water supply and use.* Geophysics Research Board, National Academy of Science, Washington D.C. (Report in press).
16. Stockton, C. W. (1977). Interpretation of past climatic variability from paleoenvironmental indicators. In *Climatic variability and the impact of possible future climatic changes on water supply and use.* Geophysics Research Board, National Academy of Science, Washington D.C. (Report in press).
17. Suomi, V. E. (1975). Atmospheric research for the nation's energy program. *Bull. Am. met. Soc.,* **56,** 1060–8. The winter of 1977 in Eastern USA is a dramatic example of weather-induced stresses on energy demand, which coupled with insufficient natural gas stockpiles, led to severe shortages and perhaps billions of dollars in economic losses.
18. CONAES (1977). Committee on Nuclear and Alternative Energy Systems, National Academy of Sciences, Washington D.C. (Report in preparation).
19. Panel on Energy and Climate (1977). *Energy and climate: outer limits to growth?* Geophysics Research Board, National Academy of Sciences, Washington D.C. (Report in press).

20. One of the leading proponents for lifestyle change is E. F. Schumacher. (1973). *Small is beautiful.* Harper & Row, New York; a similar viewpoint is expressed by P. R. Ehrlich & A. H. Ehrlich, (1974). *The end of affluence.* Ballantine, New York. The most vociferous spokesmen for accelerated affluence are H. Kahn, W. Brown, & L. Martel, (1976). *The next two hundred years: a scenario for America and the world.* Morrow, New York. 241 pp.

21. Budnitz, R. J. & Holdren, J. P. (1976). Social and environmental costs of energy systems. *A. Rev. Energy*, **1**, 553–80.

22. Lowrance, W. M. (1976). *Of acceptable risk.* Kaufmann, Los Altos, California. 182 pp.

23. Holdren, J. P. & Ehrlich, P. R. (1974). Human population and the global environment *Am. Scient.*, **62**, 282–92.

24. Lorenz, E. (1975). Climatic predictability, Appendix 2.1 in *GARP Publ. Ser.* No. 16. WMO Geneva. 265 pp.

25. Madden, R. A. (1976). Estimates of the natural variability of time-averaged sea-level pressure. *Mon. Weath. Rev.*, **104**, 942–52.

26. Schneider, S. H. & Dickinson, R. E. (1974). Climate modeling. *Rev. Geophys. Space Phys.*, **12**, 447–93.

27. Kellogg, W. W. & Schneider, S. H. (1974). Climate stabilization: for better or for worse? *Science*, **186**, 1163–72.

28. See Chapter 7 of reference 11 for more details and additional references.

29. Nordhaus, W. D. (1977). Strategies for the control of carbon dioxide. *Discussion Paper* No. 443 for the Cowles Foundation for Research in Economics at Yale University, New Haven, Connecticut; and Marchetti, C. (1977). On geoengineering and the CO_2 problem. *Climatic Change*, **1**, 59–68.

30. Toronto Conference Workshop (1975). *Living with climatic change.* Science Council of Canada, Ottawa, Ontario, Canada. 105 pp.

31. Kellogg, W. W. & Schneider, S. H. (1977). In *Desertification: environmental degradation in and around arid lands*, ed. M. H. Glantz. Westview Press, Boulder, Colorado.

32. The planetary bargain is a concept advocated by Harlan Cleveland, Aspen Institute of Humanistic Studies, Aspen, Colorado, as described briefly in Chapter 9 of reference 11.

33. An interesting study on the political constraints on the utility of long-term forecasts is given by M. H. Glantz (1977). The value of a long-term weather forecast for the West African Sahel. *Bull. Am. met. Soc.*, **58**, 150–8.

APPENDIX

This article first appeared in *Quaternary Research*, **4**, 385–404 (1974) and is reproduced here by permission of the publishers and author. It is included in the present volume to provide one example of how the pieces of the climatic puzzle can be put together to provide an overall model of a major climatic change, the initiation of a glaciation. It is not suggested that this is the only model available for such a change, nor that Professor Flohn's model is necessarily the 'best', although it is one of the most complete. New insights into the processes of climatic change are certain to lead to the development of new models of this kind, incorporating the best features from current models; it is certain that such models must draw extensively on the available information about past climates and the improving understanding of the present climate, as outlined in this book. I hope that the example of this particular detailed model will point the way and serve to encourage others who are tempted to venture into the dangerous realms of climatic model building.

John Gribbin

Background of a geophysical model of the initiation of the next glaciation*

HERMANN FLOHN

Evidence of (at least) five rapid hemispheric coolings of about 5 °C during the last 10^5 yr has been found, each event spread over not more than about a century, as examples of a global-scale climatic intransitivity. Only some of them lead to a complete glaciation at the northern continents, others ended after a few centuries by a sudden warming ('abortive glaciation'). Starting from a modified version of Wilson's hypothesis of Antarctic ice surges, an air–sea interaction model with realistic geophysical parameters is outlined to interpret the sudden initiation of the North American ice sheet. Special attention is given to the Atlantic section, where the climatic anomalies during the last glaciation appear to have been significantly larger than in other sections.

I. Introduction

In a recent symposium (Kukla *et al.*, 1972; Kukla & Matthews, 1972), several well-known Pleistocene specialists discussed the question: when will the present interglacial end? A few results will be considered here from a meteorological viewpoint.

(*a*) Since about 1945 global cooling, on a scale of ~ 0.1 °C yr^{-1}, has reversed the warming trend of the first decades of our century. The bulk of these changes is most probably not man-made, but of natural origin. Evidence exists for several short cool periods during the last 5000 yr, as well as for catastrophic dry periods in subtropical areas lasting a few decades. None of these variations are comparable in scale with the Allerød fluctuations.

(*b*) The climatic optimum of the present interglacial was reached 6–7000 B.P. Evidence from

northern Germany and England shows that the last interglacial (Eem = Sangamon) lasted little more than 10 000 yr; it was slightly warmer and wetter than the present interglacial, with a quite similar climatic time sequence.

(*c*) Based on more than 800 measurements of the $^{18}O/^{16}O$ ratio from fossil foraminifera, it has been concluded (Emiliani, 1972) that the tropical ocean surface temperatures were as high as or higher than today for only 10 % of the last 400 000 yr. Considering the length of a glacial–interglacial cycle to be nearly 10^5 yr (Broecker & Van Donk, 1970) the average duration of a warm epoch cannot have been longer than 10^4 yr.

(*d*) A large majority of the participants of the symposium concluded that the present warm epoch has reached its final phase, and that – disregarding possible man-made effects – the natural end of this interglacial epoch is 'undoubtedly near'. The time-scale of this transition may be a few millenia, perhaps only centuries.

If this is correct, earth scientists are confronted with a hitherto neglected question: what are the initial stages of a glaciation? How can we imagine the triggering of the formation of ice sheets on the northern continents eventually covering about 17×10^6 km^2 in North America and nearly 11×10^6 km^2 in Eurasia, with ice domes up to 3 or 4 km in height, thus reducing the ocean volume by about 4 %, with a eustatic sinking of the sea-level to -100 m, sometimes to -130 m? This question is not only of academic interest: its immediacy will be demonstrated in the following sections.

II. Stability or instability of climate?

E. Lorenz (1968) has recently raised a quite deep-rooted question: how stable is our climate?

Table 1. *Surface albedo* (a_S) *and equilibrium temperature* (T^*) *deviations (areas in 10^6 km^2)*

Albedo	Oceans		Continents			Average albedo a_S	Deviation ΔT^* (°K)	Remarks
	Open 0.05	Ice 0.70	Open 0.12	Ice 0.75	Snow 0.30			
N. hemisphere	145	10	70	3	27	0.1294	—	⎫
S. hemisphere	190	16	33	13	3	0.1384	—	⎬ Actual (1901–50)
Earth (E)	335	26	103	16	30	0.1339	—	⎭
Model NH 0	142	13	70	3	27	0.1373	−0.95	N. hemis. *ca.* 1890
Model SH 0	188	18	33	13	3	0.1434	−0.60	S. hemis. *ca* 1850
Model E 4	317	30 ⎫	108	49	6	⎰ 0.1731	−4.6	⎱ Ice age
Model E 5	307	40 ⎬				⎱ 0.1860	−6.2	⎰ (sea-level −100 m)
Model SH 3	165	41 ⎭	33	13	3	0.2022	−7.6	Wilson surge

Considering a complete set of basic equations with fixed external parameters – such as the solar 'constant', the rotation rate and radius of the Earth, and the chemical composition of the atmosphere – as a base for simulating the climate defined as a time-averaged state of the atmosphere, he discusses the number of possible solutions. Climate is then defined as *transitive* if only one solution exists. One may also perceive several more or less quasi-stationary different solutions, which may transform from one state into another by a sort of flip-flop mechanism '(vacillation)': this situation is defined as *intransitive*. Without discussing at length evidence for and against, Lorenz considers our climate as *semi-intransitive*.

If this is true, we cannot expect to obtain from mathematical modeling unambiguous forecasts of climatic patterns. Simplified models with a crude parameterization of synoptic-scale meridional exchange processes – such as the models developed by Budyko (1969) and Sellers (1969, 1973) – showed either a great sensitivity to comparatively small changes of external conditions or (worse than that) distinct intransitivity under exactly the same conditions. In contrast to this, Washington (1972) demonstrated on the base of the much more advanced NCAR circulation model that the response of the model to different and even contrasting externally induced disturbances was nearly identical. This state of affairs – incomplete as it stands now – is seriously disquieting. Therefore, one of the most urgent tasks is a careful and critical search for evidence of climatic instability on a hemispheric or, better, global scale.

Examples of partial (regional) instability have been given elsewhere (Flohn, 1973). The best example is known from the equatorial Pacific, where the oceanic Ekman drift causes either equatorial upwelling or downwelling, depending on the surface wind distribution, and causing in the atmosphere either a stable, cloudless and dry equatorial zone or in instability near the equator with high convective activity. Apart from short transition periods, any intermediate state cannot remain stable. Because of the large attendant differences in oceanic evaporation and precipitation, these contrasting patterns are correlated with many teleconnections over wide areas of the globe (Bjerknes, 1969; Flohn, 1972; Rowntree, 1972).

III. The role of surface albedo

Within the heat budget of the Earth's surface, the high albedo of ice and snow (0.70–0.80) – in contrast to all other surfaces, except clouds, (0.05–0.35) – dominates most other terms. During winter the tropospheric baroclinic zones have a tendency to follow the margins of the seasonal continental snow cover; this has been experienced by the author during the European winters between 1938 and 1948. In spite of all vagaries of weather, such a pattern remains superimposed in a statistical sense.

There exists (Manabe & Wetherald, 1967) a direct relation between surface albedo and an equilibrium temperature (Table 1), assuming constant relative humidity and an average cloud distribution. This relation can be checked against data on the varying extent of the Arctic and Antarctic sea-ice during the 19th and 20th centuries (Flohn, 1973). The observed decrease of the Arctic sea-ice from about 1880 to 1940 and the estimated increase of the Antarctic sea-ice during the 19th century (Lamb, 1967), both of the order of nearly 2 or 3 × 10^6 km^2, should correlate with changes in the hemispheric equilibrium temperature of the order of 0.6–0.9 °C, in good agreement with the observed data (Table 1, Models SH 0 and NH 0).

A further check can be derived from the last glaciation, with a glaciated continental area of 49×10^6 km², accompanied by a eustatic drop of the sea-level to -100 m, increasing the land area of the Earth from 149 to about 163×10^6 km². In this case (Model E 4) the equilibrium temperature of the whole Earth should be 4–5 °C lower than today, once more in agreement with the observed data. There is sufficient evidence that the Atlantic sea-ice reached, during the maximum of the last glaciation, an average latitude of about 43° N (McIntyre *et al.*, 1972). Its extension into the Bay of Biscay must also be assumed when interpreting the exceptional cooling of the adjacent territories (from northern Spain to southern Ireland) by about 12 °C, compared with only 5 °C at the same latitudes at the Pacific coast of North America (Flohn, 1969). In this case the area covered by sea-ice increases to about 40×10^6 km² with a simultaneous global temperature drop of 6 °C (Model E 5).

The good agreement between the predicted and observed equilibrium temperatures convinces us that the role of the albedo in the long-term climatic oscillations during the Pleistocene is certainly greater than that of variations of solar radiation due to the Earth's orbital elements (Hoinkes, 1971). It should be remembered that this point was raised as early as 1938 (Wund, 1938); numerical model computations of the climatic effects of the Milankovich mechanism neglecting the positive feedback effect of albedo changes are incomplete. The role of surface albedo has also been demonstrated (Kukla & Kukla, 1972) from seasonal and inter-annual changes of the snow cover. The existence of a quasi-equilibrium between area-averaged surface temperature and area-averaged surface albedo (Flohn, 1969) leads to a serious consequence: if a climate-independent mechanism producing variations of the extension of ice exists – as suggested by A. T. Wilson (1964) – the usual chain of cause and effect may be reversed: large-scale Antarctic surges will produce immediate hemispheric cooling (Model SH 3). This necessitates a critical investigation of the real time-scale of Pleistocene coolings, which appears to be inconsistent with the Milankovich time-scale.

IV. Time-scale of global coolings

Based on investigations of ice cores, ocean bottom cores, and fossil peat bogs, several drastic coolings during the last 10^5 yr have recently been revealed. Some of these, with multiple evidence, will be reviewed in stratigraphic sequence, ignoring some inevitable minor differences of time and time sequence interpretation in the literature.

(1) During the recession of the last glaciation, the well-known sequence Bølling interstadial (warm) – Older Dryas (cold) – Allerød interstadial (warm) – Younger Dryas (cold) covered less than 2000 yr, with variation in the annual temperature of up to 6 °C (Mercer, 1969). In the Mediterranean and at other subtropical and tropical sites only the second half of the sequence was marked, and the Older Dryas period was insignificant (Van der Hammen *et al.*, 1971, Fig. 2). The Allerød warming period coincides with the abrupt global environmental change after the Würm–Wisconsin Glaciation, occurring at about 11 000 B.P. in the space of a few centuries, while the melting of the ice domes – reflected in the global eustatic sea-level rise – lasted some 8000 yr.

This time sequence has been derived mainly from palynological evidence, hampered by the limited migration speed of biotopes. On the other hand, the isotopic changes preserved in the Greenland ice-cap represent largely – disregarding here some systematic sources of error (Johnsen *et al.*, 1972; Dansgaard *et al.*, 1971) – the temperature of formation of precipitation particles in clouds, i.e. the regional climate. Here (Johnsen *et al.*, 1972, Fig. 6) the cooling prior to the younger Dryas lasted less than 350 yr; the following warming, 300 yr. However, simultaneity of the climatic changes on both sides of the Atlantic has been doubted (Mercer, 1969).

(2) Before the last long warm interstadial within the Würm–Wisconsin Glaciation (Fliri, 1970) – known as Stillfried B or Plum Point – a marked cold period of not more than about 2000 yr duration occurred at about 38 000 B.P. It has been found in the Greenland ice core (Johnsen *et al.*, 1972) as well as in the Indian Ocean off the Somali Coast (4–8° N) (Olaussen *et al.*, 1971); here the time between the beginning of the event and the temperature minimum is estimated to be not more than about 500 yr.

(3) Another cold period of this magnitude is also found in the Somali Current Area,

251

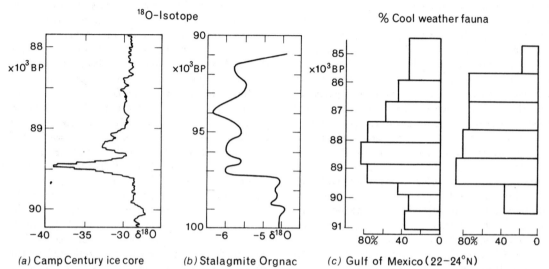

Fig. 1. Temperature variations at about 90 000 B.P. (see text). (a) $^{18}O/^{16}O$ ratio in Greenland ice core (70°N). (b) $^{18}O/^{16}O$ in a cave near Orgnac (South France). (c) Percentage of cool weather foraminifera from the Gulf of Mexico.

interpreted as the beginning of the Würm I Glaciation at about 55 000 B.P. It coincides well with the marked cooling in the Greenland core after the Odderate Interstadial and in Macedonia (van der Hammen, 1971) around 59 000 B.P.

(4) The short cooling between the Brørup and Odderade Interstadials, near 70 000 B.P., is quite dramatic: in Macedonia the vegetation changed from oak forest into steppe in much less than 1000 yr. It coincides with a marked cold period in the Greenland area, while in the Indian Ocean (Olaussen *et al.*, 1971) only a hint has been found. At the same time, a short intense cooling has been observed (Sancetta *et al.*, 1972) in an Atlantic deep-sea core at 52° N 22° W.

(5) The most dramatic, short-lived cooling event was observed (Fig. 1) in the Greenland ice at about 89 000 B.P. (all Greenland dates before 12 000 B.P. are slight uncertain). Here the climate changed within 100 yr ('almost instantaneously') from warmer than today into full glacial severity (Dansgaard *et al.*, 1972). This event has also been found in a stalagmite in a French cave (Duplessy *et al.*, 1971) at 97 000 B.P. with a cooling of the cave (!) by 3 °C in a few centuries and an extremely rapid cooling (in less than 350 yr) has been described in many cores from the Gulf of Mexico at 90 000 B.P. (Kennett & Huddleston; 1972.) At the same time the first strong cooling after the Eem Interglacial was observed in Macedonia and in the Netherlands (van der Hammen, 1971); and multiple evidence exists for a sudden sea-level rise at the eastern coast of North America and at Bermuda, possibly caused by an Antarctic surge (Hollin, 1972).

These five events show coolings of the order of up to 5 °C per century in contrast to not more than 1 °C per century in recent fluctuations. This rate is in fact a minimum value because of the smoothing role of molecular diffusion processes (Johnsen *et al.*, 1972). Of particular interest are the events in the area of the Somali Current (Olaussen *et al.*, 1971) which are far too short-lived to be interpreted as caused by orbital changes. Some other peaks, especially in the Greenland ice core (Fig. 2), may be added to this

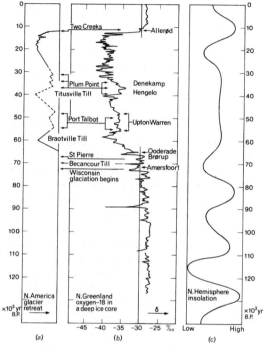

Fig. 2. Greenland ice core vs. corrected time-scale (Dansgaard *et al.*, 1971), plotted in 200-yr intervals with tentative interpretation in European (right) and American (left) terminology.

list but (up to now) without supporting evidence from other sites.

Only one of these selected cases initiated, in the northern hemisphere, a continental glaciation: this is the beginning of Würm I (case 3). If we assume a maximum glaciation of the northern continents (with a volume of 47.4×10^6 km^3 and an area of 29.5×10^6 km^2, i.e. with an average thickness of 1610 m) as resulting from an average annual accumulation of 40 cm, the growth period lasts about 4000 yr, and the minimum duration of a full glaciation is still of the order of 10 000 yr (Lamb & Woodroffe, 1970). During that time the local increase of albedo favors the persistence of glaciogenic conditions. The other four cases represent only short-lived events, with a glaciogenic anomaly of the atmospheric and/or oceanic circulation lasting 'only' a few centuries (case 5), certainly less than 2000–3000 yr. From the viewpoint of a meteorologist, these incomplete or 'abortive glaciations' are by no means less interesting: they reveal a very remarkable instability of the atmosphere–ocean–ice system repeating non-periodically over a time-scale of the order of 2×10^4 yr.

In view of the rapidity of development the initial stages must have lasted less than a century, only a few decades. What kind of atmosphere/ocean circulation anomalies are able to produce such catastrophic events? Any answer to this question can only be more or less speculative; however, it should never be unsound from the viewpoint of an experienced meteorologist specializing in physical and theoretical climatology.

V. Climatic conditions of the initial stages of a glaciation

The evidence of such dramatic global coolings with a quite short time-scale is of high importance when discussing the initiation of the glaciations of the northern continents, i.e. of the Laurentide and Scandinavian ice sheets. This initiation is in any case a problem: while in Scandinavia a spreading of the existing mountain glaciers (Svartisen, Jötunheimen) could be caused by a temperature drop of 5–6 °C, without any necessity for a substantial increase of precipitation, this is not the case in North America. Here the Rocky Mountain Ice, which expanded from the (at present heavily glaciated) high mountains in western Canada and southern Alaska, remained of medium size and never did extend far to the east. The major glaciation of North America was due to the Laurentide ice sheet, which formed on the presently unglaciated Ungava Plateau of Labrador and Quebec (now at alt. of 600–800 m, between 53° and 60° N) and even on the low-lying Keewatin country west of Hudson Bay, between 58° and 65° N. Only during the maximum phase of the last glaciation were the two ice sheets joined. Loewe (1971) has discussed the climatic conditions in Ungava and Keewatin, and concluded that a 6 °C temperature drop alone would be insufficient to cause a permanent snow cover without an increase in total precipitation (snowfall). Because of the large extent of the Laurentian ice dome (which contained more than 62 % of the total increase in ice volume during a glaciation), its formation must have a key position in the sequence of events.

The recent climate in Labrador–Ungava and

253

Keewatin is characterized by summer temperatures (June–August) of 11–12 °C and by annual precipitation near 75–80 cm in Ungava but only about 35 cm in Keewatin.

Barry (1959) and Brinkmann & Barry (1972) have investigated, by the methods of synoptic climatology, the meteorological conditions associated with high precipitation in the Labrador–Ungava area as well as in the Keewatin area, with different results. The formation of a nucleus of the ice dome on the Labrador highlands, with a temperature drop of about 6 °C together with an increase of precipitation, is only possible for a semi-permanent upper cold cyclone centered near 55° N 72° W or a deep upper trough extending from Baffin Bay or Ellesmere Island into the area near Boston. In this situation low-level flow from the north or northwest would permanently cool the area, between about longitude 85° and 70° W, on the southern flank of low-level cyclones which themselves extended even farther east. Above 700 mbar, however, relatively warm and moist air from the south would flow northward above the Labrador Peninsula, forced to ascend along a more or less stationary frontal surface and causing abundant precipitation, most frequently as snow.

In a boreal subpolar climate, the snow-melting process is not finished before late May or early June, and the first snowfall may occur as early as the end of August or early September. Each glaciation must start with a permanent snow cover lasting during the summer; its high albedo (even in a half-melted stage) prevents the soil from storing heat. If a snow cover can survive one summer with its high sunshine duration, the probability of a much higher snow cover in the next year rises substantially; this is the beginning of a positive feedback process. A few consecutive years of this type would be sufficient to build up a snow cover of several meters over the whole Ungava plateau with an area of about 60 000 km² above 600 m: then the high surface albedo during summer will prevent easy destruction, even if the large-scale flow-type changes.

Let us assume a 20 % increase of precipitation to 90 cm yr⁻¹, a snowfall fraction of 80 % of precipitation and a (high) snow cover density of 0.3; we obtain an annual snow accumulation of 240 cm (or 72 g cm⁻²). In a cloudy summer, the incoming global radiation should be slightly reduced to about 350 Ly day⁻¹ (1 Langley = 1 gcal

em⁻²); with a surface albedo of 55 % (50) and an effective terrestrial radiation of 120 Ly day⁻¹ (latitudinal average) we obtain a net radiation of 38 (55) Ly day⁻¹. During a melting season of 80 days, this would melt 38 (55) g cm⁻² snow; the remaining snow cover would reach, with a density of 0.5, a height of 68 (34) cm at the beginning of next winter. With an average (!) albedo of 45 % or less, snow cover would completely melt. Under such assumptions, a 15-m tall forest could be completely covered by snow after some 22 yr (45), with a further rise of surface albedo. Cold air will then be permanently produced near the surface, due to the combined effects of high albedo and long-wave radiation from the snow cover. This will lead, once a synoptic-scale diameter (300–500 km) is reached, to the formation of a superimposed cold low, which will force the upper flow to curve cyclonically and will further enhance snowfall. Such a powerful positive feedback mechanism is well known to the experienced meteorologist; it was discussed half a century ago by C. E. P. Brooks (1926). It can be qualitatively interpreted with the aid of the heat balance equation for an atmospheric column:

$$H + LP - \operatorname{div} Q \downarrow - \operatorname{div} \vec{A}_h - \varDelta T = 0$$

where (H = flux of sensible heat into air, LP = release of latent heat by precipitation, $\operatorname{div} Q \downarrow$ = divergence of radiative fluxes, $\operatorname{div} \vec{A}_h$ = divergence of advective heat transport, and $\varDelta T$ = heat storage in the column). In high latitudes LP is not as predominant as in the tropics; above a snow cover H is usually negative (from air to surface) and $\operatorname{div} Q \downarrow$ is strongly negative, thus $\varDelta T$ is likely to represent a heat sink.

The geophysical causes of this initial anomaly of the atmospheric circulation will be discussed in Section VI. Here it may be useful to outline the large-scale pattern connected with a deep, semi-permanent trough along 70° W, with a cold cyclonic center above western Labrador. This causes a meridionalization of the upper tropospheric flow, with anticyclonic ridges (and frequent blocking highs) near 125° W (Canadian Northwest Territories) and 20° W (Iceland) (Flohn, 1969; Lamb & Woodroffe, 1970). A secondary trough over Scandinavia and central Europe will develop near 15° E, together with a warm ridge in 50–70° E, including the mountains of central

Asia. The occurrence of a blocking high just east of Alaska leads to southerly flow over Alaska itself, locally reducing the rate of cooling. The frequent occurrence of a blocking high between Iceland and Scotland causes northerly flow over Scandinavia and central Europe, increasing snow fall and cooling. This in turn causes a quasi-stationary pattern above eastern Europe (25–40° E) corresponding to that above Labrador, with similar consequences, starting on the eastern flank of the Scandinavian mountains and in Finland. Such a pattern is nowadays frequent in cold winters and springs; here it is visualized – quite differently from today – as existing during the climax of the warm season.

VI. Antarctic surges and their geophysical consequences

Since the observed short time-scale of global cooling events (Section IV) is inconsistent with orbital effects, we ought to consider quite seriously the unorthodox Antarctic Surge hypothesis of A. T. Wilson (1964, 1966, 1969), based on the idea of a large-scale instability of the Antarctic ice dome. Since many recent examples of mountain glacier surges are known, particularly in the Alaskan mountains, and since some physical properties of glacier ice are subject to marked changes in the vicinity of the melting point, this hypothesis appears to be generally consistent with glaciological knowledge. Budd *et al.* (1970) have developed a geophysical model of the Antarctic ice, mapping such quantities as ice-cap streamlines, balance flow velocities, strain and basal heating rates, temperatures and melt rates. One of the prerequisites of a surge is basal melting, which has been found at Byrd Station, 80° S 120° W at a depth of 2164 m (Gow *et al.*, 1968). However, according to this model less than 10 % of the recent Antarctic ice-cap is now subject to melting processes near the ground. According to Oswald & Robin (1973) 17 sub-ice lakes have been discovered in East Antarctica by radioecho sounding flights, in regions of high ice thickness (2800–4200 m) and minimum velocity. Because of their small size (diameter along flight path between 2 and 15 km only), they cover, in the area of maximum frequency (around 75° S 125° E), only 0.5 % of the surface.

Wilson's hypothesis of simultaneous circumpolar Antarctic surges in its original form is hardly consistent with the roughness of the subglacial topography and with the results of this model. We should expect surges – not necessarily simultaneous – concentrated around the present ice shelves: the Weddell and Ross ice shelves, and, to a lesser degree, the Amery ice shelf (near 70° E). In analogy to present conditions the Weddell area should always have been the most productive. From the viewpoint of the heat budget and of the weather conditions in southern oceans, Wilson's assumption of a continuous quasi-permanent ice shelf around Antarctica with a size of 20–30 × 10^6 km² is unrealistic and unnecessary.

However, a surge spreading one-fourth or one-third of the present mass of the Antarctic ice dome – that means a volume of 6–10 × 10^6 km³ – more or less disintegrated from the existing shelf zones into the ocean, during a timespan of a few decades or even centuries, does not seem too unrealistic. Assuming an average thickness of 200 m for tabular icebergs, a nearly simultaneous outbreak of 6 × 10^6 km³ would produce an ice-covered ocean area of 30 × 10^6 km² (Table 1, Model SH 3). One of the prerequisites of such an event should be that the height of the ice dome approaches the highest mark on the ice-free mountains (Hoinkes, 1961), which indicates a further growth of 200–300 m is required before the next surge. (If the average positive mass budget is assumed to be – at a maximum! – 4 cm yr^{-1} (Schwerdtfeger, 1970) a growth of 200 m would need another 5000 yr; both figures, however, are crude estimates.)

Independently of the duration of the surge, each outbreak of continental ice into the ocean must lead to a significant rise of the sea-level; some evidence for such glacioeustatic rises has been found (Hollin, 1972). Assuming a mean density of 0.88 g cm^{-3}, each surging volume of 10^6 km³ should produce a eustatic rise of 2.44 m. Then the first serious consequence of a surge of the size assumed by Wilson (1964) and Hollin (1972) would be a sea-level rise of the order of 15–25 m, most probably spread over several decades; at any rate, it would be catastrophic for the densely populated coastal area, including all seaports. Denton *et al.* (1971) have given a critical survey of the climatic and glaciological history of the Antarctic ice sheets. Evidence for and against the former occurrence of large-scale surges is presented; no really conclusive proof exists at this time. Several height fluctuations of the ice of East Antarctica are quite conspicuous; they do not coincide with the northern hemisphere glaciation

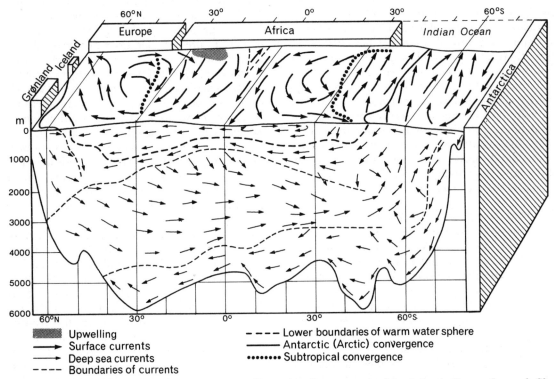

Fig. 3. Meridional cross section of abyssal circulation and surface currents of the Atlantic Ocean, Legend: Sk–subtropical convergence; P–polar front and Atlantic Convergence; dashed–boundary between warm and cold layers; dotted–layer of zero current (simplified after G. Wüst).

(cf. Hughes as quoted in the Addendum, p. 263), in spite of the nearly parallel trend of temperature in both hemispheres. If this is real, it can be taken as a suggestion that short-lived, large surges generally produce brief coolings, while the large-scale climatic fluctuations are controlled by the long-lasting glaciations of the northern hemisphere and their role in the atmosphere–ocean heat budget.

The effect of an Antarctic surge of this size on the oceanic heat budget depends mainly on ice volume and temperature. Even more important is the effect on surface albedo and thus on the atmospheric heat budget: this depends on the albedo of and the area covered by ice, regardless of the degree of its disintegration. From the Manabe–Wetherald Model (1967) it can be concluded that a surge of the size indicated by Wilson's hypothesis, with a sea-ice area increase of

30×10^6 km², is equivalent to a southern hemisphere temperature drop of 7–8 °C (Model SH 3).

South of the oceanic Antarctic Convergence – now situated at 49–50° S in the Atlantic and in the adjacent Indian Ocean, and near 60° S in the Pacific – melting of the surged ice will be quite slow, due to the low surface temperatures of the sub-antarctic ocean (between 0 and +3 °C). Here the latent heat of melting plays only an insignificant role; Antarctic cold water is permanently sinking and disappearing at the Antarctic Convergence, feeding the subantarctic intermediate water at a depth of 500–2000 m (Fig. 3). Let us assume that 40 % of the injected ice (10^{22} g water equivalent) melts in this belt, consuming 32×10^{22} gcal in latent heat distributed over a 1000 m deep ocean layer ($\sim 3 \times 10^{22}$ cm³) during a period of 100 yr. If all other terms of the heat budget re-

Initiation of a glacial period (*proposed scheme*)

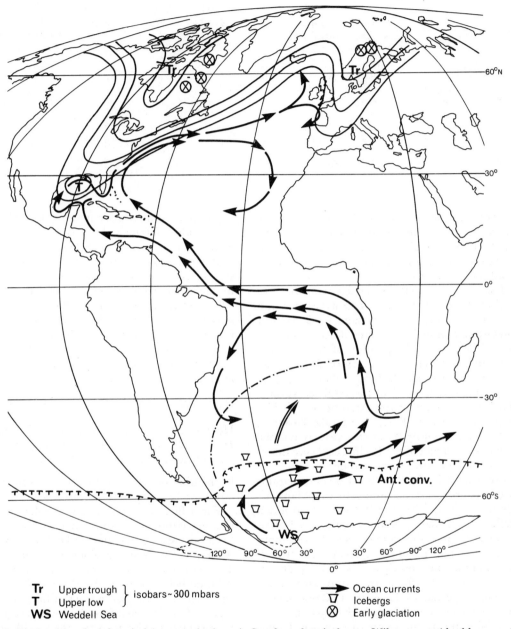

Tr	Upper trough	} isobars ~ 300 mbars
T	Upper low	
WS	Weddell Sea	

→ Ocean currents
▽ Icebergs
⊗ Early glaciation

Fig. 4. Initiation of a glacial period (proposed scheme). Southern hemisphere – Wilson surge (double arrow) across Atlantic Convergence (actual position) up to about 25° S (dash–dot line). Ocean currents – actual situation. Northern hemisphere during melting period–cold pool in Gulf of Mexico, upper troughs along long. 80° W and W. Europe, centers of early continental glaciation.

Table 2. *Estimated Heat Budget: mixing layer during surge melting*

$$Q_{sf} - (H_a + LE) - H_m + \text{div } A_t - \Delta T = 0 \text{ (in } 10^{20} \text{ cal/a)}$$

Atlantic zone	Area 10^6 km²	Transport 10^6 m³ s⁻¹	Q_{sf}	$H_a + LE$	H_m	div A_t 10^{20} cal a⁻¹	ΔT	T_{exit} °C
A. 25–50°S, 50 yr	18.6	8	16.7	0	56	−34	+5.4	0
25–50°S, 100 yr	18.6	8	16.7	0	28	−8.6	+2.7	0
B. 5–25°S (E only)								
Benguela Current	6.0	8 →	66	38	0	−28 →	+28	11.2
C. 5°S—Carib Isl.		8 } 25	51	31	0	−20 }	+42	15.8
Guayana Current	4.3	17 ↑				−22 ↑		
D. Caribbean + Gulf of Mexico	4.2	25	50	31	0	−28	+28	18.2

main constant (which is certainly unrealistic), this melting would produce an annual cooling of about 0.1 °C. Thus the regional effect of an ice surge on the heat budget of the subantarctic ocean is only small and short-lived.

The other 60 % of the injected ice is assumed to be driven across the Antarctic Convergence into the warmer water on its northern flank. Here it will be disintegrated and melted much faster than before. Because of the position of the Weddell Sea – and in agreement with observations during the 19th century (Lamb, 1967; Schott, 1942) we may assume that 35 % of the total ice volume penetrates into the narrow Atlantic sector (Fig. 4) (i.e. between 20° E and the Drake Passage), and the remaining 25 % into the vast areas of the Indian Ocean and of the Pacific, covering three-fourths of the Earth's circumference.

Over the Atlantic sector the drop of the equilibrium temperature according to the Manabe–Wetherald model (1967) will then extend much farther north than over the Pacific and Indian sectors. One may therefore expect a broad, more or less permanent upper tropospheric trough extending to (and partly across) the equatorial region, thus disturbing the subtropical anticyclonic ridge and displacing the ITC region even more to the northern hemisphere than at present, especially during the southern summer. Under present conditions a similar (but weaker) pattern is frequent only during the southern winter. The annual distribution of winds and water temperature anomalies in the equatorial region will then resemble the present northern summer situation: e.g. a southerly flow across the equator and, con-

sequently, prevalence of equatorial upwelling (cf. the results obtained by Henning for July–September, see Flohn, 1972, Figs. 3 and 4).

While the ice floating into the Indo-Pacific section is of lesser, only regional importance – e.g. in the narrow belt along the eastern coast of Africa (Olaussen *et al.*, 1971) – the ice transported into the Atlantic sector plays a key role. As a basis for discussion we may use the heat balance equation for an upper mixed oceanic layer with constant depth, above the thermocline, in the following form:

$$Q_{sf} - (H_a + LE) - H_m - \text{div } \vec{A_t} - \Delta T = 0$$

Here Q_{sf} is the net radiation at the surface, $H_a + LE$ is the turbulent flux of sensible and latent heat from sea into air, H_m is the heat used for melting of ice, and div $\vec{A_t}$ = divergence of advective heat flux of the ocean layer.

For a first-order estimate of the heat budget changes within the Atlantic current system (Table 2), let us assume that the ice floating across the Antarctic Convergence Zone melts in a short period (of a few decades or centuries) in the south Atlantic between 25° and 50° S (—18.6 × 10¹⁶ cm²). In this area, just in front of the main surge region of the Weddell Sea, the ice coverage must be assumed to rise to 0.50; then, with an ice albedo of 0.75, the average albedo is 0.40. With a global radiation of about 118 kLy yr⁻¹ (Sellers, 1965) and an atmospheric counterradiation of 63 kLy yr⁻¹ (after Brunt's formula with $T = 273$ K), Q_{sf} will be drastically reduced from its present value of 74 kLy yr⁻¹ to 9 kLy yr⁻¹ for a total energy loss of 16.7×10^{20} gcal yr^{-1}. The average water temperature of about $14\ ^\circ$C should drop to $0\ ^\circ$C: this is equivalent – assuming a mixing layer of 100 m – to $\Delta T = 260 \times 10^{20}$ gcal. The melting of 35×10^{20} g ice (water equivalent) needs 28×10^{22} gcal or the equivalent of 168 yr net radiation. During the melting period, the fluxes of H_a and LE will be small and are thus neglected, as a first-order approximation. Assuming durations of the melting period in the zone A of 50 or 100 yr, the estimated annual heat budgets of the mixing layer are given in Table 2 (T_{exit} is the temperature of the resulting mixing layer at the northern boundary of the zone).

In the following and in Table 2 we consider only this oceanic current system: Benguela current–South Equatorial current and its branch north of the South American continent (Guayana Current)–Gulf Stream. It must be assumed that during the melting period many icebergs reach 25° S – even in the 19th century some reached 35° S; the intensity of the cold Benguela Current is therefore assumed to increase from 6 to 8×10^6 m^3 s^{-1} (disregarding the contribution of layers below 100 m). In the zones B, C, and D (Table 2) it is assumed that the extremely cold water increases atmospheric stability, and that the turbulent fluxes $H_a + LE$ are reduced by 40 %. If we select a reduction rate of 30 % or less (Flohn, 1969) the water temperature (T_{exit}) remains too low; this supports the idea of a drastic reduction of the oceanic evaporation and therefore of tropical rainfall (Section VII).

A crucial point is the crossing of the equator: here we may assume that under nearly constant southerly winds (triggered by the difference of ocean temperatures in the two hemispheres) constant upwelling occurs. Its order of magnitude may be estimated by the use of Stommel's (1964) figures: he obtains, averaged between 5° N and 5° S, an upward velocity of 4×10^{-4} cm s^{-1} (about 35 cm day^{-1}). According to Henning's data (Flohn, 1972) we restrict upwelling to the latitudinal belt 5° S to 1° N; then the upward flow reaches, above an area of 4.3×10^6 km^2, a value of about 17×10^6 m^3 s^{-1}. This value is of the same order as the mass transport of the powerful South Equatorial Current of the Atlantic. In calculating the heat budget, we assume a temperature of $14\ ^\circ$C for the ascending flow – i.e. the permanent upwelling is assumed to have the minimum temperature observed near the Galapagos.

The assumption of a higher ascending mass transport would lead to unreasonably low temperatures (which should have destroyed the coral reefs in the Caribbean). It should also be mentioned that, due to the semi-permanent trough situation above the South Atlantic, the splitting of the South Equatorial Current should be displaced to the north, leading to a higher mass transport along the northern coast of South America and reducing the southward Falkland current.

A final value of T_{exit} as low as $18\ ^\circ$C in the Gulf of Mexico is not presently supported by any evidence. However, if we take into account that the branch of the North Equatorial Current entering the area as the Caribbean Current, with a similar transport of 26×10^6 km^{3}s^{-1} and a temperature of $26\ ^\circ$C, flows in a baroclinic pattern parallel to the course of the chilled waters from the southern hemisphere, we come to the result that the observed surface temperature of 21–$22\ ^\circ$C during the ice ages (Emiliani, 1970) (i.e. averaged over a much longer period of some 10^4 yr) are consistent with an area-averaged value near $22\ ^\circ$C.

It would go beyond the purpose of this article to estimate the heat budget of the Gulf Stream during the melting period. It is sufficient to mention that its temperature at Florida Strait should be perhaps $5\ ^\circ$C lower than today.

Such an – admittedly crude – consideration of the oceanic heat budget is necessary since the heat capacity of a 3-m water column is the same as that of the whole atmospheric column. If the ocean maintained its temperature and the cooling of the atmosphere above the greatest part of the globe occurred only due to advective processes in the air, the global climatic effect of an Antarctic ice outbreak would be rather insignificant. In this case it would be impossible to understand in what way the albedo-produced cooling of the subantarctic atmosphere could expand into the northern hemisphere, across the vast area of the warm tropical oceans covering – between 30° S and 15° N – about 78 % of the surface.

It has been shown in Section V that the key to the initiation of a northern glaciation (either complete or incomplete) is the summer climate of Labrador–Ungava, including Keewatin and probably Baffin Island. A quasi-permanent pattern of low-level polar air advection together with upgliding warm moist air can only be conceived together with a quasi-permanent east coast trough reaching from Labrador to Florida (Fig. 4). Such a situation must be produced, sustained, and fixed by a pool of cool water in the Gulf of Mexico and Caribbean, with surface temperatures around 22 instead of 27–28 °C. The effect of this situation in summer (with the strongly heated continent in the north) is much stronger than in winter; during the warm season a permanent high-tropospheric vortex will be maintained, and an upper trough will be fixed along the east coast by a slightly cooler Gulf Stream.

The geophysical model which has been outlined here has one advantage over others: it needs only a quite short time-span of 50 or 100 yr to produce a nucleus for glaciation growing by positive feedback, and is therefore apparently consistent with the evidence presented in Section IV. A crucial test of this model, however, lies in the magnitude of the unavoidable sea-level rise (Hollin, 1972). The evidence presented to date is hardly sufficiently conclusive; however, it is certainly difficult to find convincing traces of a marine transgression with a lifetime of the order of only 10^3 yr. Immediately after the establishment of this circulation anomaly, the storage of water in the form of ice above the northern continents begins and increases rapidly by the above-mentioned feedback process.

VII. Time and space correlations

Interpreting global coolings as caused by large-scale surges of the Antarctic ice, concentrated mainly in the Weddell Sea–Atlantic section, we may also comprehend some correlations with other evidence which remained hitherto hardly understandable.

(*a*) In all tropical continents, definite signs of a marked desiccation simultaneously with the northern glaciations have been found (Fairbridge, 1972). This is even true for the equatorial rain forests in central Africa and South America (Vuilleumier, 1972; van Zinderen Bakker & Coetzee, 1972); in both areas semi-arid, dry forest prevailed, with only a few islands of humid forest. If under the present radiation regime the oceans are advectively cooled, the evaporation from the tropical oceans as the main source of the global hydrological cycle (68 %) must have been substantially lower; a rough estimate based on advective processes alone (Flohn, 1969) gave a decrease of about 30 %. Under such conditions the intensity and extension of the tropical Hadley cell must

have decreased in comparison with present conditions.

(*b*) In several parts of the Atlantic sector the lowering of the snow line and vegetation limits during glacials was much greater than usual (equatorial zone some 800 m, midlatitudes about 1200 m). Here we mention the Itatiaya near Rio de Janeiro (Mortensen, 1957), the Costa Rica Volcanoes (Weyl, 1956) and, at least in some times, the Sabana of Bogota (van der Hammen et al., 1971), each with a glacial cooling of up to 8 °C (in earlier glaciations even 11 °C) instead of 5 °C in other areas (East Africa, Indonesia, New Guinea). Similar data have been collected from the humid Andes·Mountains of Colombia and Venezuela (Wilhelmy, 1957). Of special interest is the somewhat controversial evidence for a widespread low-land glaciation of Eastern Patagonia (Czajka, 1957).

(*c*) According to new data collected in a interdisciplinary German–Mexican Project (Heine, 1973*a*, *b*), the ^{14}C timescale of climatic events in the last 40 000 yr in the high volcanoes in central Mexico deviates in a characteristic way from the usual sequence; the greatest glacial advances occurred at about 32 000 and (during 400–500 yr only) 12 000 B.P., i.e. within the Allerød oscillation.

(*d*) The striking contrast between the glacial temperature anomalies in the area around the Bay of Biscay, from Ireland to Northern Spain (− 12 °C) or even more), and at the same latitude at the Pacific coast of North and South America (about −5 °C) has been stressed earlier (Flohn, 1969); it is consistent with abrupt progressions of polar water masses to 42° N in the Atlantic (McIntyre et al., 1972) i.e. more than 20° south of their present position. From a comparison between micropaleontological evidence and isotopic temperatures, Emiliani (1971) concludes that glacial surface temperatures were about 7–8 °C lower in the Caribbean, 5–6 °C in the equatorial Atlantic, but only 3–4 °C in the equatorial Pacific. The contrast of the large extension of ice sheets on both sides of the Atlantic with the comparatively small glaciation on both sides of the Pacific requires also a geophysical interpretation (Flohn, 1969). According to this version of the Antarctic Surge model, cooling of the Pacific may have occurred only as a secondary effect.

After the revised calculation of the solar radiation fluctuations due to orbital variations (Vernekar, 1972) the equal severity of the most recent and earlier glaciations is difficult to understand. In contrast to the solar variations, the glacial and climatic history of at least the last 20 000 yr shows a clear coincidence at both hemispheres, instead of a time lag of the order of 10 000 yr. Together with the discrepancies in the time-scale involved, this seems to be one of the strongest arguments against a primary role of the extraterrestrial 'Milankovich effect', which appears to have mesmerized nearly two generations of earth scientists, in spite of many sober and critical voices.

According to heat transport considerations (Newell, 1973) the initiation of a glaciation needs a heat deficit of 1 to 3 × 10^{19} gcal day^{-1}. A Wilson surge (Model SH 3) yields, with an average global radiation of 350 Ly day^{-1}, a heat deficit of about 9 × 10^{19} gcal day^{-1} or about 4500 TW lasting about a century. If the melting process of the ice lasts the same period, an additional loss of nearly 2 × 10^{19} gcal day^{-1} is to be added.

In contrast to this, the peak-to-peak variation of summer insolation due to the Milankovich effect – i.e. spread over a period of several 10^4 yr! – is 35 × 10^{20} gcal (Broecker, 1968) or little less than 2 × 10^{19} gcal day^{-1}, equivalent to 3.8 W m^{-2} or 970 TW.

The mass of new evidence, consisting of quantitative determinations of temperature and age (certainly not without sources of systematic error!) has shown that the time-scale of the glacial-interglacial sequence is much more complex than the classical one (Emiliani, 1972). Obviously a series of hemispheric-scale coolings occurred, some followed by a glaciation on the northern continents, others not. Such 'abortive' coolings, with a time-scale of a few centuries, are of vital interest to the meteorologist: in the human time-scale they are 'irreversible', i.e. from the viewpoint of living mankind, of the economist, and the politician. They indicate that Lorenz's (1968) unorthodox suggestion of a potential instability of our climate is quite realistic and must be taken as a serious challenge of utmost significance.

The powerful feedback between the strong albedo gradient on the outer boundary of the snow and ice-covered region and the baroclinic frontal zones (acting as cyclone tracks) contributes strongly to the development and maintenance of continental glaciations in the northern hemisphere. It is much more difficult to understand the total interruption of this process, which leads to disintegration and finally to deglaciation. In other words: what physical causes are responsible for the transition from an ice-accumulating to an ice-destroying pattern of the atmospheric circulation, right at the culmination of each glacial? It has been argued (Hoinkes, 1961; Bloch, 1964), that the aridity of northern continents causes less dust from the barren fluvioglacial deposits around the ice (with most particles well below 2 μm) to be blown on to the ice, resulting in a lowering of the surface albedo. However, the role of a dust-laden atmosphere – as it can be observed now during summer above Pakistan, Turkestan, and Sinkiang – is more complex: low-level dust absorbs solar and long-wave radiation and heats the atmosphere and the surface substantially. A most remarkable example of this effect has been observed recently in the Martian atmosphere (Gierash & Goody, 1972). Direct heating and atmospheric infra-red radiation (at high temperatures) are more powerful melting agents than the decrease of the albedo alone. Time variations of the calcium content of the Greenland ice (Hamilton & Seligo, 1972) are apparently consistent with this hypothesis: the occurrence of calcium maxima after the beginning of cooling supports our view.

VIII. Summary

In a time-scale of 10^4–10^6 yr the climate of the Earth–atmosphere–hydrosphere–cryosphere system is in fact unstable. This complex, self-regulating system – with nearly constant energy input – is in a delicate state of equilibrium, with its energy budget depending on the variable area of its subsystems. During the last 10^5 yr, evidence of at least five rapid hemispheric or global coolings has been found with temperature changes of about 5 °C (i.e. the full difference between present and ice age climate), occurring during a time-span of the order of a century. Only some of them led to a complete glaciation of the northern continents, others ended after a few centuries with a sudden warming. These facts are not compatible with the widely accepted orbital variations with a time-scale of some 10^4 yr ('Milankovich effect').

Starting from a modified version of A. T. Wilson's hypothesis of large-scale surges of the Antarctic ice dome, a purely geophysical model of such rapid coolings can be outlined:

(1) After a period of slow accumulation, the Antarctic ice dome surges – not necessarily simultaneously – mainly in the existing shelf areas, especially in the Weddell Sea. The amount of ice calving into the ocean is estimated to be $6-10 \times 10^6$ km^3, causing a eustatic sea-level rise of $15-25$ m, spread over several decades or centuries.

(2) The disintegrated ice spreads, in enormous tabular icebergs, over an area of $20-30 \times 10^6$ km^2; due to the increase of the surface albedo the average temperature of the southern hemisphere drops about 7 °C. Since the tropical zone is, at the very beginning, only weakly affected, the mid-latitude circulation intensifies significantly.

(3) During the melting period, a considerable part of the ice crosses the oceanic Antarctic Convergence, notably in the Weddell Sea–Atlantic sector. According to an estimate of the heat budget of the upper mixed layer, the melting process causes advective cooling of the system Benguela Current–South Equatorial Current–Gulf Stream of the order of at least $6-8$ °C during a timespan of about $50-100$ yr, in addition to the albedo-produced cooling. This cooling should be accompanied by a broad upper trough in the South Atlantic and a marked cold–arid phase in the neighboring continents.

(4) Advective cooling of the surface of the Caribbean and the Gulf of Mexico, together with the Gulf Stream, causes, during the warm season, a permanent high-trophospheric vortex together with a deep trough along the eastern coast of North America. This circulation anomaly produces cooling and increased snowfall in the Labrador–Ungava region and enables a survival of the snow cover during summer, as a potential nucleus of a continental glaciation.

(5) When the permanent snow cover has reached a diameter of several hundred kilometers, a positive feedback mechanism (snow surface with high albedo–tropospheric cooling–cold upper vortex–enhanced snowfall) leads to a fast-growing ice sheet. This localized heat sink maintains a quasi-permament trough–ridge pattern with blocking highs east of Alaska and east of Iceland, the latter producing a second ice center in the mountains of Fennoscandia.

(6) Further development of these ice centers either to a complete glaciation or to a reversal may perhaps be controlled by the actual state of the orbital elements; further studies are needed.

From the meterological point of view, an incomplete ('abortive') glaciation with a duration of only a few centuries is as important as a complete glaciation with a period of 10 000 yr.

(7) The final disintegration of the ice domes of the northern continents – in spite of the powerful feedback referred to in no. 5 – can be understood as caused by frequent dust storms, lowering the surface albedo of the glacier, and heating the lower atmosphere through absorption of solar and long-wave radiation.

Such a geophysical model may serve as a background for a complete physicomathematical model of the complex multiphase system designed to simulate these most dramatic events in the climatic history of the Earth. Admittedly, our knowledge of the complex interaction of processes in our geophysical system is at this time rather inadequate. Several highly interesting model experiments try to simulate the climate of a fully developed glacial epoch (Alyea, 1972; Williams *et al.*, 1973). Here we propose a much more difficult, but also more rewarding future task: the simulation of the initiation of a new glaciation. If – as stated in the introduction – a new glaciation should be expected to begin during the next, say, 5000 yr, one should expect, perhaps, a probability near 1 %, that this may happen during the next 50 yr.

Man can hardly interfere with the mass budget and the intensity of the Antarctic ice. Denton *et al.* (1971) have pointed out that a minor increase of temperature – as we expect as a result of man-made effects – may lead to disintegration of the smaller West Antarctic ice sheet; both effects should raise the accumulation rate in East Antarctica. At any rate, this problem is not only of pure academic interest: it refers to our own planet Earth, to our habitat now and in the future. Within the lifetime of our generation, it deserves a much higher priority.

Acknowledgments

The author wishes to express his gratitude to many colleagues, with whom he had the privilege to discuss some of the ideas presented here: M. R. Bloch (Beer-Sheba), K. Brunnacker (Köln), M. I. Budyko (Leningrad), H. Hoinkes (Innsbruck), F. Loewe (Melbourne), U. Radok (Melbourne), M. Schwarzbach (Köln), C. Troll (Bonn), A. T. Wilson (Wairakei, N.Z.), and E.

M. van Zinderen Bakker (Bloemfontein). A preliminary draft of this paper was first presented in a
lecture at the University of Melbourne, August
1972.

The paper shall be devoted to the memory of
the late F. Loewe (deceased March 1974). The
careful revision of the original manuscript by
Prof. Sue Bowling (Fairbanks, Alaska) is gratefully acknowledged, as well as some comments
from the reviewers.

Addendum

In a very thoughtful paper, T. Hughes (*J.
Geophys. Res.* 78, 1973, 7884–7916) evaluated a
non-equilibrium profile along a flow-line of the
Ross ice shelf suggesting an instability of the
smaller West Antarctic ice sheet. The slow retreat
of the grounding line of the Ross ice shelf is
greater than that caused by the post-glacial
eustatic sea-level rise (which was, however, of the
order of $100 \text{ m} (8000 \text{ yr})^{-1}$ or 12 mm yr^{-1}); further
retreat means therefore an increase of floating
shelf ice and of the present eustatic rise near
1 mm yr^{-1}. Since the Weddell Sea drainage area is
much more unknown than the Ross Sea area,
similar investigations there are badly needed.

The physical background of the ice core data
from Greenland and Antarctica has been recently
discussed by W. Dansgaard, S. J. Johnsen, H. B.
Clausen, and N. Gundestrup in a monograph
(*Meddel. om Grønland* **197**, No. 2, 1973, 53 pp).
The use of the hypothetical cycles of constant
length as a base for a time-scale has been seriously
doubted by N. A. Mörner (*Boreas* 2, 1973, 33–53)
and (*Geol. Magaz.* 111, 1974, in press) as well as
the use of a logarithmic scale based on a flow
model with constant velocity. Certainly any substantiated improvement of the chronology would
be highly welcome; due to the relatively high
degree of coincidence with many independent
results (cf. Section IV) it is not expected that the
results presented here would be significantly
changed by an improved chronology.

In addition to Section VII paragraph (*b*) reference should be made to the isolated and short
glacier advance in southern Chile (J. H. Mercer
& C. A. Laugenie, *Science* 182, 1973, 1017–1019)
around 36 000 B.P. and to the evidence of frost
weathering at the South African Cape Coast
indicating a winter temperature depression of
about 10 °C (K. W. Butzer, *Boreas* 2, 1972, 1–11).
Such examples of short-living episodes deserve

more attention; the highlands of Angola and
southwest Africa should be of special interest.

References

Alyea, J. (1972). Numerical simulation of an ice age
paleoclimate. *Atmos. Sci. Pap.* No. 193. Dept.
Atmos. Sci., Colorado State Univ., Fort Collins.
120 pp.
Barry, R. G. (1959). A synoptic climatology for
Labrador–Ungava. *Publ. in Meteor.* No. 17. Arct.
Meteor. Res. Group, McGill Univ., Montreal.
168 pp., Append.
——— (1966). Meteorological aspects of the glacial
history of Labrador–Ungava with special reference
to atmospheric vapour transport. *Geographical
Bulletin*, **8,** 319–40.
Bjerknes, J. (1969). Atmospheric teleconnections
from the equatorial Pacific. *Monthly Weather
Review,* **97,** 163–72.
Bloch, M. R. (1964). Die Beeinflussung der Albedo
von Eisflächen durch Staub und ihre Wirkung auf
Ozeanhöhe und Klima. *Geologische Rundschau*, **54,**
515–22.
Brinkmann, R. & Barry, R. G. (1972). Palaeo-
climatological aspects of the synoptic climatology
of Keewatin, Northwest territories, Canada.
Palaeogeography, Palaeoclimatology, Palaeoecology, **11,**
77–91.
Broecker, W. S. (1968). In defense of the astronomical
theory of glaciation. *Amer. Meteor. Soc. Meteor.
Monogr.,* **8,** No. 30, 139–41.
Broecker, W. S. & Van Donk, J. (1970). Insolation
changes, ice volumes and the O^{18} record in deep-sea
cores. *Reviews of Geophysics Space Physics,* **8,** 160–98.
Brooks, C. E. P. (1926, 1949). *Climate through the ages,*
2nd ed. London.
Budd, M., Jenssen, D. & Radok, U. (1970). The
extent of basal melting in Antarctica. *Polarforschung*
6, 293–306.
Budyko, M. (1969). The effect of solar radiation
variations on the climate of the earth. *Tellus,* **21,**
611–19.
——— (1972). The future climate, *Transactions of the
American Geophysical Union (Eos),* **54,** 868–74.
Bull, C. & Webb, P. N. (1973). Some recent developments in the investigation of the glacial history and
glaciology of Antarctica. In (Van Zinderen
Bakker, E. M. Ed.) *Palaeoecology of Africa,* vol. 8, pp.
55–84. Balkema, Cape Town.
Butzer, K. W. (1973). Pleistocene 'periglacial'
phenomena in southern Africa. *Boreas,* **2,** 1–11.
Czajka, W. (1957). Die Reichweite der pleistozanen
Vereisung Patagoniens. *Geologische Rundschau,* **45,**
634–86.
Dansgaard, W., Johnsen, S. J., Clausen, H. B. &
Langway, C. C., Jr (1971). Climatic record

revealed by the camp century ice core. In (Ture-kian, K., Ed). *Late Cenozoic glacial ages,* pp. 37–56. Yale Univ. Press, New Haven.

Dansgaard, W., Johnsen, S. J., Clausen, H. B., & Langway, C. C., Jr (1972). Speculations about the next glaciation. *Quaternary Research,* **2,** 396–8.

Denton, G. H., Armstrong, R. L. & Stuiver, M. (1971). The late cenozoic glacial history of Antarctica. In (K. Turekian, Ed.), *Late Cenozoic glacial ages,* pp. 267–306. Yale Univ. Press, New Haven.

Duplessy, J. C., Labeyrie, J., Lalou, C. & and Nguyen, H. V. (1971). La mesure des variations climatiques continentales. Application à la période comprise entre 130 000 et 90 000 ans B.P. *Quater-nary Research,* **1,** 162–74.

Emiliani, C. (1970). Pleistocene paleotemperatures. *Science,* **168,** 822–5.

——— (1971). The amplitude of pleistocene climatic cycles at low latitudes and the isotopic composition of glacial ice. In (K. Turekian, Ed.), *Late Cenozoic glacial ages,* pp. 183–97. Yale Univ. Press, New Haven.

——— (1972). Quaternary paleotemperatures and the duration of the high-temperature intervals. *Science,* **178,** 398–401.

Fairbridge, R. W. (1972). Climatology of a glacial cycle. *Quaternary Research,* **2,** 283–302.

Fliri, F., Bortenschlager, S. & Felber, (u.a.), H. (1970). Der Banderton von Baumkirchen (Inntal, Tirol). Eine neue Schlusselstellung zur Kenntnis der Würmvereisung der Alpen. *Zeitschrift fuer Z. Gletscherkunde und Glazialgeologie,* **6,** 5–35.

———, Hilscher, H. & Markgraf, V. (1971). Weitere Untersuchungen zur Chronologie der alpinen Vereisung (Bänderton von Baumkirchen, Inntal, Nordtirol). *Zeitschrift fuer Gletscherkunde und Glazialgeologie,* **7,** 5–24.

Flohn, H. (1969). Ein geophysikalisches Eiszeit-Modell. *Eiszeitalter und Gegenwart,* **20,** 204–231.

——— (1972). Investigation of equatorial upwelling and its climatic role. In (Gordon, A. H., Ed.), *Studies in physical oceanography,* vol. 1, pp. 93–102. A Tribute to Georg Wüst on his eightieth Birthday. New York, London, Gordon & Breach.

——— (1973a). Globale Energiebilanz und Klima-schwankungen. *Vorträge Rhein.-Westf. Akad. Wiss.,* N 234, 75–117.

——— (1973b). Antarctica and the global cenozoic evolution: A geophysical model. In (Van Zinderen Bakker, E. M., Ed.), *Palaeoecology of Africa,* vol. 8, pp. 37–53. Balkema, Cape Town.

Gierasch, P. J. & Goody, R. M. (1972). The effect of dust on the temperature of the Martian atmosphere. *Journal of Atmospheric Science,* **29,** 400–402.

Gow, A. J., Ueda, H. T. & Garfield, D. E. (1968).

Antarctic ice sheet: Preliminary results of first core hole to Bedrock. *Science,* **161,** 1011–13.

Hamilton, W. L. & Seliga, Th. A. (1872). Atmo-spheric turbidity and surface temperature on the polar ice sheets. *Nature (London),* **235,** 320–2.

Heine, K. (1973a). Die jungpleistozänen und holo-zänen Gletschervorstöße am Malinche-Vulkan, Mexiko. *Eiszeitalter und Gegenwart,* **23,** 46–62.

——— (1973b). Variaciones más importantes del clima durante los ultimos 40 000 años en México. *Communicationes Puebla Mexico,* **7,** 51–8.

Hoinkes, H. (1961). Die Antarktis und die geo-physikalische Erforschung der Erde. *Naturwis-senschaften,* **48,** 354–74.

——— (1971). Neue Ergebnisse und Gedanken zur Eiszeitforschung. *Jahrb. Akad. Wiss. Lit. Mainz,* 102–3.

Hollin, J. T. (1972). Interglacial climate and Antarctic ice surges. *Quaternary Research,* **2,** 401–8.

Johnsen, S. J., Dansgaard, W., Clausen, H. B., & Langway, C. C., Jr (1972). Oxygen isotope profiles through the Antarctic and Greenland ice sheets. *Nature (London),* **235,** 429–34.

Kennett, J. P. & Huddlestun, P. (1972). Abrupt climatic change at 90 000 yr B.P.: faunal evidence from Gulf of Mexico cores. *Quaternary Research,* **2,** 384–95.

Kukla, G. J., Matthews, R. K. & Mitchell, J. M., Jr. (1972a). The end of the present interglacial, *Quaternary Research,* **2,** 261–9.

——— & Matthews, R. K. (1972b). When will the present interglacial end? *Science,* **178,** 190–1.

——— & Kukla, H. J. (1972c). Insolation regime of interglacials. *Quaternary Research,* **2,** 412–24.

Lamb, H. H. (1967). On climatic variations affect-ing the Far South. *WMO Tech. Note* No. 87, 428–53.

——— & Woodroffe, A. (1970). Atmospheric circula-tion during the last ice-age. *Quaternary Research,* **1,** 29–58.

Loewe, F. (1971). Considerations on the origin of the quaternary ice sheet of North America. *Arctic and Alpine Research,* **3,** 331–44.

Lorenz, E. N. (1968). Climatic determinism. *Meteor. Monogr.,* **8,** No. 30, *Am. Meteo. Soc.* 1–3.

——— (1970). Climatic change as a mathematical problem. *Journal of Applied Meteorology,* **9,** 325–9.

McIntyre, A., Ruddiman, W. F., & Jantzen, R. (1972). Southward penetration of the North Atlantic polar front: faunal and floral evidence of the large-scale surface movements over the last 225 000 years. *Deep Sea Research,* **19,** 61–77.

Manabe, S. & Wetherald, R. T. (1967). Thermal equilibrium of the atmosphere with a given distribution of relative humidity. *Journal of Atmos-pheric Science,* **24,** 241–59.

Mercer, J. H. (1969). The Allerød oscillation: A

European climatic anomaly? *Arctic and Alpine Research*, **1**, 227–34.

——— (1970). A former ice sheet in the Arctic Ocean? *Palaeogeography, Palaeoclimatology, Palaeoecology*, **8**, 19–27.

——— & Laugenie, C. A. (1973). Glacier in Chile ended a major readvance about 36 000 years ago: some global comparisons. *Science*, **182**, 1017–19.

Mitchell, J. M., Jr. (1972). The natural breakdown of the present interglacial and its possible intervention by human activities. *Quaternary Research*, **2**, 436–45.

Mortensen, H. (1957). Temperaturgradient und Eiszeitklima am Beispiel der pleistozänen Schneegrenzdepression in den Rand- und Subtropen. *Zeitschrift fuer Geomorpholog*, **1**, 44–56.

Newell, J. M., Jr (1973). *Bull. Am. Meteor. Soc.*, **54**, 428 (Abstr.).

Olaussen, E., Bilal Ul Haq, U. I., Karlson, G. B. & Olsson, J. N. (1971). Evidence in Indian Ocean cores of late pleistocene changes in oceanic and atmospheric circulation. *Geologiska Foereningens, Stockholm Foerhandlingar*, **93**, 51–84.

Oswald, G. K. A. & de Q. Robin, G. (1973). Lakes beneath the Antarctic ice sheet. *Nature (London)*, **245**, 251–4.

Rowntree, P. R. (1972). The influence of tropical East Pacific Ocean temperatures on the atmosphere. *Quarterly Journal of the Royal Meteorological Society*, **98**, 290–321.

Sancetta, C., Imbrie, J., Kipp, N. G., McIntyre, A. & Ruddiman, W. F. (1972). Climatic record in North Atlantic deep-sea cores V 23–82: comparison of the last and present interglacials based on quantitative time series. *Quaternary Research*, **2**, 363–7.

Schott, G. (1944). *Geographie des Atlantischen Ozeans*, 3. Aufl. Hamburg, Boysen, 438 S.

Schwerdtfeger, W. (1970). The climate of the Antarctic. In *World survey of climatology*, vol. 14, pp. 253–355. Elsevier, Amsterdam, London, New York.

Sellers, W. D. (1965). *Physical climatology*. Chicago, University of Chicago Press. 272 S.

——— (1969). A global climatic model based on the energy balance of the earth-atmosphere system. *Journal of Applied Meteorology*, **8**, 392–400.

——— (1973). A new global climatic model. *Journal of Applied Meteorology*, **12**, 241–54.

Stommel, H. (1964). Summary charts of the mean dynamic topography and current field at the surface of the ocean, and related fields of the mean wind-stress. In (K. Yoshida, Ed.), *Studies on oceanography*, pp. 53–8. University of Tokyo Press, Tokyo.

Van Der Hammen, T., Wijmstra, T. A. & Zagwijn, W. H. (1971). Floral record of the late cenozoic of Europe. In (K. Turekian, Ed.), *Late Cenozoic glacial ages*, pp. 381–91. Yale Univ. Press, New Haven.

Van Zinderen Bakker, E. M. & Coetzee, J. A. (1972). A re-appraisal of late-quaternary climatic evidence from tropical Africa. In (Van Zinderen Bakker, E. M., Ed.), *Paleoecology of Africa*, vol. 7, pp. 151–82. Balkema, Cape Town.

Vernekar, A. D. (1972). Long-period global variations of incoming solar radiation. *Amer. Meteor. Soc., Meteor. Monogr.*, **12**, No. 34.

Vuilleumier, B. S. (1971). Pleistocene changes in the flora and fauna of South America. *Science*, **173**, 771–80.

Washington, W. W. (1972). Numerical climatic-change experiments: the effect of man's production of thermal energy. *Journal of Applied Meteorology*, **11**, 768–2.

Weyl, R. (1956). Spuren eiszeitlicher Vergletscherung in der Cordillera de Calamanca Costa Rica (Mittelamerika). *N. Jahrb. Geol. Paläont. Abh.*, **102**, 283–94.

Wilhelmy, H. (1957). Eiszeit und Eiszeitklima in den feuchttropischen Anden. *Petermanns Geogr. Mitt Ergänz.* Heft No. 262, 281–310.

Williams, J., Barry, R. G. & Washingon, W. W. (1973). Simulation of the climate at the last glacial maximum using the NCAR global circulation model. Occas. Pap. No. 5, Inst. of Alpine and Arctic Res., Univ. of Colorado, Boulder, Colo.

Wilson, A. T. (1964). Origin of ice ages: an ice shelf theory for pleistocene glaciation. *Nature (London)*, **201**, 147–9.

——— (1966). Variation in solar insolation to the south polar region as a trigger which induces instability in the Antarctic ice-sheet. *Nature (London)*, **210**, 477–8.

——— (1969). The climatic effects of large-scale surges of ice sheets. *Canadian Journal of Earth Science*, **6**, 911–18.

Wundt, W. (1938). Das Reflexionsvermögen der Erde zur Eiszeit. *Meteorologische Zeitschrift*, **55**, 81–7.

Author Index

Page numbers in italic type are bibliographical references.
References in brackets indicate the actual reference number under which an author may be found in the non-alphabetical bibliographies.

Subject Index